Craig's Soil Mechanics

Elise Saint-Martin
3rd year
837-5547

Craig's Soil Mechanics

Seventh edition

R.F. Craig

Formerly
Department of Civil Engineering
University of Dundee UK

Spon Press
Taylor & Francis Group

LONDON AND NEW YORK

First published 1974
by E & FN Spon, an imprint of Chapman & Hall
Second edition 1978
Third edition 1983
Fourth edition 1987
Fifth edition 1992
Sixth edition 1997
Seventh edition 2004

11 New Fetter Lane, London EC4P 4EE

Simultaneously published in the USA and Canada
by Spon Press
29 West 35th Street, New York, NY 10001

Spon Press is an imprint of the Taylor & Francis Group

© 1974, 1978, 1983, 1987, 1992, 1997, 2004 R.F. Craig

Typeset in 10/12pt Times by
Integra Software Services Pvt. Ltd, Pondicherry, India
Printed and bound in Great Britain by
TJ International Ltd, Padstow, Cornwall

British Library Cataloguing in Publication Data
A catalogue record for this book is available
from the British Library

Library of Congress Cataloging in Publication Data
Craig, R.F. (Robert F.)
  [Soil mechanics]
  Craig's soil mechanics / R.F. Craig. — 7th ed.
    p.   cm.
  Second ed.: Soil mechanics / R.F. Craig. Van Nostrand Reinhold, 1978.
  Includes bibliographical references and index.
  ISBN 0–415–32702–4 (hb.: alk. paper) — ISBN 0–415–32703–2 (pbk.: alk.
  paper)
  1. Soil mechanics. I. Title: Soil mechanics. II. Craig, R.F. (Robert F.) Soil
  mechanics. III. Title.
TA710.C685 2004
624.1′5136—dc22
                                                          2003061302

ISBN 0–415–32702–4 (hbk)
ISBN 0–415–32703–2 (pbk)

Contents

Preface

This book is intended primarily to serve the needs of the undergraduate civil engineering student and aims at the clear explanation, in adequate depth, of the fundamental principles of soil mechanics. The understanding of these principles is considered to be an essential foundation upon which future practical experience in geotechnical engineering can be built. The choice of material involves an element of personal opinion but the contents of this book should cover the requirements of most undergraduate courses to honours level as well as parts of some Masters courses.

It is assumed that the reader has no prior knowledge of the subject but has a good understanding of basic mechanics. The book includes a comprehensive range of worked examples and problems set for solution by the student to consolidate understanding of the fundamental principles and illustrate their application in simple practical situations. Both the traditional and limit state methods of design are included and some of the concepts of geotechnical engineering are introduced. The different types of field instrumentation are described and a number of case studies are included in which the differences between prediction and performance are discussed. References are included as an aid to the more advanced study of any particular topic. It is intended that the book will serve also as a useful source of reference for the practising engineer.

The author wishes to record his thanks to the various publishers, organizations and individuals who have given permission for the use of figures and tables of data, and to acknowledge his dependence on those authors whose works provided sources of material. Extracts from BS 8004: 1986 (Code of Practice for Foundations) and BS 5930: 1999 (Code of Practice for Site Investigations) are reproduced by permission of BSI. Complete copies of these codes can be obtained from BSI, Linford Wood, Milton Keynes, MK14 6LE.

Robert F. Craig
Dundee
March 2003

The unit for stress and pressure used in this book is kN/m^2 (kilonewton per square metre) or, where appropriate, MN/m^2 (meganewton per square metre). In SI the special name for the unit of stress or pressure is the *pascal* (Pa) equal to $1 N/m^2$ (newton per square metre). Thus:

$1 kN/m^2 = 1 kPa$ (kilopascal)
$1 MN/m^2 = 1 MPa$ (megapascal)

Chapter 1

Basic characteristics of soils

1.1 THE NATURE OF SOILS

To the civil engineer, soil is any uncemented or weakly cemented accumulation of mineral particles formed by the weathering of rocks, the void space between the particles containing water and/or air. Weak cementation can be due to carbonates or oxides precipitated between the particles or due to organic matter. If the products of weathering remain at their original location they constitute a residual soil. If the products are transported and deposited in a different location they constitute a transported soil, the agents of transportation being gravity, wind, water and glaciers. During transportation the size and shape of particles can undergo change and the particles can be sorted into size ranges.

The destructive process in the formation of soil from rock may be either physical or chemical. The physical process may be erosion by the action of wind, water or glaciers, or disintegration caused by alternate freezing and thawing in cracks in the rock. The resultant soil particles retain the same composition as that of the parent rock. Particles of this type are described as being of 'bulky' form and their shape can be indicated by terms such as angular, rounded, flat and elongated. The particles occur in a wide range of sizes, from boulders down to the fine rock flour formed by the grinding action of glaciers. The structural arrangement of bulky particles (Figure 1.1) is described as *single grain*, each particle being in direct contact with adjoining particles without there being any bond between them. The state of the particles can be described as dense, medium dense or loose, depending on how they are packed together.

The chemical process results in changes in the mineral form of the parent rock due to the action of water (especially if it contains traces of acid or alkali), oxygen and carbon dioxide. Chemical weathering results in the formation of groups of crystalline particles of colloidal size ($<0.002\,\text{mm}$) known as clay minerals. The clay mineral kaolinite, for example, is formed by the breakdown of feldspar by the action of water and carbon dioxide. Most clay mineral particles are of 'plate-like' form having a high specific surface (i.e. a high surface area to mass ratio) with the result that their structure is influenced significantly by surface forces. Long 'needle-shaped' particles can also occur but are comparatively rare.

The basic structural units of most clay minerals are a silicon–oxygen tetrahedron and an aluminium–hydroxyl octahedron, as illustrated in Figure 1.2(a). There are valency imbalances in both units, resulting in net negative charges. The basic units, therefore, do not exist in isolation but combine to form sheet structures. The

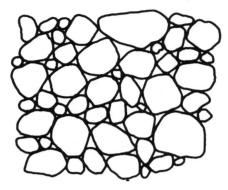

Figure 1.1 Single grain structure.

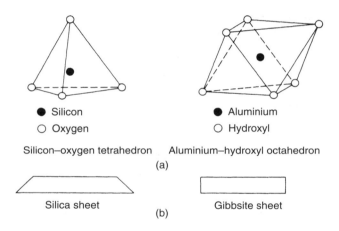

Silicon–oxygen tetrahedron   Aluminium–hydroxyl octahedron

(a)

Silica sheet

(b)

Gibbsite sheet

Figure 1.2 Clay minerals: basic units.

tetrahedral units combine by the sharing of oxygen ions to form a *silica* sheet. The octahedral units combine through shared hydroxyl ions to form a *gibbsite* sheet. The silica sheet retains a net negative charge but the gibbsite sheet is electrically neutral. Silicon and aluminium may be partially replaced by other elements, this being known as isomorphous substitution, resulting in further charge imbalance. The sheet structures are represented symbolically in Figure 1.2(b). Layer structures then form by the bonding of a silica sheet with either one or two gibbsite sheets. Clay mineral particles consist of stacks of these layers, with different forms of bonding between the layers. The structures of the principal clay minerals are represented in Figure 1.3.

Kaolinite consists of a structure based on a single sheet of silica combined with a single sheet of gibbsite. There is very limited isomorphous substitution. The combined silica–gibbsite sheets are held together relatively strongly by hydrogen bonding. A kaolinite particle may consist of over 100 stacks. *Illite* has a basic structure consisting of a sheet of gibbsite between and combined with two sheets of silica. In the silica sheet

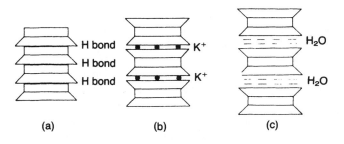

Figure 1.3 Clay minerals: (a) kaolinite, (b) illite and (c) montmorillonite.

there is partial substitution of silicon by aluminium. The combined sheets are linked together by relatively weak bonding due to non-exchangeable potassium ions held between them. *Montmorillonite* has the same basic structure as illite. In the gibbsite sheet there is partial substitution of aluminium by magnesium and iron, and in the silica sheet there is again partial substitution of silicon by aluminium. The space between the combined sheets is occupied by water molecules and exchangeable cations other than potassium, resulting in a very weak bond. Considerable swelling of montmorillonite can occur due to additional water being adsorbed between the combined sheets.

The surfaces of clay mineral particles carry residual negative charges, mainly as a result of the isomorphous substitution of silicon or aluminium by ions of lower valency but also due to disassociation of hydroxyl ions. Unsatisfied charges due to 'broken bonds' at the edges of particles also occur. The negative charges result in cations present in the water in the void space being attracted to the particles. The cations are not held strongly and, if the nature of the water changes, can be replaced by other cations, a phenomenon referred to as *base exchange*.

Cations are attracted to a clay mineral particle because of the negatively charged surface but at the same time they tend to move away from each other because of their thermal energy. The net effect is that the cations form a dispersed layer adjacent to the particle, the cation concentration decreasing with increasing distance from the surface until the concentration becomes equal to that in the general mass of water in the void space of the soil as a whole. The term *double layer* describes the negatively charged particle surface and the dispersed layer of cations. For a given particle the thickness of the cation layer depends mainly on the valency and concentration of the cations: an increase in valency (due to cation exchange) or an increase in concentration will result in a decrease in layer thickness. Temperature also affects cation layer thickness, an increase in temperature resulting in a decrease in layer thickness.

Layers of water molecules are held around a clay mineral particle by hydrogen bonding and (because water molecules are dipolar) by attraction to the negatively charged surfaces. In addition the exchangeable cations attract water (i.e. they become hydrated). The particle is thus surrounded by a layer of *adsorbed water*. The water nearest to the particle is strongly held and appears to have a high viscosity, but the viscosity decreases with increasing distance from the particle surface to that of 'free' water at the boundary of the adsorbed layer. Adsorbed water molecules can move relatively freely parallel to the particle surface but movement perpendicular to the surface is restricted.

Forces of repulsion and attraction act between adjacent clay mineral particles. Repulsion occurs between the like charges of the double layers, the force of repulsion depending on the characteristics of the layers. An increase in cation valency or concentration will result in a decrease in repulsive force and vice versa. Attraction between particles is due to short-range van der Waals forces (electrical forces of attraction between neutral molecules), which are independent of the double-layer characteristics, that decrease rapidly with increasing distance between particles. The net interparticle forces influence the structural form of clay mineral particles on deposition. If there is net repulsion the particles tend to assume a face-to-face orientation, this being referred to as a *dispersed* structure. If, on the other hand, there is net attraction the orientation of the particles tends to be edge-to-face or edge-to-edge, this being referred to as a *flocculated* structure. These structures, involving interaction between single clay mineral particles, are illustrated in Figures 1.4(a) and (b).

In natural clays, which normally contain a significant proportion of larger, bulky particles, the structural arrangement can be extremely complex. Interaction between single clay mineral particles is rare, the tendency being for the formation of elementary aggregations of particles (also referred to as domains) with a face-to-face orientation. In turn these elementary aggregations combine to form larger assemblages, the structure of which is influenced by the depositional environment. Two possible forms of particle assemblage, known as the bookhouse and turbostratic structures, are illustrated in Figures 1.4(c) and (d). Assemblages can also occur in the form of connectors or a matrix between larger particles. An example of the structure of a natural clay, in diagrammatical form, is shown in Figure 1.4(e). A secondary electron image of Errol Clay is shown in Figure 1.5, the solid bar at the bottom right of the image representing a length of 10 μm.

Particle sizes in soils can vary from over 100 mm to less than 0.001 mm. In British Standards the size ranges detailed in Figure 1.6 are specified. In Figure 1.6 the terms 'clay', 'silt', etc. are used to describe only the *sizes* of particles between specified limits. However, the same terms are also used to describe particular *types* of soil. Most soils consist of a graded mixture of particles from two or more size ranges. For example, clay is a type of soil possessing cohesion and plasticity which normally consists of particles in both the *clay size* and *silt size* ranges. Cohesion is the term used to describe the strength of a clay sample when it is unconfined, being due to negative pressure in

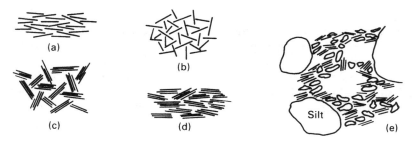

Figure 1.4 Clay structures: (a) dispersed, (b) flocculated, (c) bookhouse and (d) turbostratic; (e) example of a natural clay.

Figure 1.5 Structure of Errol Clay.

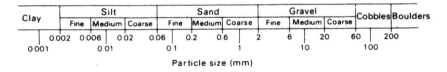

Clay	Silt			Sand			Gravel			Cobbles	Boulders
	Fine	Medium	Coarse	Fine	Medium	Coarse	Fine	Medium	Coarse		

```
        0.002  0.006   0.02    0.06    0.2    0.6      2     6    20      60    200
0.001              0.01           0.1            1          10        100
```

Particle size (mm)

Figure 1.6 Particle size ranges.

the water filling the void space, of very small size, between particles. This strength would be lost if the clay were immersed in a body of water. It should be appreciated that all clay-size particles are not necessarily clay mineral particles: the finest rock flour particles may be of clay size. If clay mineral particles are present they usually exert a considerable influence on the properties of a soil, an influence out of all proportion to their percentage by weight in the soil. Soils whose properties are influenced mainly by clay and silt size particles are referred to as *fine* soils. Those whose properties are influenced mainly by sand and gravel size particles are referred to as *coarse* soils.

1.2 PARTICLE SIZE ANALYSIS

The particle size analysis of a soil sample involves determining the percentage by mass of particles within the different size ranges. The particle size distribution of a coarse soil can be determined by the method of *sieving*. The soil sample is passed through a series of standard test sieves having successively smaller mesh sizes. The mass of soil retained in each sieve is determined and the cumulative percentage by mass passing each sieve is calculated. If fine particles are present in the soil, the sample should be treated with a deflocculating agent and washed through the sieves.

The particle size distribution of a fine soil or the fine fraction of a coarse soil can be determined by the method of *sedimentation*. This method is based on Stokes' law which governs the velocity at which spherical particles settle in a suspension: the larger the particles the greater is the settling velocity and vice versa. The law does not apply to particles smaller than 0.0002 mm, the settlement of which is influenced by Brownian movement. The size of a particle is given as the diameter of a sphere which would settle at the same velocity as the particle. Initially the soil sample is pretreated with hydrogen peroxide to remove any organic material. The sample is then made up as a suspension in distilled water to which a deflocculating agent has been added to ensure that all particles settle individually. The suspension is placed in a sedimentation tube. From Stokes' law it is possible to calculate the time, t, for particles of a certain 'size', D (the equivalent settling diameter), to settle to a specified depth in the suspension. If, after the calculated time t, a sample of the suspension is drawn off with a pipette at the specified depth below the surface, the sample will contain only particles smaller than the size D at a concentration unchanged from that at the start of sedimentation. If pipette samples are taken at the specified depth at times corresponding to other chosen particle sizes the particle size distribution can be determined from the masses of the residues. An alternative procedure to pipette sampling is the measurement of the specific gravity of the suspension by means of a special hydrometer, the specific gravity depending on the mass of soil particles in the suspension at the time of measurement. Full details of the determination of particle size distribution by both the sieving and sedimentation methods are given in BS 1377 (Part 2) [2].

The particle size distribution of a soil is presented as a curve on a semilogarithmic plot, the ordinates being the percentage by mass of particles smaller than the size given by the abscissa. The flatter the distribution curve the larger the range of particle sizes in the soil; the steeper the curve the smaller the size range. A coarse soil is described as *well graded* if there is no excess of particles in any size range and if no intermediate sizes are lacking. In general, a well-graded soil is represented by a smooth, concave distribution curve. A coarse soil is described as *poorly graded* (a) if a high proportion of the particles have sizes within narrow limits (a *uniform* soil) or (b) if particles of both large and small sizes are present but with a relatively low proportion of particles of intermediate size (a *gap-graded* or *step-graded* soil). Particle size is represented on a logarithmic scale so that two soils having the same degree of uniformity are represented by curves of the same shape regardless of their positions on the particle size distribution plot. Examples of particle size distribution curves appear in Figure 1.8. The particle size corresponding to any specified value on the 'percentage smaller' scale can be read from the particle size distribution curve. The size such that 10% of the particles are smaller than that size is denoted by D_{10}. Other sizes such as D_{30} and D_{60}

can be defined in a similar way. The size D_{10} is defined as the *effective size*. The general slope and shape of the distribution curve can be described by means of the *coefficient of uniformity* (C_U) and the *coefficient of curvature* (C_Z), defined as follows:

$$C_U = \frac{D_{60}}{D_{10}} \tag{1.1}$$

$$C_Z = \frac{D_{30}^2}{D_{60} D_{10}} \tag{1.2}$$

The higher the value of the coefficient of uniformity the larger the range of particle sizes in the soil. A well-graded soil has a coefficient of curvature between 1 and 3.

1.3   PLASTICITY OF FINE SOILS

Plasticity is an important characteristic in the case of fine soils, the term plasticity describing the ability of a soil to undergo unrecoverable deformation without cracking or crumbling. In general, depending on its water content (defined as the ratio of the mass of water in the soil to the mass of solid particles), a soil may exist in one of the liquid, plastic, semi-solid and solid states. If the water content of a soil initially in the liquid state is gradually reduced, the state will change from liquid through plastic and semi-solid, accompanied by gradually reducing volume, until the solid state is reached. The water contents at which the transitions between states occur differ from soil to soil. In the ground, most fine soils exist in the plastic state. Plasticity is due to the presence of a significant content of clay mineral particles (or organic material) in the soil. The void space between such particles is generally very small in size with the result that water is held at negative pressure by capillary tension. This produces a degree of cohesion between the particles, allowing the soil to be deformed or moulded. Adsorption of water due to the surface forces on clay mineral particles may contribute to plastic behaviour. Any decrease in water content results in a decrease in cation layer thickness and an increase in the net attractive forces between particles.

The upper and lower limits of the range of water content over which the soil exhibits plastic behaviour are defined as the *liquid limit* (w_L) and the *plastic limit* (w_P), respectively. The water content range itself is defined as the *plasticity index* (I_P), i.e.:

$$I_P = w_L - w_P$$

However, the transitions between the different states are gradual and the liquid and plastic limits must be defined arbitrarily. The natural water content (w) of a soil (adjusted to an equivalent water content of the fraction passing the 425-μm sieve) relative to the liquid and plastic limits can be represented by means of the *liquidity index* (I_L), where

$$I_L = \frac{w - w_P}{I_P}$$

The degree of plasticity of the clay-size fraction of a soil is expressed by the ratio of the plasticity index to the percentage of clay-size particles in the soil: this ratio is called the *activity*.

The transition between the semi-solid and solid states occurs at the *shrinkage limit*, defined as the water content at which the volume of the soil reaches its lowest value as it dries out.

The liquid and plastic limits are determined by means of arbitrary test procedures, fully detailed in BS 1377 (Part 2) [2]. The soil sample is dried sufficiently to enable it to be crumbled and broken up, using a mortar and a rubber pestle, without crushing individual particles; only material passing a 425-μm BS sieve is used in the tests.

The apparatus for the liquid limit test consists of a penetrometer fitted with a 30° cone of stainless steel, 35 mm long: the cone and the sliding shaft to which it is attached have a mass of 80 g. The test soil is mixed with distilled water to form a thick homogeneous paste and stored for 24 h. Some of the paste is then placed in a cylindrical metal cup, 55 mm internal diameter by 40 mm deep, and levelled off at the rim of the cup to give a smooth surface. The cone is lowered so that it just touches the surface of the soil in the cup, the cone being locked in its support at this stage. The cone is then released for a period of 5 s and its depth of penetration into the soil is measured. A little more of the soil paste is added to the cup and the test is repeated until a consistent value of penetration has been obtained. (The average of two values within 0.5 mm or of three values within 1.0 mm is taken.) The entire test procedure is repeated at least four times using the same soil sample but increasing the water content each time by adding distilled water. The penetration values should cover the range of approximately 15–25 mm, the tests proceeding from the drier to the wetter state of the soil. Cone penetration is plotted against water content and the best straight line fitting the plotted points is drawn. An example appears in Figure 1.9. The liquid limit is defined as the percentage water content (to the nearest integer) corresponding to a cone penetration of 20 mm. Also given in BS 1377 are details of the determination of liquid limit based on a single test (the one-point method), provided the cone penetration is between 15 and 25 mm.

In an alternative test for liquid limit the apparatus consists of a flat metal cup, mounted on an edge pivot: the cup rests initially on a hard rubber base. A mechanism enables the cup to be lifted to a height of 10 mm and dropped onto the base. Some of the soil paste is placed in the cup, levelled off horizontally and divided by cutting a groove, on the diameter through the pivot of the cup, using a standard grooving tool. The two halves of the soil gradually flow together as the cup is repeatedly dropped onto the base at a rate of two drops per second. The number of drops, or blows, required to close the bottom of the groove over a distance of 13 mm is recorded. Repeat determinations should be made until two successive determinations give the same number of blows. The water content of the soil in the cup is then determined. This test is also repeated at least four times, the water content of the soil paste being increased for each test; the number of blows should be within the limits of 50 and 10. Water content is plotted against the logarithm of the number of blows and the best straight line fitting the plotted points is drawn. For this test the liquid limit is defined as the water content at which 25 blows are required to close the bottom of the groove over a distance of 13 mm. Also given in BS 1377 are details of the determination of liquid limit based on a single test, provided the number of blows is between 35 and 15.

For the determination of the plastic limit the test soil is mixed with distilled water until it becomes sufficiently plastic to be moulded into a ball. Part of the soil sample (approximately 2.5 g) is formed into a thread, approximately 6 mm in diameter, between the first finger and thumb of each hand. The thread is then placed on a glass plate and rolled with the tips of the fingers of one hand until its diameter is reduced to approximately 3 mm: the rolling pressure must be uniform throughout the test. The thread is then remoulded between the fingers (the water content being reduced by the heat of the fingers) and the procedure is repeated until the thread of soil shears both longitudinally and transversely when it has been rolled to a diameter of 3 mm. The procedure is repeated using three more parts of the sample and the percentage water content of all the crumbled soil is determined as a whole. This water content (to the nearest integer) is defined as the plastic limit of the soil. The entire test is repeated using four other sub-samples and the average taken of the two values of plastic limit: the tests must be repeated if the two values differ by more than 0.5%.

1.4 SOIL DESCRIPTION AND CLASSIFICATION

It is essential that a standard language should exist for the description of soils. A comprehensive description should include the characteristics of both the soil material and the *in-situ* soil mass. Material characteristics can be determined from disturbed samples of the soil, i.e. samples having the same particle size distribution as the *in-situ* soil but in which the *in-situ* structure has not been preserved. The principal material characteristics are particle size distribution (or grading) and plasticity, from which the soil name can be deduced. Particle size distribution and plasticity properties can be determined either by standard laboratory tests or by simple visual and manual procedures. Secondary material characteristics are the colour of the soil and the shape, texture and composition of the particles. Mass characteristics should ideally be determined in the field but in many cases they can be detected in undisturbed samples, i.e. samples in which the *in-situ* soil structure has been essentially preserved. A description of mass characteristics should include an assessment of *in-situ* compactive state (coarse soils) or stiffness (fine soils) and details of any bedding, discontinuities and weathering. The arrangement of minor geological details, referred to as the soil macro-fabric, should be carefully described, as this can influence the engineering behaviour of the *in-situ* soil to a considerable extent. Examples of macro-fabric features are thin layers of fine sand and silt in clay, silt-filled fissures in clay, small lenses of clay in sand, organic inclusions and root holes. The name of the geological formation, if definitely known, should be included in the description; in addition, the type of deposit may be stated (e.g. till, alluvium, river terrace), as this can indicate, in a general way, the likely behaviour of the soil.

It is important to distinguish between soil description and soil classification. Soil description includes details of both material and mass characteristics, and therefore it is unlikely that any two soils will have identical descriptions. In soil classification, on the other hand, a soil is allocated to one of a limited number of groups on the basis of material characteristics only. Soil classification is thus independent of the *in-situ* condition of the soil mass. If the soil is to be employed in its undisturbed condition, for example to support a foundation, a full soil description will be adequate and the addition of the soil classification is discretionary. However, classification is particularly useful if the soil in question is to be used as a construction material, for example in an embankment. Engineers can also draw on past experience of the behaviour of soils of similar classification.

Rapid assessment procedures

Both soil description and classification require a knowledge of grading and plasticity. This can be determined by the full laboratory procedure using standard tests, as described in Sections 1.2 and 1.3, in which values defining the particle size distribution and the liquid and plastic limits are obtained for the soil in question. Alternatively, grading and plasticity can be assessed using a rapid procedure which involves personal judgements based on the appearance and feel of the soil. The rapid procedure can be used in the field and in other situations where the use of the laboratory procedure is not possible or not justified. In the rapid procedure the following indicators should be used.

Particles of 0.06 mm, the lower size limit for coarse soils, are just visible to the naked eye and feel harsh but not gritty when rubbed between the fingers; finer material feels smooth to the touch. The size boundary between sand and gravel is 2 mm and this represents the largest size of particles which will hold together by capillary attraction when moist. A purely visual judgement must be made as to whether the sample is well graded or poorly graded, this being more difficult for sands than for gravels.

If a predominantly coarse soil contains a significant proportion of fine material it is important to know whether the fines are essentially plastic or non-plastic (i.e. whether the fines are predominantly clay or silt respectively). This can be judged by the extent to which the soil exhibits cohesion and plasticity. A small quantity of the soil, with the largest particles removed, should be moulded together in the hands, adding water if necessary. Cohesion is indicated if the soil, at an appropriate water content, can be moulded into a relatively firm mass. Plasticity is indicated if the soil can be deformed without cracking or crumbling, i.e. without losing cohesion. If cohesion and plasticity are pronounced then the fines are plastic. If cohesion and plasticity are absent or only weakly indicated then the fines are essentially non-plastic.

The plasticity of fine soils can be assessed by means of the toughness and dilatancy tests, described below. An assessment of dry strength may also be useful. Any coarse particles, if present, are first removed, and then a small sample of the soil is moulded in the hand to a consistency judged to be just above the plastic limit; water is added or the soil is allowed to dry as necessary. The procedures are then as follows.

Toughness test

A small piece of soil is rolled out into a thread on a flat surface or on the palm of the hand, moulded together and rolled out again until it has dried sufficiently to break into lumps at a diameter of around 3 mm. In this condition, inorganic clays of high liquid limit are fairly stiff and tough; those of low liquid limit are softer and crumble more easily. Inorganic silts produce a weak and often soft thread which may be difficult to form and readily breaks and crumbles.

Dilatancy test

A pat of soil, with sufficient water added to make it soft but not sticky, is placed in the open (horizontal) palm of the hand. The side of the hand is then struck against the other hand several times. Dilatancy is indicated by the appearance of a shiny film of water on the surface of the pat; if the pat is then squeezed or pressed with the fingers the surface becomes dull as the pat stiffens and eventually crumbles. These reactions are pronounced only for predominantly silt size material and for very fine sands. Plastic clays give no reaction.

Dry strength test

A pat of soil about 6 mm thick is allowed to dry completely, either naturally or in an oven. The strength of the dry soil is then assessed by breaking and crumbling between the fingers. Inorganic clays have relatively high dry strength; the greater the strength the higher the liquid limit. Inorganic silts of low liquid limit have little or no dry strength, crumbling easily between the fingers.

Organic soils contain a significant proportion of dispersed vegetable matter which usually produces a distinctive odour and often a dark brown, dark grey or bluish grey colour. Peats consist predominantly of plant remains, usually dark brown or black in colour and with a distinctive odour. If the plant remains are recognizable and retain some strength the peat is described as fibrous. If the plant remains are recognizable but their strength has been lost they are pseudo-fibrous. If recognizable plant remains are absent, the peat is described as amorphous.

Soil description details

A detailed guide to soil description is given in BS 5930 [3]. According to this standard the basic soil types are boulders, cobbles, gravel, sand, silt and clay, defined in terms of the particle size ranges shown in Figure 1.6; added to these are organic clay, silt or sand, and peat. These names are always written in capital letters in a soil description. Mixtures of the basic soil types are referred to as composite types.

A soil is of basic type sand or gravel (these being termed coarse soils) if, after the removal of any cobbles or boulders, over 65% of the material is of sand and gravel sizes. A soil is of basic type silt or clay (termed fine soils) if, after the removal of any cobbles or boulders, over 35% of the material is of silt and clay sizes. However, these percentages should be considered as approximate guidelines, not forming a rigid boundary. Sand and gravel may each be subdivided into coarse, medium and fine fractions as defined in Figure 1.6. The state of sand and gravel can be described as well graded, poorly graded, uniform or gap graded, as defined in Section 1.2. In the case of gravels, particle shape (angular, sub-angular, sub-rounded, rounded, flat, elongated) and surface texture (rough, smooth, polished) can be described if necessary. Particle composition can also be stated. Gravel particles are usually rock fragments (e.g. sandstone, schist). Sand particles usually consist of individual mineral grains (e.g. quartz, feldspar). Fine soils should be described as either silt or clay: terms such as silty clay should not be used.

Composite types of coarse soil are named in Table 1.1, the predominant component being written in capital letters. Fine soils containing 35–65% coarse material are described as sandy and/or gravelly SILT (or CLAY). Deposits containing over 50% of boulders and cobbles are referred to as very coarse and normally can be described only in excavations and exposures. Mixes of very coarse material with finer soils can be described by combining the descriptions of the two components, e.g. COBBLES with some FINER MATERIAL (sand); gravelly SAND with occasional BOULDERS.

Table 1.1 Composite types of coarse soil

Slightly sandy GRAVEL	Up to 5% sand
Sandy GRAVEL	5–20% sand
Very sandy GRAVEL	Over 20% sand
SAND and GRAVEL	About equal proportions
Very gravelly SAND	Over 20% gravel
Gravelly SAND	5–20% gravel
Slightly gravelly SAND	Up to 5% gravel
Slightly silty SAND (and/or GRAVEL)	Up to 5% silt
Silty SAND (and/or GRAVEL)	5–20% silt
Very silty SAND (and/or GRAVEL)	Over 20% silt
Slightly clayey SAND (and/or GRAVEL)	Up to 5% clay
Clayey SAND (and/or GRAVEL)	5–20% clay
Very clayey SAND (and/or GRAVEL)	Over 20% clay

Notes
Terms such as 'Slightly clayey gravelly SAND' (having less than 5% clay and gravel)
and 'Silty sandy GRAVEL' (having 5–20% silt and sand) can be used, based on the
above proportions of secondary constituents.

The state of compaction or stiffness of the *in-situ* soil can be assessed by means of
the tests or indications detailed in Table 1.2.

Discontinuities such as fissures and shear planes, including their spacings, should be
indicated. Bedding features, including their thickness, should be detailed. Alternating
layers of varying soil types or with bands or lenses of other materials are described as
interstratified. Layers of different soil types are described as interbedded or inter-
laminated, their thickness being stated. Bedding surfaces that separate easily are referred
to as partings. If partings incorporate other material, this should be described.

Table 1.2 Compactive state and stiffness of soils

Soil group	Term	Field test or indication
Coarse soils	Very loose	Assessed on basis of N value determined by
	Loose	means of standard penetration test – see
	Medium dense	Chapter 8 and Table 8.3
	Dense	
	Very dense	
	Slightly cemented	Visual examination: pick removes soil in lumps which can be abraded
Fine soils	Uncompact	Easily moulded or crushed by the fingers
	Compact	Can be moulded or crushed by strong finger pressure
	Very soft	Finger can easily be pushed in up to 25 mm
	Soft	Finger can be pushed in up to 10 mm
	Firm	Thumb can make impression easily
	Stiff	Thumb can make slight indentation
	Very stiff	Thumb nail can make indentation
	Hard	Thumb nail can make surface scratch
Organic soils	Firm	Fibres already pressed together
	Spongy	Very compressible and open structure
	Plastic	Can be moulded in the hand and smears fingers

Some examples of soil description are as follows.

> Dense, reddish-brown, sub-angular, well-graded SAND
> Firm, grey, laminated CLAY with occasional silt partings 0.5–2.0 mm (Alluvium)
> Dense, brown, well graded, very silty SAND and GRAVEL with some
>   COBBLES (Till)
> Stiff, brown, closely fissured CLAY (London Clay)
> Spongy, dark brown, fibrous PEAT (Recent Deposits).

Soil classification systems

General classification systems in which soils are placed into groups on the basis of grading and plasticity have been used for many years. The feature of these systems is that each soil group is denoted by a letter symbol representing main and qualifying terms. The terms and letters used in the UK are detailed in Table 1.3. The boundary between coarse and fine soils is generally taken to be 35% fines (i.e. particles smaller than 0.06 mm). The liquid and plastic limits are used to classify fine soils, employing the plasticity chart shown in Figure 1.7. The axes of the plasticity chart are plasticity index and liquid limit; therefore, the plasticity characteristics of a particular soil can be represented by a point on the chart. Classification letters are allotted to the soil according to the zone within which the point lies. The chart is divided into five ranges of liquid limit. The four ranges I, H, V and E can be combined as an upper range (U) if closer designation is not required or if the rapid assessment procedure has been used to assess plasticity. The diagonal line on the chart, known as the *A-line*, should not be regarded as a rigid boundary between clay and silt for purposes of soil *description*, as opposed to classification. A similar classification system was developed in the US [10] but with less detailed subdivisions.

The letter denoting the dominant size fraction is placed first in the group symbol. If a soil has a significant content of organic matter the suffix O is added as the last letter of the group symbol. A group symbol may consist of two or more letters, for example:

> SW – well-graded SAND
> SCL – very clayey SAND (clay of low plasticity)
> CIS – sandy CLAY of intermediate plasticity
> MHSO – organic sandy SILT of high plasticity.

The name of the soil group should always be given, as above, in addition to the symbol, the extent of subdivision depending on the particular situation. If the rapid procedure has been used to assess grading and plasticity the group symbol should be enclosed in brackets to indicate the lower degree of accuracy associated with this procedure.

The term FINE SOIL or FINES (F) is used when it is not required, or not possible, to differentiate between SILT (M) and CLAY (C). SILT (M) plots below the A-line and CLAY (C) above the A-line on the plasticity chart, i.e. silts exhibit plastic properties over a lower range of water content than clays having the same liquid limit. SILT or CLAY is qualified as gravelly if more than 50% of the coarse fraction is of gravel

Table 1.3

Main terms		Qualifying terms	
GRAVEL	G	Well graded	W
SAND	S	Poorly graded	P
		Uniform	Pu
		Gap graded	Pg
FINE SOIL, FINES	F	Of low plasticity ($w_L < 35$)	L
SILT (M-SOIL)	M	Of intermediate plasticity (w_L: 35–50)	I
CLAY	C	Of high plasticity (w_L: 50–70)	H
		Of very high plasticity (w_L: 70–90)	V
		Of extremely high plasticity ($w_L > 90$)	E
		Of upper plasticity range ($w_L > 35$)	U
PEAT	Pt	Organic (may be a suffix to any group)	O

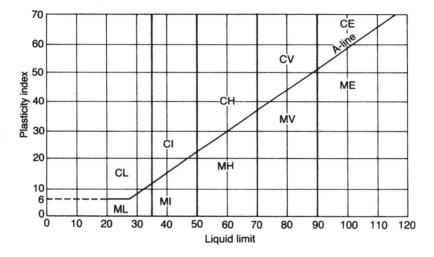

Figure 1.7 Plasticity chart: British system (BS 5930: 1999).

size and as sandy if more than 50% of the coarse fraction is of sand size. The alternative term M-SOIL is introduced to describe material which, regardless of its particle size distribution, plots below the A-line on the plasticity chart: the use of this term avoids confusion with soils of predominantly silt size (but with a significant proportion of clay-size particles) which plot above the A-line. Fine soils containing significant amounts of organic matter usually have high to extremely high liquid limits and plot below the A-line as organic silt. Peats usually have very high or extremely high liquid limits.

Any cobbles or boulders (particles retained on a 63-mm BS sieve) are removed from the soil before the classification tests are carried out but their percentages in the total sample should be determined or estimated. Mixtures of soil and cobbles or boulders can be indicated by using the letters Cb (COBBLES) or B (BOULDERS) joined by a + sign to the group symbol for the soil, the dominant component being given first, for example:

GW + Cb – well-graded GRAVEL with COBBLES
B + CL – BOULDERS with CLAY of low plasticity.

A general classification system is useful in developing an understanding of the nature of different soil types. However, in practice it is more appropriate to use systems based on properties related to the suitability of soils for use in specific construction situations. For example, the UK Department of Transport [6] has detailed a classification system for the acceptability of soils for use in earthworks in highway construction. In this system, soils are allocated to classes based mainly on grading and plasticity but, for some classes, chemical, compaction and strength characteristics are also specified.

Example 1.1

The results of particle size analyses of four soils A, B, C and D are shown in Table 1.4. The results of limit tests on soil D are:

Liquid limit:
Cone penetration (mm)   15.5   18.0   19.4   22.2   24.9
Water content (%)   39.3   40.8   42.1   44.6   45.6
Plastic limit:
Water content (%)   23.9   24.3

The fine fraction of soil C has a liquid limit of 26 and a plasticity index of 9. (a) Determine the coefficients of uniformity and curvature for soils A, B and C. (b) Allot group symbols, with main and qualifying terms to each soil.

Table 1.4

BS sieve	Particle size*	Percentage smaller			
		Soil A	Soil B	Soil C	Soil D
63 mm		100		100	
20 mm		64		76	
6.3 mm		39	100	65	
2 mm		24	98	59	
600 μm		12	90	54	
212 μm		5	9	47	100
63 μm		0	3	34	95
	0.020 mm			23	69
	0.006 mm			14	46
	0.002 mm			7	31

Note
* From sedimentation test.

The particle size distribution curves are plotted in Figure 1.8. For soils A, B and C the sizes D_{10}, D_{30} and D_{60} are read from the curves and the values of C_U and C_Z are calculated:

Soil	D_{10}	D_{30}	D_{60}	C_U	C_Z
A	0.47	3.5	16	34	1.6
B	0.23	0.30	0.41	1.8	0.95
C	0.003	0.042	2.4	800	0.25

For soil D the liquid limit is obtained from Figure 1.9, in which cone penetration is plotted against water content. The percentage water content, to the nearest integer, corresponding to a penetration of 20 mm is the liquid limit and is 42. The plastic limit is the average of the two percentage water contents, again to the nearest integer, i.e. 24. The plasticity index is the difference between the liquid and plastic limits, i.e. 18.

Soil A consists of 100% coarse material (76% gravel size; 24% sand size) and is classified as GW: well-graded, very sandy GRAVEL.

Soil B consists of 97% coarse material (95% sand size; 2% gravel size) and 3% fines. It is classified as SPu: uniform, slightly silty, medium SAND.

Soil C comprises 66% coarse material (41% gravel size; 25% sand size) and 34% fines ($w_L = 26$, $I_P = 9$, plotting in the CL zone on the plasticity chart). The classification is GCL: very clayey GRAVEL (clay of low plasticity). This is a till, a glacial deposit having a large range of particle sizes.

Soil D contains 95% fine material: the liquid limit is 42 and the plasticity index is 18, plotting just above the A-line in the CI zone on the plasticity chart. The classification is thus CI: CLAY of intermediate plasticity.

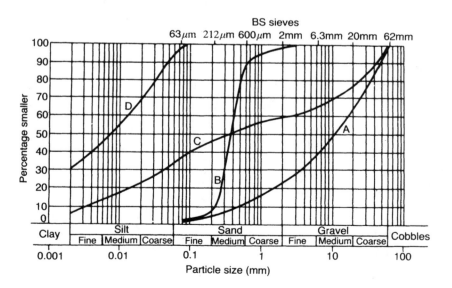

Figure 1.8 Particle size distribution curves (Example 1.1).

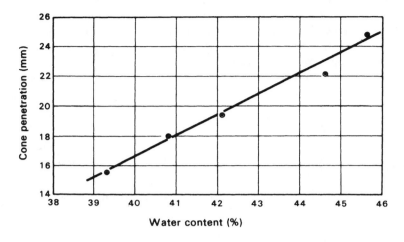

Figure 1.9 Determination of liquid limit.

1.5 PHASE RELATIONSHIPS

Soils can be of either two-phase or three-phase composition. In a completely dry soil there are two phases, namely the solid soil particles and pore air. A fully saturated soil is also two phase, being composed of solid soil particles and pore water. A partially saturated soil is three phase, being composed of solid soil particles, pore water and pore air. The components of a soil can be represented by a phase diagram as shown in Figure 1.10(a). The following relationships are defined with reference to Figure 1.10(a).

The *water content* (*w*), or *moisture content* (*m*), is the ratio of the mass of water to the mass of solids in the soil, i.e.

$$w = \frac{M_{\mathrm{w}}}{M_{\mathrm{s}}} \tag{1.3}$$

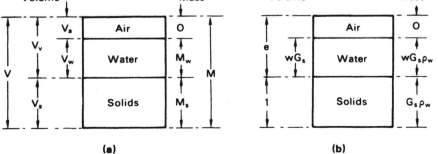

Figure 1.10 Phase diagrams.

The water content is determined by weighing a sample of the soil and then drying the sample in an oven at a temperature of 105–110 °C and reweighing. Drying should continue until the differences between successive weighings at four-hourly intervals are not greater than 0.1% of the original mass of the sample. A drying period of 24 h is normally adequate for most soils. (See BS 1377.)

The *degree of saturation* (S_r) is the ratio of the volume of water to the total volume of void space, i.e.

$$S_r = \frac{V_w}{V_v} \qquad (1.4)$$

The degree of saturation can range between the limits of zero for a completely dry soil and 1 (or 100%) for a fully saturated soil.

The *void ratio* (*e*) is the ratio of the volume of voids to the volume of solids, i.e.

$$e = \frac{V_v}{V_s} \qquad (1.5)$$

The *porosity* (*n*) is the ratio of the volume of voids to the total volume of the soil, i.e.

$$n = \frac{V_v}{V} \qquad (1.6)$$

The void ratio and the porosity are inter-related as follows:

$$e = \frac{n}{1-n} \qquad (1.7)$$

$$n = \frac{e}{1+e} \qquad (1.8)$$

The *specific volume* (*v*) is the total volume of soil which contains unit volume of solids, i.e.

$$v = 1 + e \qquad (1.9)$$

The *air content* or *air voids* (*A*) is the ratio of the volume of air to the total volume of the soil, i.e.

$$A = \frac{V_a}{V} \qquad (1.10)$$

The *bulk density* (ρ) of a soil is the ratio of the total mass to the total volume, i.e.

$$\rho = \frac{M}{V} \qquad (1.11)$$

Convenient units for density are kg/m^3 or Mg/m^3. The density of water (1000 kg/m^3 or 1.00 Mg/m^3) is denoted by ρ_w.

The specific gravity of the *soil particles* (G_s) is given by

$$G_s = \frac{M_s}{V_s \rho_w} = \frac{\rho_s}{\rho_w} \tag{1.12}$$

where ρ_s is the *particle density*. Procedures for determining the value of particle density are detailed in BS 1377 (Part 2) [2]. If the units of ρ_s are Mg/m^3 then ρ_s and G_s are numerically equal. Particle density is used in preference to G_s in British Standards but it is advantageous to use G_s (which is dimensionless) in deriving relationships from the phase diagram.

From the definition of void ratio, if the volume of solids is 1 unit then the volume of voids is e units. The mass of solids is then $G_s \rho_w$ and, from the definition of water content, the mass of water is $w G_s \rho_w$. The volume of water is thus $w G_s$. These volumes and masses are represented in Figure 1.10(b). The following relationships can now be obtained.

The degree of saturation can be expressed as

$$S_r = \frac{w G_s}{e} \tag{1.13}$$

In the case of a fully saturated soil, $S_r = 1$; hence

$$e = w G_s \tag{1.14}$$

The air content can be expressed as

$$A = \frac{e - w G_s}{1 + e} \tag{1.15}$$

or, from Equations 1.8 and 1.13,

$$A = n(1 - S_r) \tag{1.16}$$

The bulk density of a soil can be expressed as

$$\rho = \frac{G_s(1 + w)}{1 + e} \rho_w \tag{1.17}$$

or, from Equation 1.13,

$$\rho = \frac{G_s + S_r e}{1 + e} \rho_w \tag{1.18}$$

For a fully saturated soil ($S_r = 1$)

$$\rho_{sat} = \frac{G_s + e}{1 + e} \rho_w \tag{1.19}$$

For a completely dry soil ($S_r = 0$)

$$\rho_d = \frac{G_s}{1 + e} \rho_w \tag{1.20}$$

The *unit weight* (γ) of a soil is the ratio of the total weight (a force) to the total volume, i.e.

$$\gamma = \frac{W}{V} = \frac{Mg}{V}$$

Equations similar to 1.17–1.20 apply in the case of unit weights, for example

$$\gamma = \frac{G_s(1 + w)}{1 + e} \gamma_w \tag{1.17a}$$

$$\gamma = \frac{G_s + S_r e}{1 + e} \gamma_w \tag{1.18a}$$

where γ_w is the unit weight of water. Convenient units are kN/m^3, the unit weight of water being $9.8\,kN/m^3$ (or $10.0\,kN/m^3$ in the case of sea water).

When a soil *in situ* is fully saturated the solid soil particles (volume: 1 unit, weight: $G_s\gamma_w$) are subjected to upthrust (γ_w). Hence, the *buoyant unit weight* (γ') is given by

$$\gamma' = \frac{G_s\gamma_w - \gamma_w}{1 + e} = \frac{G_s - 1}{1 + e} \gamma_w \tag{1.21}$$

i.e.

$$\gamma' = \gamma_{sat} - \gamma_w \tag{1.22}$$

In the case of sands and gravels the *density index* (I_D) is used to express the relationship between the *in-situ* void ratio (e), or the void ratio of a sample, and the limiting values e_{max} and e_{min}. The density index (the term 'relative density' is also used) is defined as

$$I_D = \frac{e_{max} - e}{e_{max} - e_{min}} \tag{1.23}$$

Thus, the density index of a soil in its densest possible state ($e = e_{min}$) is 1 (or 100%) and the density index in its loosest possible state ($e = e_{max}$) is 0.

The maximum density is determined by compacting a sample underwater in a mould, using a circular steel tamper attached to a vibrating hammer: a 1-1 mould is used for sands and a 2.3-1 mould for gravels. The soil from the mould is then dried in an oven, enabling the dry density to be determined. The minimum dry density can be determined by one of the following procedures. In the case of sands, a 1-1 measuring cylinder is partially filled with a dry sample of mass 1000 g and the top of the cylinder closed with a rubber stopper. The minimum density is achieved by shaking and inverting the cylinder several times, the resulting volume being read from the graduations on the cylinder. In the case of gravels, and sandy gravels, a sample is poured from a height of about 0.5 m into a 2.3-1 mould and the resulting dry density determined. Full details of the above tests are given in BS 1377 (Part 4) [2]. Void ratio can be calculated from a value of dry density using Equation 1.20. However, the density index can be calculated directly from the maximum, minimum and *in-situ* values of dry density, avoiding the need to know the value of G_s.

Example 1.2

In its natural condition a soil sample has a mass of 2290 g and a volume of $1.15 \times 10^{-3}\,\mathrm{m^3}$. After being completely dried in an oven the mass of the sample is 2035 g. The value of G_s for the soil is 2.68. Determine the bulk density, unit weight, water content, void ratio, porosity, degree of saturation and air content.

$$\text{Bulk density,}\quad \rho = \frac{M}{V} = \frac{2.290}{1.15 \times 10^{-3}} = 1990\,\mathrm{kg/m^3} \quad (1.99\,\mathrm{Mg/m^3})$$

$$\text{Unit weight,}\quad \gamma = \frac{Mg}{V} = 1990 \times 9.8 = 19\,500\,\mathrm{N/m^3}$$

$$= 19.5\,\mathrm{kN/m^3}$$

$$\text{Water content,}\quad w = \frac{M_w}{M_s} = \frac{2290 - 2035}{2035} = 0.125 \text{ or } 12.5\%$$

From Equation 1.17,

$$\text{Void ratio,}\quad e = G_s(1 + w)\frac{\rho_w}{\rho} - 1$$

$$= \left(2.68 \times 1.125 \times \frac{1000}{1990}\right) - 1$$

$$= 1.52 - 1$$

$$= 0.52$$

$$\text{Porosity,}\quad n = \frac{e}{1+e} = \frac{0.52}{1.52} = 0.34 \text{ or } 34\%$$

$$\text{Degree of saturation,}\quad S_r = \frac{wG_s}{e} = \frac{0.125 \times 2.68}{0.52} = 0.645 \text{ or } 64.5\%$$

$$\text{Air content,}\quad A = n(1 - S_r) = 0.34 \times 0.355$$

$$= 0.121 \text{ or } 12.1\%$$

1.6 SOIL COMPACTION

Compaction is the process of increasing the density of a soil by packing the particles closer together with a reduction in the volume of *air*; there is no significant change in the volume of water in the soil. In the construction of fills and embankments, loose soil is placed in layers ranging between 75 and 450 mm in thickness, each layer being compacted to a specified standard by means of rollers, vibrators or rammers. In general, the higher the degree of compaction the higher will be the shear strength and the lower will be the compressibility of the soil. An *engineered fill* is one in which the soil has been selected, placed and compacted to an appropriate specification with the object of achieving a particular engineering performance, generally based on past experience. The aim is to ensure that the resulting fill possesses properties that are adequate for the function of the fill. This is in contrast to non-engineered fills which have been placed without regard to a subsequent engineering function.

The degree of compaction of a soil is measured in terms of dry density, i.e. the mass of solids only per unit volume of soil. If the bulk density of the soil is ρ and the water content w, then from Equations 1.17 and 1.20 it is apparent that the dry density is given by

$$\rho_{\mathrm{d}} = \frac{\rho}{1 + w} \tag{1.24}$$

The dry density of a given soil after compaction depends on the water content and the energy supplied by the compaction equipment (referred to as the *compactive effort*).

The compaction characteristics of a soil can be assessed by means of standard laboratory tests. The soil is compacted in a cylindrical mould using a standard compactive effort. In BS 1377 (Part 4) [2] three compaction procedures are detailed. In the *Proctor* test the volume of the mould is 1 l and the soil (with all particles larger than 20 mm removed) is compacted by a rammer consisting of a 2.5-kg mass falling freely through 300 mm: the soil is compacted in three equal layers, each layer receiving 27 blows with the rammer. In the *modified AASHTO* test the mould is the same as is used in the above test but the rammer consists of a 4.5-kg mass falling 450 mm: the soil (with all particles larger than 20 mm removed) is compacted in five layers, each layer receiving 27 blows with the rammer. If the sample contains a limited proportion of particles up to 37.5 mm in size, a 2.3-1 mould should be used, each layer receiving 62 blows with either the 2.5- or 4.5-kg rammer. In the *vibrating hammer* test, the soil (with all particles larger than 37.5 mm removed) is compacted in three layers in a 2.3-1 mould, using a circular tamper fitted in the vibrating hammer, each layer being compacted for a period of 60 s.

After compaction using one of the three standard methods, the bulk density and water content of the soil are determined and the dry density calculated. For a given soil the process is repeated at least five times, the water content of the sample being increased each time. Dry density is plotted against water content and a curve of the form shown in Figure 1.11 is obtained. This curve shows that for a particular method of compaction (i.e. a particular compactive effort) there is a particular value of water content, known as the *optimum water content* (w_{opt}), at which a maximum value of dry density is obtained. At low values of water content most soils tend to be stiff and are difficult to compact. As the water content is increased the soil becomes more workable, facilitating compaction and resulting in higher dry densities. At high water contents, however, the dry density decreases with increasing water content, an increasing proportion of the soil volume being occupied by water.

If all the air in a soil could be expelled by compaction the soil would be in a state of full saturation and the dry density would be the maximum possible value for the given

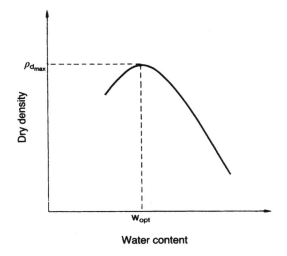

Figure 1.11 Dry density–water content relationship.

water content. However, this degree of compaction is unattainable in practice. The maximum possible value of dry density is referred to as the 'zero air voids' dry density or the saturation dry density and can be calculated from the expression:

$$\rho_d = \frac{G_s}{1 + wG_s}\rho_w \tag{1.25}$$

In general, the dry density after compaction at water content w to an air content A can be calculated from the following expression, derived from Equations 1.15 and 1.20:

$$\rho_d = \frac{G_s(1 - A)}{1 + wG_s}\rho_w \tag{1.26}$$

The calculated relationship between zero air voids dry density and water content (for $G_s = 2.65$) is shown in Figure 1.12; the curve is referred to as the zero air voids line or the saturation line. The experimental dry density–water content curve for a particular compactive effort must lie completely to the left of the zero air voids line. The curves relating dry density at air contents of 5 and 10% with water content are also shown in Figure 1.12, the values of dry density being calculated from Equation 1.26. These curves enable the air content at any point on the experimental dry density–water content curve to be determined by inspection.

For a particular soil, different dry density–water content curves are obtained for different compactive efforts. Curves representing the results of tests using the 2.5- and 4.5-kg rammers are shown in Figure 1.12. The curve for the 4.5-kg test is situated above and to the left of the curve for the 2.5-kg test. Thus, a higher compactive effort results in a higher value of maximum dry density and a lower value of optimum water content; however, the values of air content at maximum dry density are approximately equal.

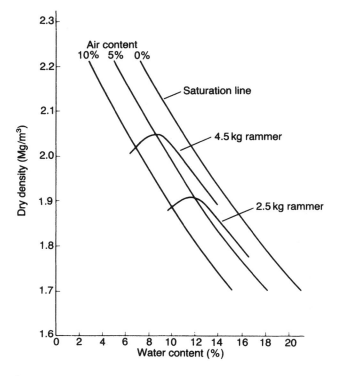

Figure 1.12 Dry density–water content curves for different compactive efforts.

The dry density–water content curves for a range of soil types using the same compactive effort (the BS 2.5-kg rammer) are shown in Figure 1.13. In general, coarse soils can be compacted to higher dry densities than fine soils.

Field compaction

The results of laboratory compaction tests are not directly applicable to field compaction because the compactive efforts in the laboratory tests are different, and are applied in a different way, from those produced by field equipment. Further, the laboratory tests are carried out only on material smaller than either 20 or 37.5 mm. However, the maximum dry densities obtained in the laboratory using the 2.5- and 4.5-kg rammers cover the range of dry density normally produced by field compaction equipment.

A minimum number of passes must be made with the chosen compaction equipment to produce the required value of dry density. This number, which depends on the type and mass of the equipment and on the thickness of the soil layer, is usually within the range 3–12. Above a certain number of passes no significant increase in dry density is obtained. In general, the thicker the soil layer the heavier the equipment required to produce an adequate degree of compaction.

There are two approaches to the achievement of a satisfactory standard of compaction in the field, known as *method* and *end-product* compaction. In method compaction

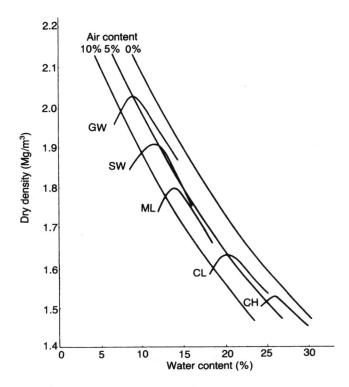

Figure 1.13 Dry density–water content curves for a range of soil types.

the type and mass of equipment, the layer depth and the number of passes are specified. In the UK these details are given, for the class of material in question, in the Specification for Highway Works [6]. In end-product compaction the required dry density is specified: the dry density of the compacted fill must be equal to or greater than a stated percentage of the maximum dry density obtained in one of the standard laboratory compaction tests. Method compaction is used in most earthworks. End-product compaction is normally restricted to pulverized fuel ash in general fill and to certain selected fills.

Field density tests can be carried out, if considered necessary, to verify the standard of compaction in earthworks, dry density or air content being calculated from measured values of bulk density and water content. A number of methods of measuring bulk density in the field are detailed in BS 1377 (Part 4) [2].

The following types of compaction equipment are used in the field.

Smooth-wheeled rollers

These consist of hollow steel drums, the mass of which can be increased by water or sand ballast. They are suitable for most types of soil except uniform sands and silty sands, provided a mixing or kneading action is not required. A smooth surface is produced on the compacted layer, encouraging the run-off of any rainfall but resulting in relatively poor bonding between successive layers; the fill as a whole will therefore tend to be laminated. Smooth-wheeled rollers, and the other types of roller described below, can be either towed or self-propelled.

Pneumatic-tyred rollers

This equipment is suitable for a wide range of coarse and fine soils but not for uniformly graded material. Wheels are mounted close together on two axles, the rear set overlapping the lines of the front set to ensure complete coverage of the soil surface. The tyres are relatively wide with a flat tread so that the soil is not displaced laterally. This type of roller is also available with a special axle which allows the wheels to wobble, thus preventing the bridging over of low spots. Pneumatic-tyred rollers impart a kneading action to the soil. The finished surface is relatively smooth, resulting in a low degree of bonding between layers. If good bonding is essential, the compacted surface must be scarified between layers. Increased compactive effort can be obtained by increasing the tyre inflation pressure or, less effectively, by adding kentledge to the body of the roller.

Sheepsfoot rollers

This type of roller consists of hollow steel drums with numerous tapered or club-shaped feet projecting from their surfaces. The mass of the drums can be increased by ballasting. The arrangement of the feet can vary but they are usually from 200 to 250 mm in length with an end area of 40–65 cm^2. The feet thus impart a relatively high pressure over a small area. Initially, when the layer of soil is loose, the drums are in contact with the soil surface. Subsequently, as the projecting feet compact below the surface and the soil becomes sufficiently dense to support the high contact pressure, the drums rise above the soil. Sheepsfoot rollers are most suitable for fine soils, both plastic and non-plastic, especially at water contents dry of optimum. They are also suitable for coarse soils with more than 20% of fines. The action of the feet causes significant mixing of the soil, improving its degree of homogeneity, and will break up lumps of stiff material. Due to the penetration of the feet, excellent bonding is produced between successive soil layers, an important requirement for water-retaining earthworks. *Tamping* rollers are similar to sheepsfoot rollers but the feet have a larger end area, usually over 100 cm^2, and the total area of the feet exceeds 15% of the surface area of the drums.

Grid rollers

These rollers have a surface consisting of a network of steel bars forming a grid with square holes. Kentledge can be added to the body of the roller. Grid rollers provide high contact pressure but little kneading action and are suitable for most coarse soils.

Vibratory rollers

These are smooth-wheeled rollers fitted with a power-driven vibration mechanism. They are used for most soil types and are more efficient if the water content of the soil is slightly wet of optimum. They are particularly effective for coarse soils with little or no fines. The mass of the roller and the frequency of vibration must be matched to the soil type and layer thickness. The lower the speed of the roller the fewer the number of passes required.

Vibrating plates

This equipment, which is suitable for most soil types, consists of a steel plate with upturned edges, or a curved plate, on which a vibrator is mounted. The unit, under manual guidance, propels itself slowly over the surface of the soil.

Power rammers

Manually controlled power rammers, generally petrol-driven, are used for the compaction of small areas where access is difficult or where the use of larger equipment would not be justified. They are also used extensively for the compaction of backfill in trenches. They do not operate effectively on uniformly graded soils.

Moisture condition test

As an alternative to standard compaction tests, the moisture condition test is widely used in the UK. This test, developed by the Transport and Road Research Laboratory [8], enables a rapid assessment to be made of the suitability of soils for use as fill materials. The test does not involve the determination of water content, a cause of delay in obtaining the results of compaction tests. In principle, the test consists of determining the effort required to compact a soil sample (normally 1.5 kg) close to its maximum density. The soil is compacted in a cylindrical mould having an internal diameter of 100 mm centred on the base plate of the apparatus. Compaction is imparted by a rammer having a diameter of 97 mm and a mass of 7 kg falling freely from a height of 250 mm. The fall of the rammer is controlled by an adjustable release mechanism and two vertical guide rods. The penetration of the rammer into the mould is measured by means of a scale on the side of the rammer. A fibre disc is placed on top of the soil to prevent extrusion between the rammer and the inside of the mould. Full details are given in BS 1377 (Part 4) [2].

The penetration is measured at various stages of compaction. For a given number of rammer blows (n) the penetration is subtracted from the penetration at four times that number of blows ($4n$). The change in penetration between n and $4n$ blows is plotted against the logarithm (to base 10) of the lesser number of blows (n). A change in penetration of 5 mm is arbitrarily chosen to represent the condition beyond which no significant increase in density occurs. The *moisture condition value* (MCV) is defined as 10 times the logarithm of the number of blows corresponding to a change in penetration of 5 mm on the above plot. An example of such a plot is shown in Figure 1.14. For a range of soil types it has been shown that the relationship between water content and MCV is linear over a substantial range of water content.

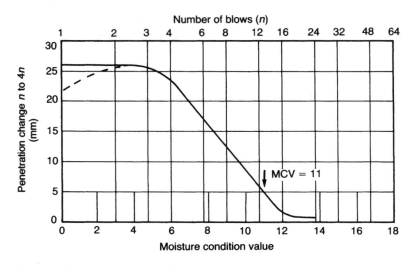

Figure 1.14 Moisture condition test.

PROBLEMS

1.1   The results of particle size analyses and, where appropriate, limit tests on samples of four soils are given in Table 1.5. Allot group symbols and give main and qualifying terms appropriate for each soil.

1.2   A soil has a bulk density of 1.91 Mg/m³ and a water content of 9.5%. The value of G_s is 2.70. Calculate the void ratio and degree of saturation of the soil. What would be the values of density and water content if the soil were fully saturated at the same void ratio?

Table 1.5

BS sieve	Particle size	Percentage smaller			
		Soil E	Soil F	Soil G	Soil H
63 mm					
20 mm		100			
6.3 mm		94	100		
2 mm		69	98		
600 μm		32	88	100	
212 μm		13	67	95	100
63 μm		2	37	73	99
	0.020 mm		22	46	88
	0.006 mm		11	25	71
	0.002 mm		4	13	58
Liquid limit			Non-plastic	32	78
Plastic limit				24	31

1.3   Calculate the dry unit weight, the saturated unit weight and the buoyant unit weight of a soil having a void ratio of 0.70 and a value of G_s of 2.72. Calculate also the unit weight and water content at a degree of saturation of 75%.

1.4   A soil specimen is 38 mm in diameter and 76 mm long and in its natural condition weighs 168.0 g. When dried completely in an oven the specimen weighs 130.5 g. The value of G_s is 2.73. What is the degree of saturation of the specimen?

1.5   Soil has been compacted in an embankment at a bulk density of 2.15 Mg/m^3 and a water content of 12%. The value of G_s is 2.65. Calculate the dry density, void ratio, degree of saturation and air content. Would it be possible to compact the above soil at a water content of 13.5% to a dry density of 2.00 Mg/m^3?

1.6   The following results were obtained from a standard compaction test on a soil:

Mass (g)	2010	2092	2114	2100	2055
Water content (%)	12.8	14.5	15.6	16.8	19.2

The value of G_s is 2.67. Plot the dry density–water content curve and give the optimum water content and maximum dry density. Plot also the curves of zero, 5 and 10% air content and give the value of air content at maximum dry density. The volume of the mould is 1000 cm^3.

1.7   The *in-situ* dry density of a sand is 1.72 Mg/m^3. The maximum and minimum dry densities, determined by standard laboratory tests, are 1.81 and 1.54 Mg/m^3, respectively. Determine the density index of the sand.

REFERENCES

1   American Society for Testing and Materials *Annual Book of ASTM Standards*, Vol. 04/08, Hitchin, Herts.

2   British Standard 1377 (1990) *Methods of Test for Soils for Civil Engineering Purposes*, British Standards Institution, London.

3   British Standard 5930 (1999) *Code of Practice for Site Investigations*, British Standards Institution, London.

4   British Standard 6031 (1981) *Code of Practice for Earthworks*, British Standards Institution, London.

5   Collins, K. and McGown, A. (1974) The form and function of microfabric features in a variety of natural soils, *Geotechnique*, **24**, 223–54.

6   Department of Transport (1993) Earthworks, in *Specification for Highway Works*, HMSO, Series 600, London.

7   Grim, R.E. (1962) *Clay Mineralogy*, McGraw-Hill, New York.

8   Parsons, A.W. and Boden, J.B. (1979) *The Moisture Condition Test and its Potential Applications in Earthworks*, TRRL Report 522, Crowthorne, Berks.

9   Rowe, P.W. (1972) The relevance of soil fabric to site investigation practice, *Geotechnique*, **22**, 193–300.

10 Wagner, A.A. (1957) The use of the unified soil classification system by the bureau of reclamation, in *Proceedings of the 4th International Conference of SMFE, London*, Vol. 1, Butterworths, London, pp. 125–34.

Chapter 2

Seepage

2.1 SOIL WATER

All soils are *permeable* materials, water being free to flow through the interconnected pores between the solid particles. The pressure of the pore water is measured relative to atmospheric pressure and the level at which the pressure is atmospheric (i.e. zero) is defined as the *water table* (WT) or the *phreatic surface*. Below the water table the soil is assumed to be fully saturated, although it is likely that, due to the presence of small volumes of entrapped air, the degree of saturation will be marginally below 100%. The level of the water table changes according to climatic conditions but the level can change also as a consequence of constructional operations. A *perched* water table can occur locally, contained by soil of low permeability, above the normal water table level. *Artesian* conditions can exist if an inclined soil layer of high permeability is confined locally by an overlying layer of low permeability; the pressure in the artesian layer is governed not by the local water table level but by a higher water table level at a distant location where the layer is unconfined.

Below the water table the pore water may be static, the hydrostatic pressure depending on the depth below the water table, or may be seeping through the soil under hydraulic gradient: this chapter is concerned with the second case. Bernoulli's theorem applies to the pore water but seepage velocities in soils are normally so small that velocity head can be neglected. Thus

$$h = \frac{u}{\gamma_w} + z \tag{2.1}$$

where h is the total head, u the pore water pressure, γ_w the unit weight of water ($9.8 \, \text{kN/m}^3$) and z the elevation head above a chosen datum.

Above the water table, water can be held at negative pressure by capillary tension; the smaller the size of the pores the higher the water can rise above the water table. The capillary rise tends to be irregular due to the random pore sizes occurring in a soil. The soil can be almost completely saturated in the lower part of the capillary zone but in general the degree of saturation decreases with height. When water percolates through the soil from the surface towards the water table some of this water can be held by surface tension around the points of contact between particles. The negative pressure of water held above the water table results in attractive forces between the particles: this attraction is referred to as soil suction and is a function of pore size and water content.

2.2 PERMEABILITY

In one dimension, water flows through a fully saturated soil in accordance with Darcy's empirical law:

$$q = Aki \qquad (2.2)$$

or

$$v = \frac{q}{A} = ki$$

where q is the volume of water flowing per unit time, A the cross-sectional area of soil corresponding to the flow q, k the coefficient of permeability, i the hydraulic gradient and v the discharge velocity. The units of the coefficient of permeability are those of velocity (m/s).

The coefficient of permeability depends primarily on the average size of the pores, which in turn is related to the distribution of particle sizes, particle shape and soil structure. In general, the smaller the particles the smaller is the average size of the pores and the lower is the coefficient of permeability. The presence of a small percentage of fines in a coarse-grained soil results in a value of k significantly lower than the value for the same soil without fines. For a given soil the coefficient of permeability is a function of void ratio. If a soil deposit is stratified the permeability for flow parallel to the direction of stratification is higher than that for flow perpendicular to the direction of stratification. The presence of fissures in a clay results in a much higher value of permeability compared with that of the unfissured material.

The coefficient of permeability also varies with temperature, upon which the viscosity of the water depends. If the value of k measured at 20 °C is taken as 100% then the values at 10 and 0 °C are 77 and 56%, respectively. The coefficient of permeability can also be represented by the equation:

$$k = \frac{\gamma_w}{\eta} K$$

where γ_w is the unit weight of water, η the viscosity of water and K (units m^2) an absolute coefficient depending only on the characteristics of the soil skeleton.

The values of k for different types of soil are typically within the ranges shown in Table 2.1. For sands, Hazen showed that the approximate value of k is given by

$$k = 10^{-2} D_{10}^2 \quad \text{(m/s)} \qquad (2.3)$$

where D_{10} is the effective size in mm.

On the microscopic scale the water seeping through a soil follows a very tortuous path between the solid particles but macroscopically the flow path (in one dimension) can be considered as a smooth line. The average velocity at which the water flows through the soil pores is obtained by dividing the volume of water flowing per unit

Table 2.1 Coefficient of permeability (m/s) (BS 8004: 1986)

I	10^{-1}	10^{-2}	10^{-3}	10^{-4}	10^{-5}	10^{-6}	10^{-7}	10^{-8}	10^{-9}	10^{-10}

Clean gravels	Clean sands and sand–gravel mixtures	Very fine sands, silts and clay-silt laminate	Unfissured clays and clay-silts (>20% clay)
	Desiccated and fissured clays		

time by the average area of voids (A_v) on a cross-section normal to the macroscopic direction of flow: this velocity is called the seepage velocity (v'). Thus

$$v' = \frac{q}{A_v}$$

The porosity of a soil is defined in terms of volume:

$$n = \frac{V_v}{V}$$

However, on average, the porosity can also be expressed as

$$n = \frac{A_v}{A}$$

Hence

$$v' = \frac{q}{nA} = \frac{v}{n}$$

or

$$v' = \frac{ki}{n} \tag{2.4}$$

Determination of coefficient of permeability

Laboratory methods

The coefficient of permeability for coarse soils can be determined by means of the *constant-head* permeability test (Figure 2.1(a)). The soil specimen, at the appropriate density, is contained in a Perspex cylinder of cross-sectional area A: the specimen rests on a coarse filter or a wire mesh. A steady vertical flow of water, under a constant total head, is maintained through the soil and the volume of water flowing per unit time (q)

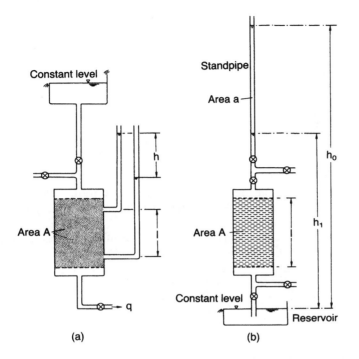

Figure 2.1 Laboratory permeability tests: (a) constant head and (b) falling head.

is measured. Tappings from the side of the cylinder enable the hydraulic gradient (h/l) to be measured. Then from Darcy's law:

$$k = \frac{ql}{Ah}$$

A series of tests should be run, each at a different rate of flow. Prior to running the test a vacuum is applied to the specimen to ensure that the degree of saturation under flow will be close to 100%. If a high degree of saturation is to be maintained the water used in the test should be de-aired.

For fine soils the *falling-head* test (Figure 2.1(b)) should be used. In the case of fine soils, undisturbed specimens are normally tested and the containing cylinder in the test may be the sampling tube itself. The length of the specimen is l and the cross-sectional area A. A coarse filter is placed at each end of the specimen and a standpipe of internal area a is connected to the top of the cylinder. The water drains into a reservoir of constant level. The standpipe is filled with water and a measurement is made of the time (t_1) for the water level (relative to the water level in the reservoir) to fall from h_0 to h_1. At any intermediate time t the water level in the standpipe is given by h and its rate of change by $-dh/dt$. At time t the difference in total head between the top and bottom of the specimen is h. Then, applying Darcy's law:

$$-a\frac{dh}{dt} = Ak\frac{h}{l}$$

$$\therefore \quad -a\int_{h_0}^{h_1}\frac{dh}{h} = \frac{Ak}{l}\int_0^{t_1}dt$$

$$\therefore \quad k = \frac{al}{At_1}\ln\frac{h_0}{h_1}$$

$$= 2.3\frac{al}{At_1}\log\frac{h_0}{h_1}$$

Again, precautions must be taken to ensure that the degree of saturation remains close to 100%. A series of tests should be run using different values of h_0 and h_1 and/or standpipes of different diameters.

The coefficient of permeability of fine soils can also be determined indirectly from the results of consolidation tests (see Chapter 7).

The reliability of laboratory methods depends on the extent to which the test specimens are representative of the soil mass as a whole. More reliable results can generally be obtained by the *in-situ* methods described below.

Well pumping test

This method is most suitable for use in homogeneous coarse soil strata. The procedure involves continuous pumping at a constant rate from a well, normally at least 300 mm in diameter, which penetrates to the bottom of the stratum under test. A screen or filter is placed in the bottom of the well to prevent ingress of soil particles. Perforated casing is normally required to support the sides of the well. Steady seepage is established, radially towards the well, resulting in the water table being drawn down to form a 'cone of depression'. Water levels are observed in a number of boreholes spaced on radial lines at various distances from the well. An unconfined stratum of uniform thickness with a (relatively) impermeable lower boundary is shown in Figure 2.2(a), the water table being below the upper surface of the stratum. A confined layer between two impermeable strata is shown in Figure 2.2(b), the original water table being within the overlying stratum. Frequent recordings are made of the water levels in the boreholes, usually by means of an electrical dipper. The test enables the average coefficient of permeability of the soil mass below the cone of depression to be determined. Full details of the test procedure are given in BS 6316.

Analysis is based on the assumption that the hydraulic gradient at any distance r from the centre of the well is constant with depth and is equal to the slope of the water table, i.e.

$$i_r = \frac{dh}{dr}$$

where h is the height of the water table at radius r. This is known as the Dupuit assumption and is reasonably accurate except at points close to the well.

In the case of an unconfined stratum (Figure 2.2(a)), consider two boreholes located on a radial line at distances r_1 and r_2 from the centre of the well, the respective water

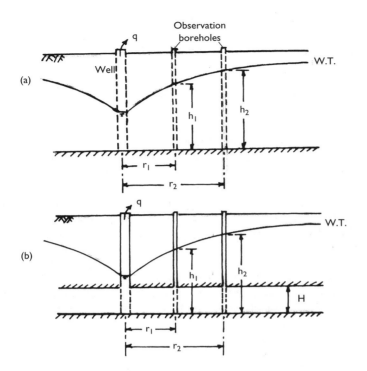

Figure 2.2 Well pumping tests: (a) unconfined stratum and (b) confined stratum.

levels relative to the bottom of the stratum being h_1 and h_2. At distance r from the well the area through which seepage takes place is $2\pi rh$, where r and h are variables. Then applying Darcy's law:

$$q = 2\pi rhk\frac{dh}{dr}$$

$$\therefore q\int_{r_1}^{r_2}\frac{dr}{r} = 2\pi k\int_{h_1}^{h_2} h\,dh$$

$$\therefore q\ln\left(\frac{r_2}{r_1}\right) = \pi k(h_2^2 - h_1^2)$$

$$\therefore k = \frac{2.3q\log(r_2/r_1)}{\pi(h_2^2 - h_1^2)}$$

For a confined stratum of thickness H (Figure 2.2(b)) the area through which seepage takes place is $2\pi rH$, where r is variable and H is constant. Then

$$q = 2\pi rHk\frac{dh}{dr}$$

$$\therefore q\int_{r_1}^{r_2}\frac{dr}{r} = 2\pi Hk\int_{h_1}^{h_2} dh$$

$$\therefore q \ln\left(\frac{r_2}{r_1}\right) = 2\pi Hk(h_2 - h_1)$$

$$\therefore k = \frac{2.3q\log(r_2/r_1)}{2\pi H(h_2 - h_1)}$$

Borehole tests

The general principle is that water is either introduced into or pumped out of a borehole which terminates within the stratum in question, the procedures being referred to as inflow and outflow tests, respectively. A hydraulic gradient is thus established, causing seepage either into or out of the soil mass surrounding the borehole and the rate of flow is measured. In a constant-head test the water level is maintained throughout at a given level (Figure 2.3(a)). In a variable-head test the water level is allowed to fall or rise from its initial position and the time taken for the level to change between two values is recorded (Figure 2.3(b)). The tests indicate the permeability of the soil within a radius of only 1–2 m from the centre of the borehole. Careful boring is essential to avoid disturbance in the soil structure.

A problem in such tests is that clogging of the soil face at the bottom of the borehole tends to occur due to the deposition of sediment from the water. To alleviate the

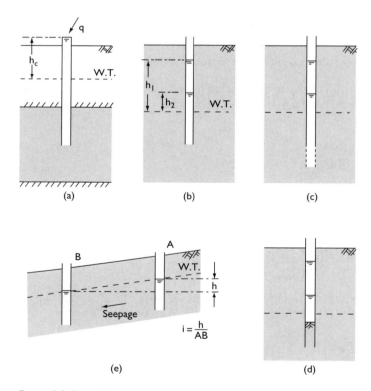

Figure 2.3 Borehole tests.

problem the borehole may be extended below the bottom of the casing, as shown in Figure 2.3(c), increasing the area through which seepage takes place. The extension may be uncased or supported by perforated casing depending on the type of soil. Another solution is to install within the casing a central tube perforated at its lower end and set within a pocket of coarser material.

Expressions for the coefficient of permeability depend on whether the stratum is unconfined or confined, the position of the bottom of the casing within the stratum and details of the drainage face in the soil. If the soil is anisotropic with respect to permeability and if the borehole extends below the bottom of the casing (Figure 2.3(c)) then the horizontal permeability tends to be measured. If, on the other hand, the casing penetrates below soil level in the bottom of the borehole (Figure 2.3(d)) then vertical permeability tends to be measured. General formulae can be written, with the above details being represented by an 'intake factor' (F). Values of intake factor F were published by Hvorslev [5] and are also given in BS 5930 [1].

For a constant-head test:

$$k = \frac{q}{Fh_{c}}$$

For a variable-head test:

$$k = \frac{2.3A}{F(t_2 - t_1)} \log \frac{h_1}{h_2}$$

where k is the coefficient of permeability, q the rate of flow, h_{c} the constant head, h_1 the variable head at time t_1, h_2 the variable head at time t_2 and A the cross-sectional area of casing or standpipe.

The coefficient of permeability for a coarse soil can also be obtained from *in-situ* measurements of seepage velocity, using Equation 2.4. The method involves excavating uncased boreholes or trial pits at two points A and B (Figure 2.3(e)), seepage taking place from A towards B. The hydraulic gradient is given by the difference in the steady-state water levels in the boreholes divided by the distance AB. Dye or any other suitable tracer is inserted into borehole A and the time taken for the dye to appear in borehole B is measured. The seepage velocity is then the distance AB divided by this time. The porosity of the soil can be determined from density tests. Then

$$k = \frac{v'n}{i}$$

2.3 SEEPAGE THEORY

The general case of seepage in two dimensions will now be considered. Initially it will be assumed that the soil is homogeneous and isotropic with respect to permeability, the coefficient of permeability being k. In the x–z plane, Darcy's law can be written in the generalized form:

$$v_x = ki_x = -k\frac{\partial h}{\partial x} \tag{2.5a}$$

$$v_z = ki_z = -k\frac{\partial h}{\partial z} \tag{2.5b}$$

with the total head h decreasing in the directions of v_x and v_z.

An element of fully saturated soil having dimensions dx, dy and dz in the x, y and z directions, respectively, with flow taking place in the x–z plane only, is shown in Figure 2.4. The components of discharge velocity of water entering the element are v_x and v_z, and the rates of change of discharge velocity in the x and z directions are $\partial v_x/\partial x$ and $\partial v_z/\partial z$, respectively. The volume of water entering the element per unit time is

$$v_x\,dy\,dz + v_z\,dx\,dy$$

and the volume of water leaving per unit time is

$$\left(v_x + \frac{\partial v_x}{\partial x}\,dx\right)dy\,dz + \left(v_z + \frac{\partial v_z}{\partial z}\,dz\right)dx\,dy$$

If the element is undergoing no volume change and if water is assumed to be incompressible, the difference between the volume of water entering the element per unit time and the volume leaving must be zero. Therefore

$$\frac{\partial v_x}{\partial x} + \frac{\partial v_z}{\partial z} = 0 \tag{2.6}$$

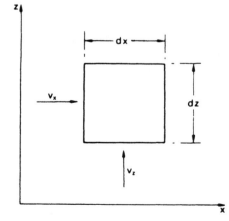

Figure 2.4 Seepage through a soil element.

Equation 2.6 is the *equation of continuity* in two dimensions. If, however, the volume of the element is undergoing change, the equation of continuity becomes

$$\left(\frac{\partial v_x}{\partial x} + \frac{\partial v_z}{\partial z}\right) dx\,dy\,dz = \frac{dV}{dt} \tag{2.7}$$

where dV/dt is the volume change per unit time.

Consider, now, the function $\phi(x, z)$, called the *potential function*, such that

$$\frac{\partial \phi}{\partial x} = v_x = -k\frac{\partial h}{\partial x} \tag{2.8a}$$

$$\frac{\partial \phi}{\partial z} = v_z = -k\frac{\partial h}{\partial z} \tag{2.8b}$$

From the Equations 2.6 and 2.8 it is apparent that

$$\frac{\partial^2 \phi}{\partial x^2} + \frac{\partial^2 \phi}{\partial z^2} = 0 \tag{2.9}$$

i.e. the function $\phi(x, z)$ satisfies the Laplace equation.

Integrating Equation 2.8:

$$\phi(x, z) = -kh(x, z) + C$$

where C is a constant. Thus, if the function $\phi(x, z)$ is given a constant value, equal to ϕ_1 (say), it will represent a curve along which the value of total head (h_1) is constant. If the function $\phi(x, z)$ is given a series of constant values, ϕ_1, ϕ_2, ϕ_3, etc., a family of curves is specified along each of which the total head is a constant value (but a different value for each curve). Such curves are called *equipotentials*.

A second function $\psi(x, z)$, called the *flow function*, is now introduced, such that

$$-\frac{\partial \psi}{\partial x} = v_z = -k\frac{\partial h}{\partial z} \tag{2.10a}$$

$$\frac{\partial \psi}{\partial z} = v_x = -k\frac{\partial h}{\partial x} \tag{2.10b}$$

It can be shown that this function also satisfies the Laplace equation.

The total differential of the function $\psi(x, z)$ is

$$dψ = \frac{\partial \psi}{\partial x} dx + \frac{\partial \psi}{\partial z} dz$$
$$= -v_z\,dx + v_x\,dz$$

If the function $\psi(x, z)$ is given a constant value ψ_1 then $d\psi = 0$ and

$$\frac{dz}{dx} = \frac{v_z}{v_x} \tag{2.11}$$

Thus, the tangent at any point on the curve represented by

$$\psi(x, z) = \psi_1$$

specifies the direction of the resultant discharge velocity at that point: the curve therefore represents the flow path. If the function $\psi(x, z)$ is given a series of constant values, ψ_1, ψ_2, ψ_3, etc., a second family of curves is specified, each representing a flow path. These curves are called *flow lines*.

Referring to Figure 2.5, the flow per unit time between two flow lines for which the values of the flow function are ψ_1 and ψ_2 is given by

$$\Delta q = \int_{\psi_1}^{\psi_2} (-v_z \, \mathrm{d}x + v_x \, \mathrm{d}z)$$

$$= \int_{\psi_1}^{\psi_2} \left(\frac{\partial \psi}{\partial x} \mathrm{d}x + \frac{\partial \psi}{\partial z} \mathrm{d}z \right) = \psi_2 - \psi_1$$

Thus, the flow through the 'channel' between the two flow lines is constant.

The total differential of the function $\phi(x, z)$ is

$$\mathrm{d}\phi = \frac{\partial \phi}{\partial x} \mathrm{d}x + \frac{\partial \phi}{\partial z} \mathrm{d}z$$

$$= v_x \, \mathrm{d}x + v_z \, \mathrm{d}z$$

If $\phi(x, z)$ is constant then $\mathrm{d}\phi = 0$ and

$$\frac{\mathrm{d}z}{\mathrm{d}x} = -\frac{v_x}{v_z} \tag{2.12}$$

Comparing Equations 2.11 and 2.12 it is apparent that the flow lines and the equipotentials intersect each other at right angles.

Consider, now, two flow lines ψ_1 and $(\psi_1 + \Delta\psi)$ separated by the distance Δn. The flow lines are intersected orthogonally by two equipotentials ϕ_1 and $(\phi_1 + \Delta\phi)$

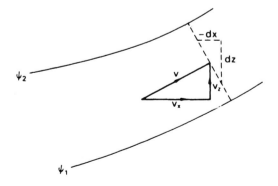

Figure 2.5 Seepage between two flow lines.

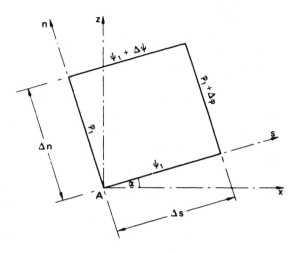

Figure 2.6 Flow lines and equipotentials.

separated by the distance Δs, as shown in Figure 2.6. The directions s and n are inclined at angle α to the x and z axes, respectively. At point A the discharge velocity (in direction s) is v_s; the components of v_s in the x and z directions, respectively, are

$$v_x = v_s \cos \alpha$$
$$v_z = v_s \sin \alpha$$

Now

$$\frac{\partial \phi}{\partial s} = \frac{\partial \phi}{\partial x}\frac{\partial x}{\partial s} + \frac{\partial \phi}{\partial z}\frac{\partial z}{\partial s}$$
$$= v_s \cos^2 \alpha + v_s \sin^2 \alpha = v_s$$

and

$$\frac{\partial \psi}{\partial n} = \frac{\partial \psi}{\partial x}\frac{\partial x}{\partial n} + \frac{\partial \psi}{\partial z}\frac{\partial z}{\partial n}$$
$$= -v_s \sin \alpha (-\sin \alpha) + v_s \cos^2 \alpha = v_s$$

Thus

$$\frac{\partial \psi}{\partial n} = \frac{\partial \phi}{\partial s}$$

or approximately

$$\frac{\Delta \psi}{\Delta n} = \frac{\Delta \phi}{\Delta s} \tag{2.13}$$

2.4 FLOW NETS

In principle, for the solution of a practical seepage problem the functions $\phi(x, z)$ and $\psi(x, z)$ must be found for the relevant boundary conditions. The solution is represented by a family of flow lines and a family of equipotentials, constituting what is referred to as a flow net. Possible methods of solution are complex variable techniques, the finite difference method, the finite element method, electrical analogy and the use of hydraulic models. Computer software based on either the finite difference or finite element methods is widely available for the solution of seepage problems. Williams *et al.* [10] described how solutions can be obtained from the finite difference form of the Laplace equation by means of a spreadsheet. Relatively simple problems can be solved by the trial and error sketching of the flow net, the general form of which can be deduced from consideration of the boundary conditions. Flow net sketching leads to a greater understanding of seepage principles. However, for problems in which the geometry becomes complex and there are zones of different permeabilities throughout the flow region, use of the finite element method is usually necessary.

The fundamental condition to be satisfied in a flow net is that every intersection between a flow line and an equipotential must be at right angles. In addition, it is *convenient* to construct the flow net such that $\Delta\psi$ has the same value between any two adjacent flow lines and $\Delta\phi$ has the same value between any two adjacent equipotentials. It is also *convenient* to make $\Delta s = \Delta n$ in Equation 2.13, i.e. the flow lines and equipotentials form 'curvilinear squares' throughout the flow net. Then for any curvilinear square

$$\Delta\psi = \Delta\phi$$

Now, $\Delta\psi = \Delta q$ and $\Delta\phi = k\Delta h$, therefore:

$$\Delta q = k\Delta h \tag{2.14}$$

For the entire flow net, h is the difference in total head between the first and last equipotentials, N_d the number of *equipotential drops*, each representing the same total head loss Δh, and N_f the number of *flow channels*, each carrying the same flow Δq. Then,

$$\Delta h = \frac{h}{N_d} \tag{2.15}$$

and

$$q = N_f \Delta q$$

Hence, from Equation 2.14

$$q = kh\frac{N_f}{N_d} \tag{2.16}$$

Equation 2.16 gives the total volume of water flowing per unit time (per unit dimension in the y direction) and is a function of the *ratio* N_f/N_d.

Between two adjacent equipotentials the hydraulic gradient is given by

$$i = \frac{\Delta h}{\Delta s} \tag{2.17}$$

Example of a flow net

As an illustration the flow net for the problem detailed in Figure 2.7(a) will be considered. The figure shows a line of sheet piling driven 6.00 m into a stratum of soil 8.60 m thick, underlain by an impermeable stratum. On one side of the piling the depth of water is 4.50 m; on the other side the depth of water (reduced by pumping) is 0.50 m.

The first step is to consider the boundary conditions of the flow region. At every point on the boundary AB the total head is constant, so AB is an equipotential; similarly CD is an equipotential. The datum to which total head is referred may be any level but in seepage problems it is convenient to select the downstream water level as datum. Then, the total head on equipotential CD is zero (pressure head 0.50 m; elevation head −0.50 m) and the total head on equipotential AB is 4.00 m (pressure head 4.50 m; elevation head −0.50 m). From point B, water must flow down the upstream face BE of the piling, round the tip E and up the down-stream face EC. Water from point F must flow along the impermeable surface FG. Thus BEC and FG are flow lines. The shapes of other flow lines must be between the extremes of BEC and FG.

The first trial sketching of the flow net (Figure 2.7(b)) can now be attempted using a procedure suggested by Casagrande [2]. The estimated line of flow (HJ) from a point on AB near the piling is lightly sketched. This line must start at right angles to equipotential AB and follow a smooth curve round the bottom of the piling. Trial equipotential lines are then drawn between the flow lines BEC and HJ, intersecting both flow lines at right angles and forming curvilinear squares. If necessary the position of HJ should be altered slightly so that a whole number of squares is obtained between BH and CJ. The procedure is continued by sketching the estimated line of flow (KL) from a second point on AB and extending the equipotentials already drawn. The flow line KL and the equipotential extensions are adjusted so that all intersections are at right angles and all areas are square. The procedure is repeated until the boundary FG is reached. At the first attempt it is almost certain that the last flow line drawn will be inconsistent with the boundary FG as, for example, in Figure 2.7(b). By studying the nature of this inconsistency the position of the first flow line (HJ) can be adjusted in a way that will tend to correct the inconsistency. The entire flow net is then adjusted and the inconsistency should now be small. After a third trial the last flow line should be consistent with the boundary FG, as shown in Figure 2.7(c). In general, the areas between the last flow line and the lower boundary will not be square but the length/breadth ratio of each area should be constant within this flow channel. In constructing a flow net it is a mistake to draw too many flow lines; typically, four to five flow channels are sufficient. Once experience has been gained and flow nets for various seepage situations have been studied, the detailed procedure described above

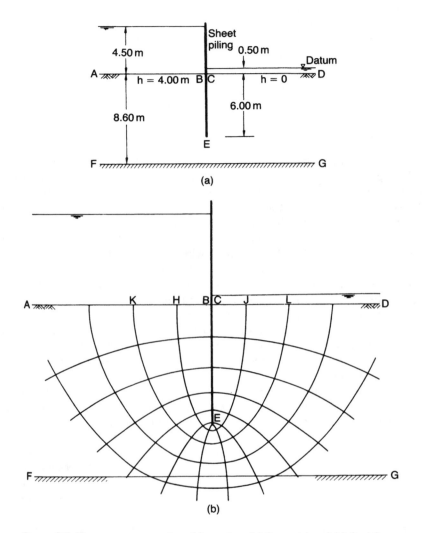

Figure 2.7 Flow net construction: (a) section, (b) first trial and (c) final flow net.

can be short-circuited. Three or four flow lines, varying in shape between the extremes of the two boundaries, can be sketched before or together with the equipotentials. Subsequent adjustments are made until a satisfactory flow net is achieved.

In the flow net in Figure 2.7(c) the number of flow channels is 4.3 and the number of equipotential drops is 12; thus the ratio N_f/N_d is 0.36. The equipotentials are numbered from zero at the downstream boundary; this number is denoted by n_d. The loss in total head between any two adjacent equipotentials is

$$\Delta h = \frac{h}{N_d} = \frac{4.00}{12} = 0.33 \, \text{m}$$

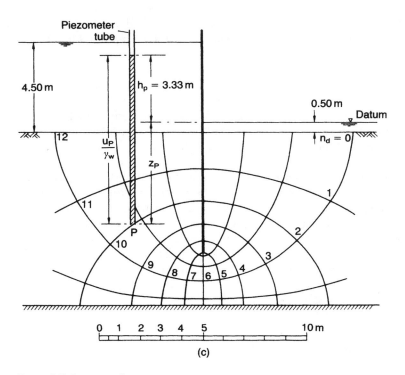

Figure 2.7 (continued)

The total head at every point on an equipotential numbered n_d is $n_d \Delta h$. The total volume of water flowing under the piling per unit time per unit length of piling is given by

$$q = kh\frac{N_f}{N_d} = k \times 4.00 \times 0.36$$
$$= 1.44k \, \text{m}^3/\text{s}$$

A piezometer tube is shown at a point P on the equipotential denoted by $n_d = 10$. The total head at P is

$$h_p = \frac{n_d}{N_d}h = \frac{10}{12} \times 4.00 = 3.33 \, \text{m}$$

i.e. the water level in the tube is 3.33 m above the datum. The point P is at a distance z_p *below* the datum, i.e. the elevation head is $-z_p$. The pore water pressure at P can then be calculated from Bernoulli's theorem:

$$u_p = \gamma_w\{h_p - (-z_p)\}$$
$$= \gamma_w(h_p + z_p)$$

The hydraulic gradient across any square in the flow net involves measuring the average dimension of the square (Equation 2.17). The highest hydraulic gradient (and hence the highest seepage velocity) occurs across the smallest square and vice versa.

Example 2.1

The section through a sheet pile wall along a tidal estuary is given in Figure 2.8. At low tide the depth of water in front of the wall is 4.00 m; the water table behind the wall lags 2.50 m behind tidal level. Plot the net distribution of water pressure on the piling.

The flow net is shown in the figure. The water level in front of the piling is selected as datum. The total head at water table level (the upstream equipotential) is 2.50 m (pressure head zero; elevation head +2.50 m). The total head on the soil surface in front of the piling (the downstream equipotential) is zero (pressure head 4.00 m; elevation head −4.00 m). There are 12 equipotential drops in the flow net.

The water pressures are calculated on both sides of the piling at selected levels numbered 1–7. For example, at level 4 the total head on the back of the piling is

$$h_b = \frac{8.8}{12} \times 2.50 = 1.83 \, \text{m}$$

and the total head on the front is

$$h_f = \frac{1}{12} \times 2.50 = 0.21 \, \text{m}$$

The elevation head at level 4 is −5.5 m.

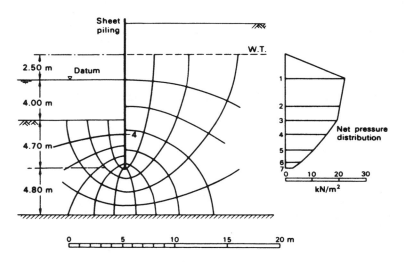

Figure 2.8 Example 2.1.

Table 2.2

Level	z (m)	h_b (m)	u_b/γ_w (m)	h_f (m)	u_f/γ_w (m)	$u_b - u_f$ (kN/m^2)
1	0	2.30	2.30	0	0	22.6
2	−2.70	2.10	4.80	0	2.70	20.6
3	−4.00	2.00	6.00	0	4.00	19.6
4	−5.50	1.83	7.33	0.21	5.71	15.9
5	−7.10	1.68	8.78	0.50	7.60	11.6
6	−8.30	1.51	9.81	0.84	9.14	6.6
7	−8.70	1.25	9.95	1.04	9.74	2.1

Therefore, the net pressure on the back of the piling is

$$u_b - u_f = 9.8(1.83 + 5.5) - 9.8(0.21 + 5.5)$$
$$= 9.8(7.33 - 5.71)$$
$$= 15.9 \, \text{kN/m}^2$$

The calculations for the selected points are shown in Table 2.2 and the net pressure diagram is plotted in Figure 2.8.

Example 2.2

The section through a dam is shown in Figure 2.9. Determine the quantity of seepage under the dam and plot the distribution of uplift pressure on the base of the dam. The coefficient of permeability of the foundation soil is 2.5×10^{-5} m/s.

The flow net is shown in the figure. The downstream water level is selected as datum. Between the upstream and downstream equipotentials the total head loss is 4.00 m. In the flow net there are 4.7 flow channels and 15 equipotential drops. The seepage is given by

$$q = kh\frac{N_f}{N_d} = 2.5 \times 10^{-5} \times 4.00 \times \frac{4.7}{15}$$
$$= 3.1 \times 10^{-5} \, \text{m}^3/\text{s} \quad \text{(per m)}$$

The pore water pressure is calculated at the points of intersection of the equipotentials with the base of the dam. The total head at each point is obtained from the flow net and the elevation head from the section. The calculations are shown in Table 2.3 and the pressure diagram is plotted in Figure 2.9.

Example 2.3

A river bed consists of a layer of sand 8.25 m thick overlying impermeable rock; the depth of water is 2.50 m. A long cofferdam 5.50 m wide is formed by driving two lines of sheet piling to a depth of 6.00 m below the level of the river bed and excavation to a depth of 2.00 m below bed level is carried out within the cofferdam. The water level within the cofferdam is kept at excavation level by pumping. If the flow of water into

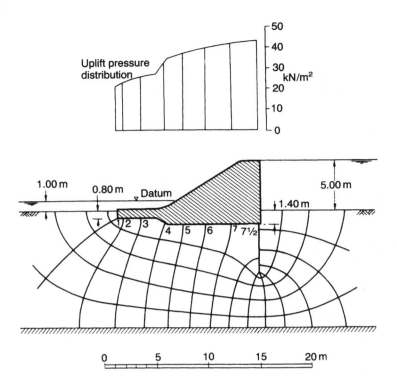

Figure 2.9 Example 2.2.

Table 2.3

Point	h (m)	z (m)	h − z (m)	$u = \gamma_w(h - z)$ (kN/m²)
I	0.27	−1.80	2.07	20.3
2	0.53	−1.80	2.33	22.9
3	0.80	−1.80	2.60	25.5
4	1.07	−2.10	3.17	31.1
5	1.33	−2.40	3.73	36.6
6	1.60	−2.40	4.00	39.2
7	1.87	−2.40	4.27	41.9
$7\frac{1}{2}$	2.00	−2.40	4.40	43.1

the cofferdam is 0.25 m³/h per unit length, what is the coefficient of permeability of the sand? What is the hydraulic gradient immediately below the excavated surface?

The section and flow net appear in Figure 2.10. In the flow net there are 6.0 flow channels and 10 equipotential drops. The total head loss is 4.50 m. The coefficient of permeability is given by

$$k = \frac{q}{h(N_f/N_d)}$$

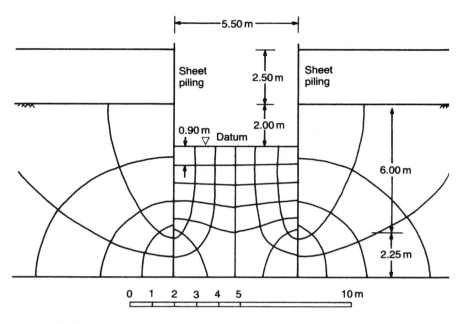

Figure 2.10 Example 2.3.

$$= \frac{0.25}{4.50 \times 6/10 \times 60^2} = 2.6 \times 10^{-5}\,\text{m/s}$$

The distance (Δs) between the last two equipotentials is measured as 0.9 m. The required hydraulic gradient is given by

$$i = \frac{\Delta h}{\Delta s}$$

$$= \frac{4.50}{10 \times 0.9} = 0.50$$

2.5 ANISOTROPIC SOIL CONDITIONS

It will now be assumed that the soil, although homogeneous, is anisotropic with respect to permeability. Most natural soil deposits are anisotropic, with the coefficient of permeability having a maximum value in the direction of stratification and a minimum value in the direction normal to that of stratification; these directions are denoted by x and z, respectively, i.e.

$$k_x = k_{\text{max}} \quad \text{and} \quad k_z = k_{\text{min}}$$

In this case the generalized form of Darcy's law is

$$v_x = k_x i_x = -k_x \frac{\partial h}{\partial x} \qquad (2.18a)$$

$$v_z = k_z i_z = -k_z \frac{\partial h}{\partial z} \qquad (2.18b)$$

Also, in any direction s, inclined at angle α to the x direction, the coefficient of permeability is defined by the equation

$$v_s = -k_s \frac{\partial h}{\partial s}$$

Now

$$\frac{\partial h}{\partial s} = \frac{\partial h}{\partial x}\frac{\partial x}{\partial s} + \frac{\partial h}{\partial z}\frac{\partial z}{\partial s}$$

i.e.

$$\frac{v_s}{k_s} = \frac{v_x}{k_x}\cos\alpha + \frac{v_z}{k_z}\sin\alpha$$

The components of discharge velocity are also related as follows:

$$v_x = v_s \cos\alpha$$
$$v_z = v_s \sin\alpha$$

Hence

$$\frac{1}{k_s} = \frac{\cos^2\alpha}{k_x} + \frac{\sin^2\alpha}{k_z}$$

or

$$\frac{s^2}{k_s} = \frac{x^2}{k_x} + \frac{z^2}{k_z} \qquad (2.19)$$

The directional variation of permeability is thus described by Equation 2.19 which represents the ellipse shown in Figure 2.11.

Given the generalized form of Darcy's law (Equation 2.18) the equation of continuity (2.6) can be written:

$$k_x \frac{\partial^2 h}{\partial x^2} + k_z \frac{\partial^2 h}{\partial z^2} = 0 \qquad (2.20)$$

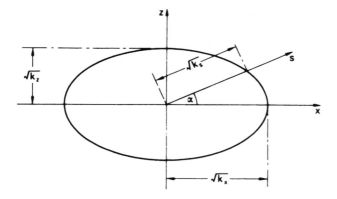

Figure 2.11 Permeability ellipse.

or

$$\frac{\partial^2 h}{(k_z/k_x)\partial x^2} + \frac{\partial^2 h}{\partial z^2} = 0$$

Substituting

$$x_t = x\sqrt{\frac{k_z}{k_x}} \qquad (2.21)$$

the equation of continuity becomes

$$\frac{\partial^2 h}{\partial x_t^2} + \frac{\partial^2 h}{\partial z^2} = 0 \qquad (2.22)$$

which is the equation of continuity for an *isotropic* soil in an x_t–z plane.

Thus, Equation 2.21 defines a scale factor which can be applied in the x direction to transform a given anisotropic flow region into a fictitious isotropic flow region in which the Laplace equation is valid. Once the flow net (representing the solution of the Laplace equation) has been drawn for the transformed section the flow net for the natural section can be obtained by applying the inverse of the scaling factor. Essential data, however, can normally be obtained from the transformed section. The necessary transformation could also be made in the z direction.

The value of coefficient of permeability applying to the transformed section, referred to as the equivalent isotropic coefficient, is

$$k' = \sqrt{(k_x k_z)} \qquad (2.23)$$

A formal proof of Equation 2.23 has been given by Vreedenburgh [9]. The validity of Equation 2.23 can be demonstrated by considering an elemental flow net field through

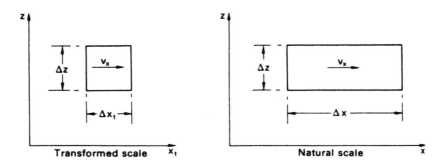

Figure 2.12 Elemental flow net field.

which flow is in the x direction. The flow net field is drawn to the transformed and natural scales in Figure 2.12, the transformation being in the x direction. The discharge velocity v_x can be expressed in terms of either k' (transformed section) or k_x (natural section), i.e.

$$v_x = -k' \frac{\partial h}{\partial x_t} = -k_x \frac{\partial h}{\partial x}$$

where

$$\frac{\partial h}{\partial x_t} = \frac{\partial h}{\sqrt{(k_z/k_x)}\partial x}$$

Thus

$$k' = k_x \sqrt{\frac{k_z}{k_x}} = \sqrt{(k_x k_z)}$$

2.6   NON-HOMOGENEOUS SOIL CONDITIONS

Two *isotropic* soil layers of thicknesses H_1 and H_2 are shown in Figure 2.13, the respective coefficients of permeability being k_1 and k_2; the boundary between the layers is horizontal. (If the layers are anisotropic, k_1 and k_2 represent the equivalent isotropic coefficients for the layers.) The two layers can be considered as a single homogeneous anisotropic layer of thickness $(H_1 + H_2)$ in which the coefficients in the directions parallel and normal to that of stratification are $\bar{k}_x$ and $\bar{k}_z$, respectively.

For one-dimensional seepage in the horizontal direction, the equipotentials in each layer are vertical. If h_1 and h_2 represent total head at any point in the respective layers, then for a common point on the boundary $h_1 = h_2$. Therefore, any vertical line through the two layers represents a common equipotential. Thus, the hydraulic gradients in the two layers, and in the equivalent single layer, are equal; the equal hydraulic gradients are denoted by i_x.

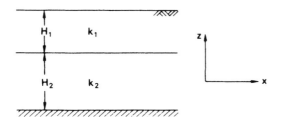

Figure 2.13 Non-homogeneous soil conditions.

The total horizontal flow per unit time is given by

$$\bar{q}_x = (H_1 + H_2)\bar{k}_x i_x = (H_1 k_1 + H_2 k_2)i_x$$

$$\therefore \bar{k}_x = \frac{H_1 k_1 + H_2 k_2}{H_1 + H_2} \tag{2.24}$$

For one-dimensional seepage in the vertical direction, the discharge velocities in each layer, and in the equivalent single layer, must be equal if the requirement of continuity is to be satisfied. Thus

$$v_z = \bar{k}_z \bar{i}_z = k_1 i_1 = k_2 i_2$$

where $\bar{i}_z$ is the average hydraulic gradient over the depth $(H_1 + H_2)$. Therefore

$$i_1 = \frac{\bar{k}_z}{k_1}\bar{i}_z \quad \text{and} \quad i_2 = \frac{\bar{k}_z}{k_2}\bar{i}_z$$

Now the loss in total head over the depth $(H_1 + H_2)$ is equal to the sum of the losses in total head in the individual layers, i.e.

$$\bar{i}_z(H_1 + H_2) = i_1 H_1 + i_2 H_2$$

$$= \bar{k}_z \bar{i}_z \left(\frac{H_1}{k_1} + \frac{H_2}{k_2}\right)$$

$$\therefore \bar{k}_z = \frac{H_1 + H_2}{\left(\dfrac{H_1}{k_1}\right) + \left(\dfrac{H_2}{k_2}\right)} \tag{2.25}$$

Similar expressions for $\bar{k}_x$ and $\bar{k}_z$ apply in the case of any number of soil layers. It can be shown that $\bar{k}_x$ must always be greater than $\bar{k}_z$, i.e. seepage can occur more readily in the direction parallel to stratification than in the direction perpendicular to stratification.

2.7 TRANSFER CONDITION

Consideration is now given to the condition which must be satisfied when seepage takes place diagonally across the boundary between two isotropic soils 1 and 2 having

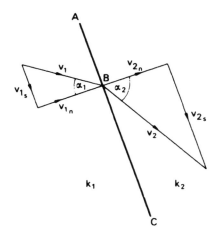

Figure 2.14 Transfer condition.

coefficients of permeability k_1 and k_2, respectively. The direction of seepage approaching a point B on the boundary ABC is at angle α_1 to the normal at B, as shown in Figure 2.14; the discharge velocity approaching B is v_1. The components of v_1 along the boundary and normal to the boundary are v_{1s} and v_{1n} respectively. The direction of seepage leaving point B is at angle α_2 to the normal, as shown; the discharge velocity leaving B is v_2. The components of v_2 are v_{2s} and v_{2n}.

For soils 1 and 2 respectively

$$\phi_1 = -k_1 h_1 \quad \text{and} \quad \phi_2 = -k_2 h_2$$

At the common point B, $h_1 = h_2$; therefore

$$\frac{\phi_1}{k_1} = \frac{\phi_2}{k_2}$$

Differentiating with respect to s, the direction along the boundary:

$$\frac{1}{k_1}\frac{\partial \phi_1}{\partial s} = \frac{1}{k_2}\frac{\partial \phi_2}{\partial s}$$

i.e.

$$\frac{v_{1s}}{k_1} = \frac{v_{2s}}{k_2}$$

For continuity of flow across the boundary the normal components of discharge velocity must be equal, i.e.

$$v_{1n} = v_{2n}$$

Therefore

$$\frac{1}{k_1}\frac{v_{1s}}{v_{1n}} = \frac{1}{k_2}\frac{v_{2s}}{v_{2n}}$$

Hence it follows that

$$\frac{\tan \alpha_1}{\tan \alpha_2} = \frac{k_1}{k_2} \tag{2.26}$$

Equation 2.26 specifies the change in direction of the flow line passing through point B. This equation must be satisfied on the boundary by every flow line crossing the boundary.

Equation 2.13 can be written as

$$\Delta\psi = \frac{\Delta n}{\Delta s}\Delta\phi$$

i.e.

$$\Delta q = \frac{\Delta n}{\Delta s}k\Delta h$$

If Δq and Δh are each to have the same values on both sides of the boundary then

$$\left(\frac{\Delta n}{\Delta s}\right)_1 k_1 = \left(\frac{\Delta n}{\Delta s}\right)_2 k_2$$

and it is clear that curvilinear squares are possible only in one soil. If

$$\left(\frac{\Delta n}{\Delta s}\right)_1 = 1$$

then

$$\left(\frac{\Delta n}{\Delta s}\right)_2 = \frac{k_1}{k_2} \tag{2.27}$$

If the permeability ratio is less than $\frac{1}{10}$ it is unlikely that the part of the flow net in the soil of higher permeability needs to be considered.

2.8 SEEPAGE THROUGH EMBANKMENT DAMS

This problem is an example of unconfined seepage, one boundary of the flow region being a phreatic surface on which the pressure is atmospheric. In section the phreatic surface constitutes the top flow line and its position must be estimated before the flow net can be drawn.

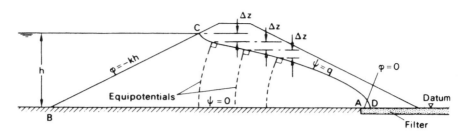

Figure 2.15 Homogeneous embankment dam section.

Consider the case of a homogeneous isotropic embankment dam on an impermeable foundation, as shown in Figure 2.15. The impermeable boundary BA is a flow line and CD is the required top flow line. At every point on the upstream slope BC the total head is constant (u/γ_w and z varying from point to point but their sum remaining constant); therefore, BC is an equipotential. If the downstream water level is taken as datum then the total head on equipotential BC is equal to h, the difference between the upstream and downstream water levels. The discharge surface AD, for the case shown in Figure 2.15 only, is the equipotential for zero total head. At every point on the top flow line the pressure is zero (atmospheric), so total head is equal to elevation head and there must be equal vertical intervals Δz between the points of intersection between successive equipotentials and the top flow line.

A suitable filter must always be constructed at the discharge surface in an embankment dam. The function of the filter is to keep the seepage entirely within the dam; water seeping out onto the downstream slope would result in the gradual erosion of the slope. A horizontal underfilter is shown in Figure 2.15. Other possible forms of filter are illustrated in Figure 2.19(a) and (b); in these two cases the discharge surface AD is neither a flow line nor an equipotential since there are components of discharge velocity both normal and tangential to AD.

The boundary conditions of the flow region ABCD in Figure 2.15 can be written as follows:

Equipotential BC: $\phi = -kh$
Equipotential AD: $\phi = 0$
Flow line CD: $\psi = q$ (also, $\phi = -kz$)
Flow line BA: $\psi = 0$

The conformal transformation $r = w^2$

Complex variable theory can be used to obtain a solution to the embankment dam problem. Let the complex number $w = \phi + i\psi$ be an analytic function of $r = x + iz$. Consider the function

$$r = w^2$$

Thus

$$(x + iz) = (\phi + i\psi)^2$$
$$= (\phi^2 + 2i\phi\psi - \psi^2)$$

Equating real and imaginary parts:

$$x = \phi^2 - \psi^2 \tag{2.28}$$

$$z = 2\phi\psi \tag{2.29}$$

Equations 2.28 and 2.29 govern the transformation of points between the r and w planes.
Consider the transformation of the straight lines $\psi = n$, where $n = 0, 1, 2, 3$ (Figure 2.16(a)). From Equation 2.29

$$\phi = \frac{z}{2n}$$

and Equation 2.28 becomes

$$x = \frac{z^2}{4n^2} - n^2 \tag{2.30}$$

Equation 2.30 represents a family of confocal parabolas. For positive values of z the parabolas for the specified values of n are plotted in Figure 2.16(b).
Consider also the transformation of the straight lines $\phi = m$, where $m = 0, 1, 2, \ldots, 6$ (Figure 2.16(a)). From Equation 2.29

$$\psi = \frac{z}{2m}$$

and Equation 2.28 becomes

$$x = m^2 - \frac{z^2}{4m^2} \tag{2.31}$$

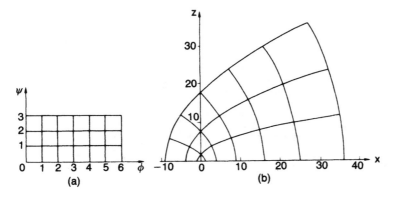

Figure 2.16 Conformal transformation $r = w^2$: (a) w plane and (b) r plane.

Equation 2.31 represents a family of confocal parabolas conjugate with the parabolas represented by Equation 2.30. For positive values of z the parabolas for the specified values of m are plotted in Figure 2.16(b). The two families of parabolas satisfy the requirements of a flow net.

Application to embankment dam section

The flow region in the w plane satisfying the boundary conditions for the section (Figure 2.15) is shown in Figure 2.17(a). In this case the transformation function

$$r = Cw^2$$

will be used, where C is a constant. Equations 2.28 and 2.29 then become

$$x = C(\phi^2 - \psi^2)$$
$$z = 2C\phi\psi$$

The equation of the top flow line can be derived by substituting the conditions

$$\psi = q$$
$$\phi = -kz$$

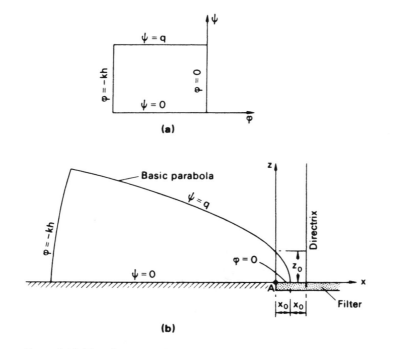

Figure 2.17 Transformation for embankment dam section: (a) w plane and (b) r plane.

Thus

$$z = -2Ckzq$$

$$\therefore C = -\frac{1}{2kq}$$

Hence

$$x = -\frac{1}{2kq}(k^2z^2 - q^2)$$

$$x = \frac{1}{2}\left(\frac{q}{k} - \frac{k}{q}z^2\right) \tag{2.32}$$

The curve represented by Equation 2.32 is referred to as Kozeny's basic parabola and is shown in Figure 2.17(b), the origin and focus both being at A.

When $z = 0$ the value of x is given by

$$x_0 = \frac{q}{2k}$$

$$\therefore q = 2kx_0 \tag{2.33}$$

where $2x_0$ is the directrix distance of the basic parabola. When $x = 0$ the value of z is given by

$$z_0 = \frac{q}{k} = 2x_0$$

Substituting Equation 2.33 in Equation 2.32 yields

$$x = x_0 - \frac{z^2}{4x_0} \tag{2.34}$$

The basic parabola can be drawn using Equation 2.34, provided the coordinates of one point on the parabola are known initially.

An inconsistency arises due to the fact that the conformal transformation of the straight line $\phi = -kh$ (representing the upstream equipotential) is a parabola, whereas the upstream equipotential in the embankment dam section is the upstream slope. Based on an extensive study of the problem, Casagrande [2] recommended that the initial point of the basic parabola should be taken at G (Figure 2.18) where GC = 0.3HC. The coordinates of point G, substituted in Equation 2.34, enable the value of x_0 to be determined; the basic parabola can then be plotted. The top flow line must intersect the upstream slope at right angles; a correction CJ must therefore be made (using personal judgement) to the basic parabola. The flow net can then be completed as shown in Figure 2.18.

If the discharge surface AD is not horizontal, as in the cases shown in Figure 2.19, a further correction KD to the basic parabola is required. The angle β is used to describe the direction of the discharge surface relative to AB. The correction can be made with

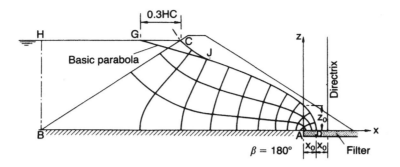

Figure 2.18 Flow net for embankment dam section.

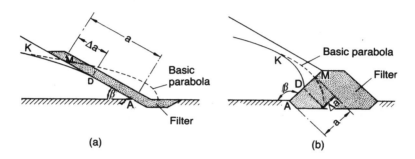

Figure 2.19 Downstream correction to basic parabola.

Table 2.4 Downstream correction to basic parabola. Reproduced from A. Casagrande (1940) 'Seepage through dams', in *Contributions to Soil Mechanics 1925–1940*, by permission of the Boston Society of Civil Engineers

β	30°	60°	90°	120°	150°	180°
$\Delta a/a$	(0.36)	0.32	0.26	0.18	0.10	0

the aid of values of the ratio $MD/MA = \Delta a/a$, given by Casagrande for the range of values of β (Table 2.4).

Seepage control in embankment dams

The design of an embankment dam section and, where possible, the choice of soils are aimed at reducing or eliminating the detrimental effects of seeping water. Where high hydraulic gradients exist there is a possibility that the seeping water may cause internal erosion within the dam, especially if the soil is poorly compacted. Erosion can work its way back into the embankment, creating voids in the form of channels or 'pipes', and thus impairing the stability of the dam. This form of erosion is referred to as *piping*.

A section with a central core of low permeability, aimed at reducing the volume of seepage, is shown in Figure 2.20(a). Practically all the total head is lost in the core and if the core is narrow, high hydraulic gradients will result. There is a particular danger of erosion at the boundary between the core and the adjacent soil (of higher permeability) under a high exit gradient from the core. Protection against this danger can be given by means of a 'chimney' drain (Figure 2.20(a)) at the downstream boundary of the core. The drain, designed as a filter to provide a barrier to soil particles from the core, also serves as an interceptor, keeping the downstream slope in an unsaturated state.

Most embankment dam sections are non-homogeneous owing to zones of different soil types, making the construction of the flow net more difficult. The basic parabola construction for the top flow line applies only to homogeneous sections but the condition that there must be equal vertical distances between the points of intersection

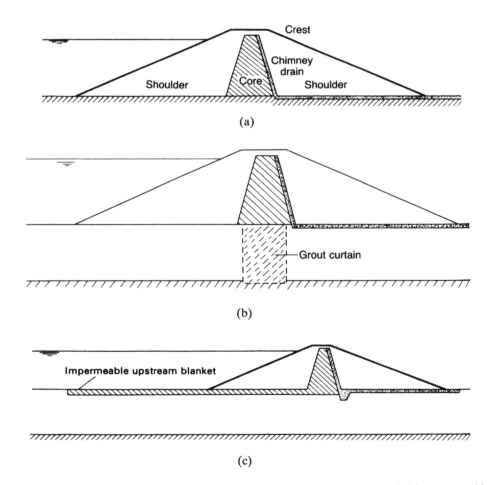

Figure 2.20 (a) Central core and chimney drain, (b) grout curtain and (c) impermeable upstream blanket.

of equipotentials with the top flow line applies equally to a non-homogeneous section. The transfer condition (Equation 2.26) must be satisfied at all zone boundaries. In the case of a section with a central core of low permeability, the application of Equation 2.26 means that the lower the permeability ratio the lower the position of the top flow line in the downstream zone (in the absence of a chimney drain).

If the foundation soil is more permeable than the dam, the control of underseepage is essential. Underseepage can be virtually eliminated by means of an 'impermeable' cut-off such as a grout curtain (Figure 2.20(b)). Another form of cut-off is the concrete diaphragm wall (Section 6.9). Any measure designed to lengthen the seepage path, such as an impermeable upstream blanket (Figure 2.20(c)), will result in a partial reduction in underseepage.

An excellent treatment of seepage control is given by Cedergren [3].

Filter design

Filters used to control seepage must satisfy certain fundamental requirements. The pores must be small enough to prevent particles from being carried in from the adjacent soil. The permeability must be high enough to ensure the free drainage of water entering the filter. The capacity of a filter should be such that it does not become fully saturated. In the case of an embankment dam, a filter placed downstream from the core should be capable of controlling and sealing any leak which develops through the core as a result of internal erosion. The filter must also remain stable under the abnormally high hydraulic gradient which is liable to develop adjacent to such a leak.

Based on extensive laboratory tests by Sherard *et al.* [7, 8] and on design experience, it has been shown that filter performance can be related to the size D_{15} obtained from the particle size distribution curve of the filter material. Average pore size, which is largely governed by the smaller particles in the filter, is well represented by D_{15}. A filter of uniform grading will trap all particles larger than around $0.11D_{15}$; particles smaller than this size will be carried through the filter in suspension in the seeping water. The characteristics of the adjacent soil, in respect of its retention by the filter, can be represented by the size D_{85} for that soil. The following criterion has been recommended for satisfactory filter performance:

$$\frac{(D_{15})_f}{(D_{85})_s} < 5 \tag{2.35}$$

where $(D_{15})_f$ and $(D_{85})_s$ refer to the filter and the adjacent (upstream) soil, respectively. However, in the case of filters for fine soils the following limit is recommended for the filter material:

$$D_{15} \leq 0.5\,\mathrm{mm}$$

Care must be taken to avoid segregation of the component particles of the filter during construction.

To ensure that the permeability of the filter is high enough to allow free drainage, it is recommended that

$$\frac{(D_{15})_f}{(D_{15})_s} > 5 \qquad\qquad (2.36)$$

Graded filters comprising two (or more) layers with different gradings can also be used, the finer layer being on the upstream side. The above criterion (Equation 2.35) would also be applied to the component layers of the filter.

Example 2.4

A homogeneous anisotropic embankment dam section is detailed in Figure 2.21(a), the coefficients of permeability in the x and z directions being 4.5×10^{-8} and 1.6×10^{-8} m/s, respectively. Construct the flow net and determine the quantity of seepage through the dam. What is the pore water pressure at point P?

The scale factor for transformation in the x direction is

$$\sqrt{\frac{k_z}{k_x}} = \sqrt{\frac{1.6}{4.5}} = 0.60$$

The equivalent isotropic permeability is

$$k' = \sqrt{(k_x k_z)}$$
$$= \sqrt{(4.5 \times 1.6)} \times 10^{-8} = 2.7 \times 10^{-8}\,\text{m/s}$$

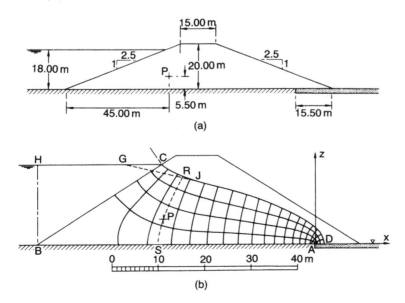

Figure 2.21 Example 2.4.

The section is drawn to the transformed scale as in Figure 2.21(b). The focus of the basic parabola is at point A. The basic parabola passes through point G such that

$$GC = 0.3HC = 0.3 \times 27.00 = 8.10 \, m$$

i.e. the coordinates of G are

$$x = -40.80, \quad z = +18.00$$

Substituting these coordinates in Equation 2.34:

$$-40.80 = x_0 - \frac{18.00^2}{4x_0}$$

Hence

$$x_0 = 1.90 \, m$$

Using Equation 2.34 the coordinates of a number of points on the basic parabola are now calculated:

x	1.90	0	−5.00	−10.00	−20.00	−30.00
z	0	3.80	7.24	9.51	12.90	15.57

The basic parabola is plotted in Figure 2.21(b). The upstream correction is made and the flow net completed, ensuring that there are equal vertical intervals between the points of intersection of successive equipotentials with the top flow line. In the flow net there are 3.8 flow channels and 18 equipotential drops. Hence, the quantity of seepage (per unit length) is

$$q = k'h\frac{N_f}{N_d}$$

$$= 2.7 \times 10^{-8} \times 18 \times \frac{3.8}{18} = 1.0 \times 10^{-7} \, m^3/s$$

The quantity of seepage can also be determined from Equation 2.33 (without the necessity of drawing the flow net):

$$q = 2k'x_0$$

$$= 2 \times 2.7 \times 10^{-8} \times 1.90 = 1.0 \times 10^{-7} \, m^3/s$$

Level AD is selected as datum. An equipotential RS is drawn through point P (transformed position). By inspection the total head at P is 15.60 m. At P the elevation head is 5.50 m, so the pressure head is 10.10 m and the pore water pressure is

$$u_p = 9.8 \times 10.10 = 99 \, kN/m^2$$

Alternatively, the pressure head at P is given directly by the vertical distance of P below the point of intersection (R) of equipotential RS with the top flow line.

Example 2.5

Draw the flow net for the non-homogeneous embankment dam section detailed in Figure 2.22 and determine the quantity of seepage through the dam. Zones 1 and 2 are isotropic, having coefficients of permeability 1.0×10^{-7} and 4.0×10^{-7} m/s, respectively.

The ratio $k_2/k_1 = 4$. The basic parabola is not applicable in this case. Three fundamental conditions must be satisfied in the flow net:

1   There must be equal vertical intervals between points of intersection of equipotentials with the top flow line.
2   If the part of the flow net in zone 1 consists of curvilinear squares then the part in zone 2 must consist of curvilinear rectangles having a length/breadth ratio of 4.
3   For each flow line the transfer condition (Equation 2.26) must be satisfied at the inter-zone boundary.

The flow net is shown in Figure 2.22. In the flow net there are 3.6 flow channels and 8 equipotential drops. The quantity of seepage per unit length is given by

$$q = k_1 h \frac{N_f}{N_d}$$

$$= 1.0 \times 10^{-7} \times 16 \times \frac{3.6}{8} = 7.2 \times 10^{-7}\,\mathrm{m^3/s}$$

(If curvilinear squares are used in zone 2, curvilinear rectangles having a length/breadth ratio of 0.25 must be used in zone 1 and k_2 must be used in the seepage equation.)

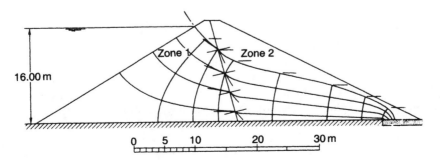

16.00 m

0   5   10   20   30 m

Figure 2.22 Example 2.5. (Reproduced from H.R. Cedergren (1989) *Seepage, Drainage and Flow Nets*, © John Wiley & Sons, Inc., New York, by permission.)

2.9 GROUTING

The permeability of coarse-grained soils can be considerably reduced by means of grouting. The process consists of injecting suitable fluids, known as grouts, into the pore space of the soil; the grout subsequently solidifies, preventing or reducing the seepage of water. Grouting also results in an increase in the strength of the soil. Fluids used for grouting include mixes of cement and water, clay suspensions, chemical solutions, such as sodium silicate or synthetic resins, and bitumen emulsion. Injection is usually effected through a pipe which is either driven into the soil or placed in a borehole and held with a packer.

The particle size distribution of the soil governs the type of grout that can be used. Particles in suspension in a grout, such as cement or clay, will only penetrate soil pores whose size is greater than a certain value; pores smaller than this size will be blocked and grout acceptability will be impaired. Cement and clay grouts are suitable only for gravels and coarse sands. For medium and fine sands, grouts of the solution or emulsion types must be used.

The extent of penetration for a given soil depends on the viscosity of the grout and the pressure under which it is injected. These factors in turn govern the required spacing of the injection points. The injection pressure must be kept below the pressure of the soil overburden, or heaving of the ground surface may occur and fissures may open within the soil. In soils having a wide variation of pore sizes it is expedient to use a primary injection of grout of relatively high viscosity to treat the larger pores, followed by a secondary injection of grout of relatively low viscosity for the smaller pores.

2.10 FROST HEAVE

Frost heave is the rise of the ground surface due to frost action. The freezing of water is accompanied by a volume increase of approximately 9%; therefore, in a saturated soil the void volume above the level of freezing will increase by the same amount, representing an overall increase in the volume of the soil of $2\frac{1}{2}-5\%$ depending on the void ratio. However, under certain circumstances, a much greater increase in volume can occur due to the formation of ice lenses within the soil.

In a soil having a high degree of saturation the pore water freezes immediately below the surface when the temperature falls below 0 °C. The soil temperature increases with depth but during a prolonged period of subzero temperatures the zone of freezing gradually extends downwards. The limit of frost penetration in the UK is normally assumed to be 0.5 m, although under exceptional conditions this depth may approach 1.0 m. The temperature at which water freezes in the pores of a soil depends on the pore size; the smaller the pores the lower the freezing temperature. Water therefore freezes initially in the larger pores, remaining unfrozen in the smaller pores. As the temperature falls below zero, higher soil suction develops and water migrates towards the ice in the larger voids where it freezes and adds to the volume of ice. Continued migration gradually results in the formation of ice lenses and a rise in the ground surface. The process continues only if the bottom of the zone of freezing is within the zone of capillary rise so that water can migrate upwards from below the water table. The magnitude of frost heave decreases as the degree of saturation of the soil decreases.

When thawing eventually takes place the soil previously frozen will contain an excess of water with the result that it will become soft and its strength will be reduced.

In the case of coarse-grained soils with little or no fines, virtually all the pores are large enough for freezing to take place throughout the soil and the only volume increase is due to the 9% increase in the volume of water on freezing. In the case of soils of very low permeability, water migration is restricted by the slow rate of flow; consequently, the development of ice lenses is restricted. However, the presence of fissures can result in an increase in the rate of migration. The worst conditions for water migration occur in soils having a high percentage of silt-size particles; such soils usually have a network of small pores, yet, at the same time, the permeability is not too low. A well-graded soil is reckoned to be frost-susceptible if more than 3% of the particles are smaller than 0.02 mm. A poorly graded soil is susceptible if more than 10% of the particles are smaller than 0.02 mm.

PROBLEMS

2.1 In a falling-head permeability test the initial head of 1.00 m dropped to 0.35 m in 3 h, the diameter of the standpipe being 5 mm. The soil specimen was 200 mm long by 100 mm in diameter. Calculate the coefficient of permeability of the soil.

2.2 A deposit of soil is 16 m deep and overlies an impermeable stratum: the coefficient of permeability is 10^{-6} m/s. A sheet pile wall is driven to a depth of 12.00 m in the deposit. The difference in water level between the two sides of the piling is 4.00 m. Draw the flow net and determine the quantity of seepage under the piling.

2.3 Draw the flow net for seepage under the structure detailed in Figure 2.23 and determine the quantity of seepage. The coefficient of permeability of the soil is 5.0×10^{-5} m/s. What is the uplift force on the base of the structure?

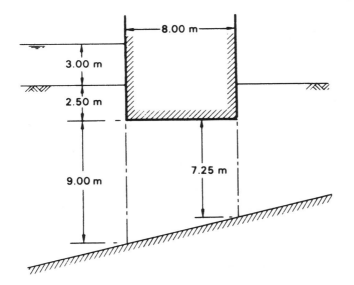

Figure 2.23

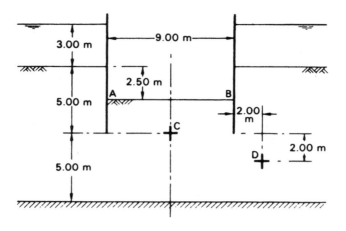

Figure 2.24

2.4 The section through a long cofferdam is shown in Figure 2.24, the coefficient of permeability of the soil being 4.0×10^{-7} m/s. Draw the flow net and determine the quantity of seepage entering the cofferdam.

2.5 The section through part of a cofferdam is shown in Figure 2.25, the coefficient of permeability of the soil being 2.0×10^{-6} m/s. Draw the flow net and determine the quantity of seepage.

2.6 The dam shown in section in Figure 2.26 is located on anisotropic soil. The coefficients of permeability in the x and z directions are 5.0×10^{-7} and 1.8×10^{-7} m/s, respectively. Determine the quantity of seepage under the dam.

2.7 An embankment dam is shown in section in Figure 2.27, the coefficients of permeability in the horizontal and vertical directions being 7.5×10^{-6} and 2.7×10^{-6} m/s, respectively. Construct the top flow line and determine the quantity of seepage through the dam.

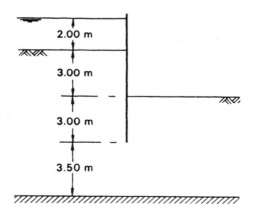

Figure 2.25

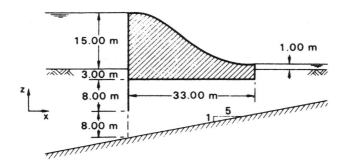

Figure 2.26

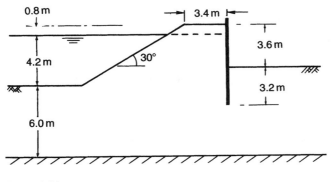

Figure 2.27

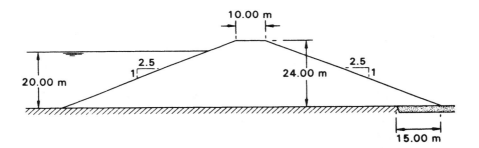

Figure 2.28

2.8 Details of an excavation adjacent to a canal are shown in Figure 2.28. Determine the quantity of seepage into the excavation if the coefficient of permeability is 4.5×10^{-5} m/s.

2.9 Determine the quantity of seepage under the dam shown in section in Figure 2.29. Both layers of soil are isotropic, the coefficients of permeability of the upper and lower layers being 2.0×10^{-6} and 1.6×10^{-5} m/s, respectively.

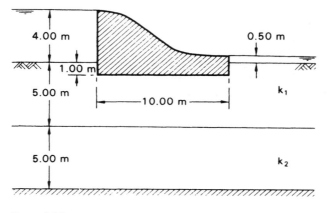

Figure 2.29

REFERENCES

1   British Standard 5930 (1981) *Code of Practice for Site Investigation*, British Standards Institution, London.
2   Casagrande, A. (1940) Seepage through dams, in *Contributions to Soil Mechanics 1925–1940*, Boston Society of Civil Engineers, Boston, MA, pp. 295–336.
3   Cedergren, H.R. (1989) *Seepage, Drainage and Flow Nets*, 3rd edn, John Wiley & Sons, New York.
4   Harr, M.E. (1962) *Groundwater and Seepage*, McGraw-Hill, New York.
5   Hvorslev, M.J. (1951) *Time Lag and Soil Permeability in Ground-Water Observations*, Bulletin No. 36, Waterways Experimental Station, US Corps of Engineers, Vicksburg, MS.
6   Ischy, E. and Glossop, R. (1962) An introduction to alluvial grouting, *Proceedings of the ICE*, **21**, 449–74.
7   Sherard, J.L., Dunnigan, L.P. and Talbot, J.R. (1984) Basic properties of sand and gravel filters, *Journal of the ASCE*, **110**, No. GT6, 684–700.
8   Sherard, J.L., Dunnigan, L.P. and Talbot, J.R. (1984) Filters for silts and clays, *Journal of the ASCE*, **110**, No. GT6, 701–18.
9   Vreedenburgh, C.G.F. (1936) On the steady flow of water percolating through soils with homogeneous-anisotropic permeability, in *Proceedings of the 1st International Conference of SMFE*, Cambridge, MA, Vol. 1.
10   Williams, B.P., Smyrell, A.G. and Lewis, P.J. (1993) Flownet diagrams – the use of finite differences and a spreadsheet to determine potential heads, *Ground Engineering*, **25** (5), 32–8.

Chapter 3

Effective stress

3.1 INTRODUCTION

A soil can be visualized as a skeleton of solid particles enclosing continuous voids which contain water and/or air. For the range of stresses usually encountered in practice the individual solid particles and water can be considered incompressible; air, on the other hand, is highly compressible. The volume of the soil skeleton as a whole can change due to rearrangement of the soil particles into new positions, mainly by rolling and sliding, with a corresponding change in the forces acting between particles. The actual compressibility of the soil skeleton will depend on the structural arrangement of the solid particles. In a fully saturated soil, since water is considered to be incompressible, a reduction in volume is possible only if some of the water can escape from the voids. In a dry or a partially saturated soil a reduction in volume is always possible due to compression of the air in the voids, provided there is scope for particle rearrangement.

Shear stress can be resisted only by the skeleton of solid particles, by means of forces developed at the interparticle contacts. Normal stress may be resisted by the soil skeleton through an increase in the interparticle forces. If the soil is fully saturated, the water filling the voids can also withstand normal stress by an increase in pressure.

3.2 THE PRINCIPLE OF EFFECTIVE STRESS

The importance of the forces transmitted through the soil skeleton from particle to particle was recognized in 1923 when Terzaghi presented the principle of effective stress, an intuitive relationship based on experimental data. The principle applies only to *fully saturated* soils and relates the following three stresses:

1   the *total normal stress* (σ) on a plane within the soil mass, being the force per unit area transmitted in a normal direction across the plane, imagining the soil to be a solid (single-phase) material;
2   the *pore water pressure* (u), being the pressure of the water filling the void space between the solid particles;
3   the *effective normal stress* (σ') on the plane, representing the stress transmitted through the soil skeleton only.

The relationship is:

$$\sigma = \sigma' + u \tag{3.1}$$

The principle can be represented by the following physical model. Consider a 'plane' XX in a fully saturated soil, passing through points of interparticle contact only, as shown in Figure 3.1. The wavy plane XX is really indistinguishable from a true plane on the mass scale due to the relatively small size of individual soil particles. A normal force P applied over an area A may be resisted partly by interparticle forces and partly by the pressure in the pore water. The interparticle forces are very random in both magnitude and direction throughout the soil mass but at every point of contact on the wavy plane may be split into components normal and tangential to the direction of the true plane to which XX approximates; the normal and tangential components are N' and T, respectively. Then, the effective normal stress is interpreted as the sum of all the components N' within the area A, divided by the area A, i.e.

$$\sigma' = \frac{\Sigma N'}{A} \tag{3.2}$$

The total normal stress is given by

$$\sigma = \frac{P}{A} \tag{3.3}$$

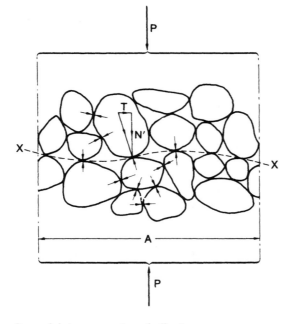

Figure 3.1 Interpretation of effective stress.

If point contact is assumed between the particles, the pore water pressure will act on the plane over the entire area A. Then, for equilibrium in the direction normal to XX

$$P = \Sigma N' + uA$$

or

$$\frac{P}{A} = \frac{\Sigma N'}{A} + u$$

i.e.

$$\sigma = \sigma' + u$$

The pore water pressure which acts equally in every direction will act on the entire surface of any particle but is assumed not to change the volume of the particle; also, the pore water pressure does not cause particles to be pressed together. The error involved in assuming point contact between particles is negligible in soils, the total contact area normally being between 1 and 3% of the cross-sectional area A. It should be understood that σ' does not represent the true contact stress between two particles, which would be the random but very much higher stress N'/a, where a is the actual contact area between the particles.

Effective vertical stress due to self-weight of soil

Consider a soil mass having a horizontal surface and with the water table at surface level. The total vertical stress (i.e. the total normal stress on a horizontal plane) at depth z is equal to the weight of all material (solids + water) per unit area above that depth, i.e.

$$\sigma_v = \gamma_{sat} z$$

The pore water pressure at any depth will be hydrostatic since the void space between the solid particles is continuous, so at depth z

$$u = \gamma_w z$$

Hence, from Equation 3.1 the effective vertical stress at depth z will be

$$\sigma'_v = \sigma_v - u$$
$$= (\gamma_{sat} - \gamma_w)z = \gamma' z$$

where γ' is the buoyant unit weight of the soil.

3.3 RESPONSE OF EFFECTIVE STRESS TO A CHANGE IN TOTAL STRESS

As an illustration of how effective stress responds to a change in total stress, consider the case of a fully saturated soil subject to an *increase* in total vertical stress and in which the lateral strain is zero, volume change being entirely due to deformation of the soil in the vertical direction. This condition may be assumed in practice when there is a change in total vertical stress over an area which is large compared with the thickness of the soil layer in question.

It is assumed initially that the pore water pressure is constant at a value governed by a constant position of the water table. This initial value is called the *static pore water pressure* (u_s). When the total vertical stress is increased, the solid particles immediately try to take up new positions closer together. However, if water is incompressible and the soil is laterally confined, no such particle rearrangement, and therefore no increase in the interparticle forces, is possible unless some of the pore water can escape. Since the pore water is resisting the particle rearrangement the pore water pressure is increased above the static value immediately the increase in total stress takes place. The increase in pore water pressure will be equal to the increase in total vertical stress, i.e. the increase in total vertical stress is carried entirely by the pore water. Note that if the lateral strain were not zero some degree of particle rearrangement would be possible, resulting in an immediate increase in effective vertical stress and the increase in pore water pressure would be less than the increase in total vertical stress.

The increase in pore water pressure causes a pressure gradient, resulting in a transient flow of pore water towards a free-draining boundary of the soil layer. This flow or *drainage* will continue until the pore water pressure again becomes equal to the value governed by the position of the water table. The component of pore water pressure above the static value is known as the *excess pore water pressure* (u_e). It is possible, however, that the position of the water table will have changed during the time necessary for drainage to take place, i.e. the datum against which excess pore water pressure is measured will have changed. In such cases the excess pore water pressure should be expressed with reference to the static value governed by the new water table position. At any time during drainage the overall pore water pressure (u) is equal to the sum of the static and excess components, i.e.

$$u = u_s + u_e \tag{3.4}$$

The reduction of excess pore water pressure as drainage takes place is described as *dissipation* and when this has been completed (i.e. when $u_e = 0$) the soil is said to be in the *drained* condition. Prior to dissipation, with the excess pore water pressure at its initial value, the soil is said to be in the *undrained* condition. It should be noted that the term 'drained' does not mean that all water has flowed out of the soil pores: it means that there is no stress-induced pressure in the pore water. The soil remains fully saturated throughout the process of dissipation.

As drainage of pore water takes place the solid particles become free to take up new positions with a resulting increase in the interparticle forces. In other words, as the excess pore water pressure dissipates, the effective vertical stress increases, accompanied by a corresponding reduction in volume. When dissipation of excess pore water pressure is complete the increment of total vertical stress will be carried entirely by the soil skeleton. The time taken for drainage to be completed depends on the permeability of the soil. In soils of low permeability, drainage will be slow, whereas in soils of high permeability, drainage will be rapid. The whole process is referred to as *consolidation*. With deformation taking place in one direction only, consolidation is described as one-dimensional.

When a soil is subject to a *reduction* in total normal stress the scope for volume increase is limited because particle rearrangement due to total stress increase is largely irreversible. As a result of increase in the interparticle forces there will be small elastic strains (normally ignored) in the solid particles, especially around the contact areas, and if clay mineral particles are present in the soil they may experience bending. In addition, the adsorbed water surrounding clay mineral particles will experience recoverable compression due to increases in interparticle forces, especially if there is face-to-face orientation of the particles. When a decrease in total normal stress takes place in a soil there will thus be a tendency for the soil skeleton to expand to a limited extent, especially in soils containing an appreciable proportion of clay mineral particles. As a result the pore water pressure will be reduced and the excess pore water pressure will be negative. The pore water pressure will gradually increase to the static value, flow taking place into the soil, accompanied by a corresponding reduction in effective normal stress and increase in volume. This process, the reverse of consolidation, is known as *swelling*.

Under seepage (as opposed to static) conditions, the excess pore water pressure is the value above or below the *steady seepage pore water pressure* (u_{ss}), which is determined, at the point in question, from the appropriate flow net.

Consolidation analogy

The mechanics of the one-dimensional consolidation process can be represented by means of a simple analogy. Figure 3.2(a) shows a spring inside a cylinder filled with water and a piston, fitted with a valve, on top of the spring. It is assumed that there can be no leakage between the piston and the cylinder and no friction. The spring

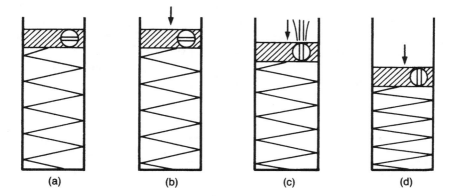

(a)   (b)   (c)   (d)

Figure 3.2 Consolidation analogy.

represents the compressible soil skeleton, the water in the cylinder the pore water and the bore diameter of the valve the permeability of the soil. The cylinder itself simulates the condition of no lateral strain in the soil.

Suppose a load is now placed on the piston with the valve closed, as in Figure 3.2(b). Assuming water to be incompressible, the piston will not move as long as the valve is closed, with the result that no load can be transmitted to the spring; the load will be carried by the water, the increase in pressure in the water being equal to the load divided by the piston area. This situation with the valve closed corresponds to the undrained condition in the soil.

If the valve is now opened, water will be forced out through the valve at a rate governed by the bore diameter. This will allow the piston to move and the spring to be compressed as load is gradually transferred to it. This situation is shown in Figure 3.2(c). At any time the increase in load on the spring will correspond to the reduction in pressure in the water. Eventually, as shown in Figure 3.2(d), all the load will be carried by the spring and the piston will come to rest, this corresponding to the drained condition in the soil. At any time, the load carried by the spring represents the effective normal stress in the soil, the pressure of the water in the cylinder the pore water pressure and the load on the piston the total normal stress. The movement of the piston represents the change in volume of the soil and is governed by the compressibility of the spring (the equivalent of the compressibility of the soil skeleton). The piston and spring analogy represents only an element of soil since the stress conditions vary from point to point throughout a soil mass.

Example 3.1

A layer of saturated clay 4 m thick is overlain by sand 5 m deep, the water table being 3 m below the surface. The saturated unit weights of the clay and sand are 19 and 20 kN/m^3, respectively; above the water table the unit weight of the sand is 17 kN/m^3. Plot the values of total vertical stress and effective vertical stress against depth. If sand to a height of 1 m above the water table is saturated with capillary water, how are the above stresses affected?

The total vertical stress is the weight of all material (solids + water) per unit area above the depth in question. Pore water pressure is the hydrostatic pressure corresponding to the depth below the water table. The effective vertical stress is the difference between the total vertical stress and the pore water pressure at the same depth. Alternatively, effective vertical stress may be calculated directly using the buoyant unit weight of the soil below the water table. The stresses need to be calculated only at depths where there is a change in unit weight (Table 3.1).

Table 3.1

Depth (m)	σ_v (kN/m^2)		u (kN/m^2)	$\sigma_v' = \sigma_v - u$ (kN/m^2)
3	3×17	$= 51.0$	0	51.0
5	$(3 \times 17) + (2 \times 20)$	$= 91.0$	$2 \times 9.8 = 19.6$	71.4
9	$(3 \times 17) + (2 \times 20) + (4 \times 19)$	$= 167.0$	$6 \times 9.8 = 58.8$	108.2

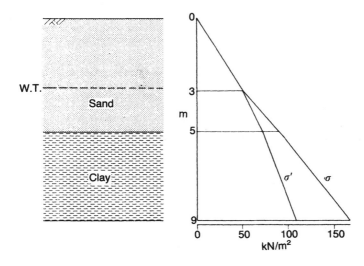

Figure 3.3 Example 3.1.

The alternative calculation of σ'_v at depths of 5 and 9 m is as follows:

Buoyant unit weight of sand $= 20 - 9.8 = 10.2\,\text{kN/m}^3$
Buoyant unit weight of clay $= 19 - 9.8 = 9.2\,\text{kN/m}^3$
At 5 m depth: $\sigma'_v = (3 \times 17) + (2 \times 10.2) = 71.4\,\text{kN/m}^2$
At 9 m depth: $\sigma'_v = (3 \times 17) + (2 \times 10.2) + (4 \times 9.2) = 108.2\,\text{kN/m}^2$

The alternative method is recommended when only the effective stress is required. In all cases the stresses would normally be rounded off to the nearest whole number. The stresses are plotted against depth in Figure 3.3.

Effect of capillary rise

The water table is the level at which pore water pressure is atmospheric (i.e. $u = 0$). Above the water table, water is held under negative pressure and, even if the soil is saturated above the water table, does not contribute to hydrostatic pressure below the water table. The only effect of the 1 m capillary rise, therefore, is to increase the total unit weight of the sand between 2 and 3 m depth from 17 to $20\,\text{kN/m}^3$, an increase of $3\,\text{kN/m}^3$. Both total and effective vertical stresses below 3 m depth are therefore increased by the constant amount $3 \times 1 = 3.0\,\text{kN/m}^2$, pore water pressures being unchanged.

Example 3.2

A 5 m depth of sand overlies a 6 m layer of clay, the water table being at the surface; the permeability of the clay is very low. The saturated unit weight of the sand is $19 \, \text{kN/m}^3$ and that of the clay is $20 \, \text{kN/m}^3$. A 4 m depth of fill material of unit weight $20 \, \text{kN/m}^3$ is placed on the surface over an extensive area. Determine the effective vertical stress at the centre of the clay layer (a) immediately after the fill has been placed, assuming this to take place rapidly and (b) many years after the fill has been placed.

The soil profile is shown in Figure 3.4. Since the fill covers an extensive area it can be assumed that the condition of zero lateral strain applies. As the permeability of the clay is very low, dissipation of excess pore water pressure will be very slow; immediately after the rapid placing of the fill, no appreciable dissipation will have taken place. Therefore, the effective vertical stress at the centre of the clay layer immediately after placing will be virtually unchanged from the original value, i.e.

$$\sigma'_v = (5 \times 9.2) + (3 \times 10.2) = 76.6 \, \text{kN/m}^2$$

(the buoyant unit weights of the sand and the clay, respectively, being 9.2 and $10.2 \, \text{kN/m}^3$).

Many years after the placing of the fill, dissipation of excess pore water pressure should be essentially complete and the effective vertical stress at the centre of the clay layer will be

$$\sigma'_v = (4 \times 20) + (5 \times 9.2) + (3 \times 10.2) = 156.6 \, \text{kN/m}^2$$

Immediately after the fill has been placed, the total vertical stress at the centre of the clay increases by $80 \, \text{kN/m}^2$ due to the weight of the fill. Since the clay is saturated and there is no lateral strain there will be a corresponding increase in pore water pressure of $80 \, \text{kN/m}^2$ (the initial excess pore water pressure). The static pore water pressure is $(8 \times 9.8) = 78.4 \, \text{kN/m}^2$. Immediately after placing, the pore water pressure increases from 78.4 to $158.4 \, \text{kN/m}^2$ and then during subsequent consolidation gradually decreases again to $78.4 \, \text{kN/m}^2$, accompanied by the gradual increase of effective vertical stress from 76.6 to $156.6 \, \text{kN/m}^2$.

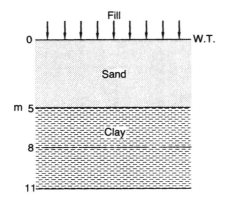

Figure 3.4 Example 3.2.

3.4   PARTIALLY SATURATED SOILS

In the case of partially saturated soils part of the void space is occupied by water and part by air. The pore water pressure (u_w) must always be less than the pore air pressure (u_a) due to surface tension. Unless the degree of saturation is close to unity the pore air will form continuous channels through the soil and the pore water will be concentrated in the regions around the interparticle contacts. The boundaries between pore water and pore air will be in the form of menisci whose radii will depend on the size of the pore spaces within the soil. Part of any wavy plane through the soil will therefore pass through water and part through air.

In 1955 Bishop proposed the following effective stress equation for partially saturated soils:

$$\sigma = \sigma' + u_a - \chi(u_a - u_w) \qquad\qquad (3.5)$$

where χ is a parameter, to be determined experimentally, related primarily to the degree of saturation of the soil. The term $(u_a - u_w)$ is a measure of the suction in the soil. For a fully saturated soil ($S_r = 1$), $\chi = 1$; and for a completely dry soil ($S_r = 0$), $\chi = 0$. Equation 3.5 thus degenerates to Equation 3.1 when $S_r = 1$. The value of χ is also influenced, to a lesser extent, by the soil structure and the way the particular degree of saturation was brought about. Equation 3.5 is not convenient for use in practice because of the presence of the parameter χ.

A physical model may be considered in which the parameter χ is interpreted as the average proportion of any cross-section which passes through water. Then, across a given section of gross area A (Figure 3.5) total force is given by the equation

$$\sigma A = \sigma' A + u_w \chi A + u_a (1 - \chi) A \qquad\qquad (3.6)$$

which leads to Equation 3.5.

If the degree of saturation of the soil is close to unity it is likely that the pore air will exist in the form of bubbles within the pore water and it is possible to draw a wavy plane through pore water only. The soil can then be considered as a fully saturated soil but with the pore water having some degree of compressibility due to the presence of

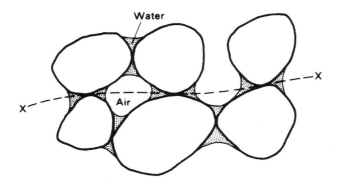

Figure 3.5 Partially saturated soil.

the air bubbles; Equation 3.1 may then represent effective stress with sufficient accuracy for most practical purposes.

3.5 INFLUENCE OF SEEPAGE ON EFFECTIVE STRESS

When water is seeping through the pores of a soil, total head is dissipated as viscous friction producing a frictional drag, acting in the direction of flow, on the solid particles. A transfer of energy thus takes place from the water to the solid particles and the force corresponding to this energy transfer is called *seepage force*. Seepage force acts on the particles of a soil in addition to gravitational force and the combination of the forces on a soil mass due to gravity and seeping water is called the resultant body force. It is the resultant body force that governs the effective normal stress on a plane within a soil mass through which seepage is taking place.

Consider a point in a soil mass where the direction of seepage is at angle θ below the horizontal. A square element ABCD of dimension b (unit dimension normal to the paper) is centred at the above point with sides parallel and normal to the direction of seepage, as shown in Figure 3.6(a), i.e. the square element can be considered as a flow net field. Let the drop in total head between the sides AD and BC be Δh. Consider the pore water pressures on the boundaries of the element, taking the value of pore water pressure at point A as u_A. The difference in pore water pressure between A and D is due only to the difference in elevation head between A and D, the total head being the same at A and D. However, the difference in pore water pressure between A and either B or C is due to the difference in elevation head and the difference in total head between A and either B or C. The pore water pressures at B, C and D are as follows:

$$u_B = u_A + \gamma_w(b\sin\theta - \Delta h)$$
$$u_C = u_A + \gamma_w(b\sin\theta + b\cos\theta - \Delta h)$$
$$u_D = u_A + \gamma_w b\cos\theta$$

The following pressure differences can now be established:

$$u_B - u_A = u_C - u_D = \gamma_w(b\sin\theta - \Delta h)$$
$$u_D - u_A = u_C - u_B = \gamma_w b\cos\theta$$

These values are plotted in Figure 3.6(b), giving the distribution diagrams of net pressure across the element in directions parallel and normal to the direction of flow.

Therefore, the *force* on BC due to pore water pressure acting on the boundaries of the element, called the *boundary water force*, is given by

$$\gamma_w(b\sin\theta - \Delta h)b$$

or

$$\gamma_w b^2 \sin\theta - \Delta h \gamma_w b$$

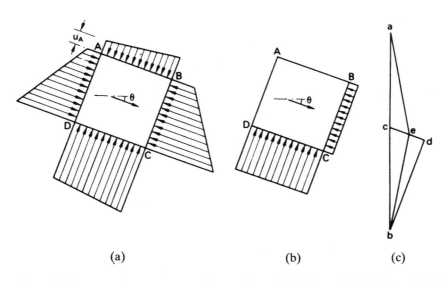

Figure 3.6 Forces under seepage conditions. (Reproduced from D.W. Taylor (1948) *Fundamentals of Soil Mechanics*, © John Wiley & Sons Inc. [2], by permission.)

and the boundary water force on CD by

$$\gamma_w b^2 \cos \theta$$

If there were no seepage, i.e. if the pore water were static, the value of Δh would be zero, the forces on BC and CD would be $\gamma_w b^2 \sin \theta$ and $\gamma_w b^2 \cos \theta$, respectively, and their resultant would be $\gamma_w b^2$ acting in the vertical direction. The force $\Delta h \gamma_w b$ represents the only difference between the static and seepage cases and is therefore called the seepage force (J), *acting in the direction of flow* (in this case normal to BC).

Now, the average hydraulic gradient across the element is given by

$$i = \frac{\Delta h}{b}$$

hence,

$$J = \Delta h \gamma_w b = \frac{\Delta h}{b} \gamma_w b^2 = i \gamma_w b^2$$

or

$$J = i \gamma_w V \qquad\qquad (3.7)$$

where V is the volume of the soil element.

The *seepage pressure* (j) is defined as the seepage force per unit volume, i.e.

$$j = i\gamma_w \tag{3.8}$$

It should be noted that j (and hence J) depends only on the value of hydraulic gradient.

All the forces, both gravitational and forces due to seeping water, acting on the element ABCD, may be represented in the vector diagram (Figure 3.6(c)). The forces are summarized below.

Total weight of the element $= \gamma_{sat}b^2 =$ vector ab
Boundary water force on CD (seepage and static cases)
$$= \gamma_w b^2 \cos\theta = \text{vector } bd$$
Boundary water force on BC (seepage case)
$$= \gamma_w b^2 \sin\theta - \Delta h \gamma_w b = \text{vector } de$$
Boundary water force on BC (static case)
$$= \gamma_w b^2 \sin\theta = \text{vector } dc$$
Resultant boundary water force (seepage case)
$$= \text{vector } be$$
Resultant boundary water force (static case)
$$= \gamma_w b^2 = \text{vector } bc$$
Seepage force $\qquad = \Delta h \gamma_w b = \text{vector } ce$
Resultant body force (seepage case)
$$= \text{vector } ae$$
Resultant body force (static case)
$$= \text{vector } ac = \gamma' b^2$$

The resultant body force can be obtained by one or other of the following force combinations:

1   Total (saturated) weight + resultant boundary water force, i.e. vector ab + vector be.
2   Effective (buoyant) weight + seepage force, i.e. vector ac + vector ce.

Only the resultant body force contributes to effective stress. A component of seepage force acting vertically upwards will therefore reduce a vertical effective stress component from the static value. A component of seepage force acting vertically downwards will increase a vertical effective stress component from the static value.

A problem may be solved using either force combination 1 or force combination 2, but it may be that one combination is more suitable than the other for a particular problem. Combination 1 involves consideration of the equilibrium of the whole soil mass (solids + water), while combination 2 involves consideration of the equilibrium of the soil skeleton only.

The quick condition

Consider the special case of seepage vertically upwards. The vector ce in Figure 3.6(c) would then be vertically upwards and if the hydraulic gradient were high enough the resultant body force would be zero. The value of hydraulic gradient corresponding to zero resultant body force is called the *critical hydraulic gradient* (i_c). For an element of soil of volume V subject to upward seepage under the critical hydraulic gradient, the seepage force is therefore equal to the effective weight of the element, i.e.

$$i_c \gamma_w V = \gamma' V$$

Therefore

$$i_c = \frac{\gamma'}{\gamma_w} = \frac{G_s - 1}{1 + e} \tag{3.9}$$

The ratio γ'/γ_w, and hence the critical hydraulic gradient, is approximately 1.0 for most soils.

When the hydraulic gradient is i_c, the effective normal stress on any plane will be zero, gravitational forces having been cancelled out by upward seepage forces. In the case of sands the contact forces between particles will be zero and the soil will have no strength. The soil is then said to be in a *quick* condition (quick meaning 'alive') and if the critical gradient is exceeded the surface will appear to be 'boiling' as the particles are moved around in the upward flow of water. It should be realized that 'quicksand' is not a special type of soil but simply sand through which there is an upward flow of water under a hydraulic gradient equal to or exceeding i_c. In the case of clays, the quick condition may not necessarily result when the hydraulic gradient reaches the critical value given by Equation 3.9.

Conditions adjacent to sheet piling

High upward hydraulic gradients may be experienced in the soil adjacent to the downstream face of a sheet pile wall. Figure 3.7 shows part of the flow net for seepage under a sheet pile wall, the embedded length on the downstream side being d. A mass of soil adjacent to the piling may become unstable and be unable to support the wall. Model tests have shown that failure is likely to occur within a soil mass of approximate dimensions $d \times d/2$ in section (ABCD in Figure 3.7). Failure first shows in the form of a rise or *heave* at the surface, associated with an expansion of the soil which results in an increase in permeability. This in turn leads to increased flow, surface 'boiling' in the case of sands and complete failure.

The variation of total head on the lower boundary CD of the soil mass can be obtained from the flow net equipotentials, but for purposes of analysis it is sufficient to determine the average total head h_m by inspection. The total head on the upper boundary AB is zero.

The average hydraulic gradient is given by

$$i_m = \frac{h_m}{d}$$

Since failure due to heaving may be expected when the hydraulic gradient becomes i_c, the factor of safety (F) against heaving may be expressed as

$$F = \frac{i_c}{i_m} \tag{3.10}$$

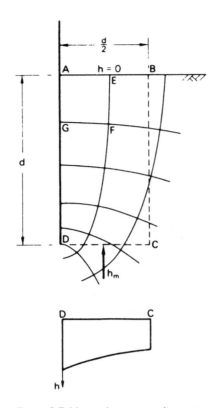

Figure 3.7 Upward seepage adjacent to sheet piling.

In the case of sands, a factor of safety can also be obtained with respect to 'boiling' at the surface. The *exit* hydraulic gradient (i_e) can be determined by measuring the dimension Δs of the flow net field AEFG adjacent to the piling:

$$i_e = \frac{\Delta h}{\Delta s}$$

where Δh is the drop in total head between equipotentials GF and AE. Then, the factor of safety is

$$F = \frac{i_c}{i_e} \tag{3.11}$$

There is unlikely to be any appreciable difference between the values of F given by Equations 3.10 and 3.11.

The sheet pile wall problem shown in Figure 3.7 can also be used to illustrate the two methods of combining gravitational and water forces.

1   Total weight of mass ABCD $= \frac{1}{2}\gamma_{sat}d^2$

Average total head on CD $= h_m$
Elevation head on CD $= -d$
Average pore water pressure on CD $= (h_m + d)\gamma_w$
Boundary water force on CD $= \frac{d}{2}(h_m + d)\gamma_w$

$$\text{Resultant body force of ABCD} = \frac{1}{2}\gamma_{sat}d^2 - \frac{d}{2}(h_m + d)\gamma_w$$
$$= \frac{1}{2}(\gamma' + \gamma_w)d^2 - \frac{1}{2}(h_m d + d^2)\gamma_w$$
$$= \frac{1}{2}\gamma'd^2 - \frac{1}{2}h_m\gamma_w d$$

2   Effective weight of mass ABCD $= \frac{1}{2}\gamma'd^2$

Average hydraulic gradient through ABCD $= \dfrac{h_m}{d}$

$$\text{Seepage force on ABCD} = \frac{h_m}{d}\gamma_w \frac{d^2}{2}$$
$$= \frac{1}{2}h_m\gamma_w d$$

Resultant body force of ABCD$= \frac{1}{2}\gamma'd^2 - \frac{1}{2}h_m\gamma_w d$ as in method 1 above.
The resultant body force will be zero, leading to heaving, when

$$\frac{1}{2}h_m\gamma_w d = \frac{1}{2}\gamma'd^2$$

The factor of safety can then be expressed as

$$F = \frac{\frac{1}{2}\gamma'd^2}{\frac{1}{2}h_m\gamma_w d} = \frac{\gamma'd}{h_m\gamma_w} = \frac{i_c}{i_m}$$

If the factor of safety against heaving is considered inadequate, the embedded length d may be increased or a surcharge load in the form of a filter may be placed on the surface AB, the filter being designed to prevent entry of soil particles. If the effective weight of the filter per unit area is w' then the factor of safety becomes

$$F = \frac{\gamma'd + w'}{h_m\gamma_w}$$

Example 3.3

The flow net for seepage under a sheet pile wall is shown in Figure 3.8(a), the saturated unit weight of the soil being 20 kN/m³. Determine the values of effective vertical stress at A and B.

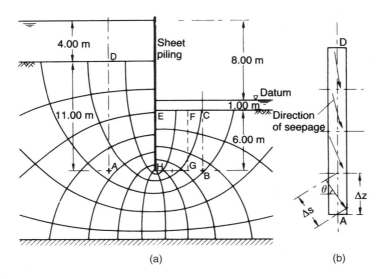

Figure 3.8 Examples 3.3 and 3.4.

1   First consider the combination of total weight and resultant boundary water force. Consider the column of saturated soil of *unit area* between A and the soil surface at D. The total weight of the column is $11\gamma_{sat}$ (220 kN). Due to the change in level of the equipotentials across the column, the boundary water forces on the sides of the column will not be equal although in this case the difference will be small. There is thus a net horizontal boundary water force on the column. However, as the effective *vertical* stress is to be calculated, only the vertical component of the resultant body force is required and the net horizontal boundary water force need not be considered. The boundary water force on the top surface of the column is only due to the depth of water above D and is $4\gamma_w$ (39 kN). The boundary water force on the bottom surface of the column must be determined from the flow net, as follows:

> Number of equipotential drops between the downstream soil surface and A = 8.2. There are 12 equipotential drops between the upstream and downstream soil surfaces, representing a loss in total head of 8 m.
>
> Total head at A, $h_A = \dfrac{8.2}{12} \times 8 = 5.5\,\text{m}$.
>
> Elevation head at A, $z_A = -7.0\,\text{m}$.
>
> Pore water pressure at A, $u_A = \gamma_w(h_A - z_A)$
> $$= 9.8(5.5 + 7.0) = 122\,\text{kN/m}^2$$
> i.e. boundary water force on bottom surface = 122 kN.
> Net vertical boundary water force = 122 − 39 = 83 kN.
> Total weight of the column = 220 kN.
> Vertical component of resultant body force = 220 − 83 = 137 kN
> i.e. effective vertical stress at A = 137 kN/m².

It should be realized that the same result would be obtained by the direct application of the effective stress equation, the total vertical stress at A being the weight of saturated soil and water, per unit area, above A. Thus

$$\sigma_A = 11\gamma_{sat} + 4\gamma_w = 220 + 39 = 259\,kN/m^2$$

$$u_A = 122\,kN/m^2$$

$$\sigma'_A = \sigma_A - u_A = 259 - 122 = 137\,kN/m^2$$

The only difference in concept is that the boundary water force per unit area on top of the column of saturated soil AD contributes to the total vertical stress at A. Similarly at B

$$\sigma_B = 6\gamma_{sat} + 1\gamma_w = 120 + 9.8 = 130\,kN/m^2$$

$$h_B = \frac{2.4}{12} \times 8 = 1.6\,m$$

$$z_B = -7.0\,m$$

$$u_B = \gamma_w(h_B - z_B) = 9.8(1.6 + 7.0) = 84\,kN/m^2$$

$$\sigma'_B = \sigma_B - u_B = 130 - 84 = 46\,kN/m^2$$

2   Now consider the combination of effective weight and seepage force. The direction of seepage alters over the depth of the column of soil AD as illustrated in Figure 3.8(b), the direction of seepage for any section of the column being determined from the flow net; the effective weight of the column must be combined with the vertical components of seepage force. More conveniently, the effective stress at A can be calculated using the algebraic sum of the buoyant unit weight of the soil and the average value of the vertical component of seepage pressure between A and D.

Between any two equipotentials the hydraulic gradient is $\Delta h/\Delta s$ (Equation 2.17). Hence, if θ is the angle between the direction of flow and the horizontal, the vertical component of seepage pressure ($j\sin\theta$) is

$$\frac{\Delta h}{\Delta s}\gamma_w\sin\theta = \frac{\Delta h}{\Delta z}\gamma_w$$

where $\Delta z\ (=\Delta s/\sin\theta)$ is the vertical distance between the same equipotentials. The calculation is as follows.

Number of equipotential drops between D and A $= 3.8$.

Loss in total head between D and A $= \dfrac{3.8}{12} \times 8 = 2.5\,m$.

Average value of vertical component of seepage pressure between D and A, acting in the same direction as gravity

$$= \frac{2.5}{11} \times 9.8 = 2.3\,kN/m^3$$

Buoyant unit weight of soil, $\gamma' = 20 - 9.8 = 10.2\,\text{kN/m}^3$.
For column AD, of unit area, resultant body force

$$= 11(10.2 + 2.3) = 137\,\text{kN}$$

i.e. effective vertical stress at A = $137\,\text{kN/m}^2$.

The calculation is now given for point B.

Loss in total head between B and $C = \dfrac{2.4}{12} \times 8 = 1.6\,\text{m}$.

Average value of vertical component of seepage pressure between B and C, acting in the opposite direction to gravity

$$= \frac{1.6}{6} \times 9.8 = 2.6\,\text{kN/m}^3$$

Hence, $\sigma'_B = 6(10.2 - 2.6) = 46\,\text{kN/m}^2$.

Example 3.4

Using the flow net in Figure 3.8(a), determine the factor of safety against failure by heaving adjacent to the downstream face of the piling. The saturated unit weight of the soil is $20\,\text{kN/m}^3$.

The stability of the soil mass EFGH in Figure 3.8(a), 6 m by 3 m in section, will be analyzed.

By inspection of the flow net, the average value of total head on the base GH is given by

$$h_m = \frac{3.5}{12} \times 8 = 2.3\,\text{m}$$

The average hydraulic gradient between GH and the soil surface EF is

$$i_m = \frac{2.3}{6} = 0.39$$

Critical hydraulic gradient, $i_c = \dfrac{\gamma'}{\gamma_w} = \dfrac{10.2}{9.8} = 1.04$

Factor of safety, $F = \dfrac{i_c}{i_m} = \dfrac{1.04}{0.39} = 2.7$

PROBLEMS

3.1   A river is 2 m deep. The river bed consists of a depth of sand of saturated unit weight $20\,\text{kN/m}^3$. What is the effective vertical stress 5 m below the top of the sand?

3.2   The North Sea is 200 m deep. The sea bed consists of a depth of sand of saturated unit weight $20\,\text{kN/m}^3$. What is the effective vertical stress 5 m below the top of the sand?

3.3 A layer of clay 4 m thick lies between two layers of sand each 4 m thick, the top of the upper layer of sand being ground level. The water table is 2 m below ground level but the lower layer of sand is under artesian pressure, the piezometric surface being 4 m above ground level. The saturated unit weight of the clay is $20 \, kN/m^3$ and that of the sand $19 \, kN/m^3$; above the water table the unit weight of the sand is $16.5 \, kN/m^3$. Calculate the effective vertical stresses at the top and bottom of the clay layer.

3.4 In a deposit of fine sand the water table is 3.5 m below the surface but sand to a height of 1.0 m above the water table is saturated by capillary water; above this height the sand may be assumed to be dry. The saturated and dry unit weights, respectively, are 20 and $16 \, kN/m^3$. Calculate the effective vertical stress in the sand 8 m below the surface.

3.5 A layer of sand extends from ground level to a depth of 9 m and overlies a layer of clay, of very low permeability, 6 m thick. The water table is 6 m below the surface of the sand. The saturated unit weight of the sand is $19 \, kN/m^3$ and that of the clay $20 \, kN/m^3$; the unit weight of the sand above the water table is $16 \, kN/m^3$. Over a short period of time the water table rises by 3 m and is expected to remain permanently at this new level. Determine the effective vertical stress at depths of 8 and 12 m below ground level (a) immediately after the rise of the water table and (b) several years after the rise of the water table.

3.6 An element of soil with sides horizontal and vertical measures 1 m in each direction. Water is seeping through the element in a direction inclined upwards at 30° above the horizontal under a hydraulic gradient of 0.35. The saturated unit weight of the soil is $21 \, kN/m^3$. Draw a force diagram to scale showing the following: total and effective weights, resultant boundary water force, seepage force. What is the magnitude and direction of the resultant body force?

3.7 For the seepage situations shown in Figure 3.9, determine the effective normal stress on plane XX in each case (a) by considering pore water pressure

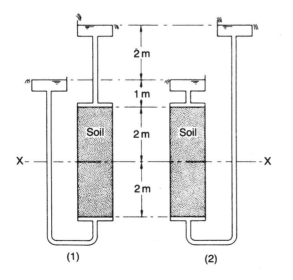

Figure 3.9

and (b) by considering seepage pressure. The saturated unit weight of the soil is $20 \, \text{kN/m}^3$.

3.8   The section through a long cofferdam is shown in Figure 2.24, the saturated unit weight of the soil being $20 \, \text{kN/m}^3$. Determine the factor of safety against 'boiling' at the surface AB and the values of effective vertical stress at C and D.

3.9   The section through part of a cofferdam is shown in Figure 2.25, the saturated unit weight of the soil being $19.5 \, \text{kN/m}^3$. Determine the factor of safety against a heave failure in the excavation adjacent to the sheet piling. What depth of filter (unit weight $21 \, \text{kN/m}^3$) would be required to ensure a factor of safety of 3.0?

REFERENCES

1   Skempton, A.W. (1961) Effective stress in soils, concrete and rocks, in *Proceedings of Conference on Pore Pressure and Suction in Soils*, Butterworths, London, pp. 4–16.
2   Taylor, D.W. (1948) *Fundamentals of Soil Mechanics*, John Wiley & Sons, New York.
3   Terzaghi, K. (1943) *Theoretical Soil Mechanics*, John Wiley & Sons, New York.

Chapter 4

Shear strength

4.1 SHEAR FAILURE

This chapter is concerned with the resistance of a soil to failure in shear, a knowledge of which is required in analysis of the stability of soil masses. If at a point on any plane within a soil mass the shear stress becomes equal to the shear strength of the soil then failure will occur at that point. Originally, prior to the postulation of the principle of effective stress, the shear strength (τ_f) of a soil at a point on a particular plane was expressed by Coulomb as a linear function of the normal stress at failure (σ_f) on the plane at the same point:

$$\tau_f = c + \sigma_f \tan \phi \tag{4.1}$$

where c and ϕ are the *shear strength parameters* referred to as the *cohesion intercept* and the *angle of shearing resistance*, respectively. However, in accordance with the principle that shear stress in a soil can be resisted only by the skeleton of solid particles, shear strength should be expressed as a function of *effective* normal stress at failure (σ_f'), the shear strength parameters being denoted c' and ϕ':

$$\tau_f = c' + \sigma_f' \tan \phi' \tag{4.2}$$

Failure will thus occur at any point in the soil where a critical combination of shear stress and effective normal stress develops. It should be appreciated that c' and ϕ' are simply mathematical constants defining a linear relationship between shear strength and effective normal stress. Shearing resistance is developed by interparticle forces; therefore, if effective normal stress is zero then shearing resistance must be zero (unless there is cementation between the particles) and the value of c' would be zero. This point is crucial to the interpretation of shear strength parameters.

States of stress in two dimensions can be represented on a plot of shear stress (τ) against effective normal stress (σ'). A stress state can be represented either by a point with coordinates τ and σ', or by a Mohr circle defined by the effective principal stresses σ_1' and σ_3'. Stress points and Mohr circles representing stress states at failure are shown in Figures 4.1(a) and (b). The line through the stress points or the line touching the Mohr circles may be straight or slightly curved and is referred to as the *failure envelope*. A state of stress represented by a stress point that plots above the failure envelope, or by a Mohr circle part of which lies above the envelope, is impossible.

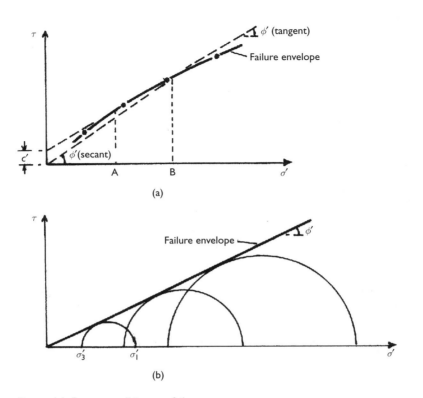

Figure 4.1 Stress conditions at failure.

There are two methods of specifying shear strength parameters. (1) The envelope is represented by the straight line defined by Equation 4.2, from which the parameters c' and ϕ' can be obtained. These are referred to as *tangent parameters* and are only valid over a limited stress range. This has been the traditional approach to representing shear strength. If the straight line passes through the origin, as in Figure 4.1(b), then, of course, c' is zero. If the failure envelope is slightly curved the parameters are obtained from a straight line approximation to the curve over the stress range of interest, e.g. between A and B in Figure 4.1(a). It should be appreciated that the use of tangent parameters does *not* infer that the shear strength is c' at zero effective normal stress. (2) A straight line is drawn between a particular stress point and the origin, as in Figure 4.1(a), or a line is drawn through the origin and tangential to a particular Mohr circle. The parameter c' is zero and the slope of the line gives ϕ', the shear strength equation being $\tau_f = \sigma'_f \tan \phi'$. The angle ϕ', determined in this way, is referred to as a *secant parameter* and is valid only for one particular stress state. Generally, the value of secant ϕ' used in practice would be that corresponding to the highest expected value of effective normal stress (i.e. the lowest value of the parameter for the stress range of interest).

The relationship between the shear strength parameters and the effective principal stresses at failure at a particular point can be deduced. The general case with $c' > 0$ is

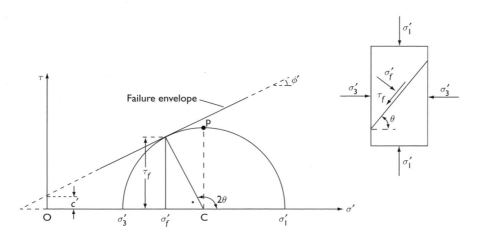

Figure 4.2 Mohr–Coulomb failure criterion.

represented in Figure 4.2, compressive stress being taken as positive. The coordinates of the tangent point are τ_f and σ'_f where

$$\tau_f = \frac{1}{2}(\sigma'_1 - \sigma'_3)\sin 2\theta \tag{4.3}$$

$$\sigma'_f = \frac{1}{2}(\sigma'_1 + \sigma'_3) + \frac{1}{2}(\sigma'_1 - \sigma'_3)\cos 2\theta \tag{4.4}$$

and θ is the theoretical angle between the major principal plane and the plane of failure. It is apparent that

$$\theta = 45° + \frac{\phi'}{2} \tag{4.5}$$

Now

$$\sin\phi' = \frac{\frac{1}{2}(\sigma'_1 - \sigma'_3)}{c'\cot\phi' + \frac{1}{2}(\sigma'_1 + \sigma'_3)}$$

Therefore

$$(\sigma'_1 - \sigma'_3) = (\sigma'_1 + \sigma'_3)\sin\phi' + 2c'\cos\phi' \tag{4.6a}$$

or

$$\sigma'_1 = \sigma'_3\tan^2\left(45° + \frac{\phi'}{2}\right) + 2c'\tan\left(45° + \frac{\phi'}{2}\right) \tag{4.6b}$$

Equation 4.6 is referred to as the Mohr–Coulomb failure criterion.

For a given state of stress it is apparent that, because $\sigma_1' = \sigma_1 - u$ and $\sigma_3' = \sigma_3 - u$, the Mohr circles for total and effective stresses have the same diameter but their centres are separated by the corresponding pore water pressure u. Similarly, total and effective stress points are separated by the value of u.

The state of stress represented in Figure 4.2 could also be defined by the coordinates of point P, rather than by the Mohr circle. The coordinates of P are $\frac{1}{2}(\sigma_1' - \sigma_3')$ and $\frac{1}{2}(\sigma_1' + \sigma_3')$, also denoted by t' and s', respectively, being the maximum shear stress and the average principal stress. The stress state could also be expressed in terms of total stress. It should be noted that

$$\frac{1}{2}(\sigma_1' - \sigma_3') = \frac{1}{2}(\sigma_1 - \sigma_3)$$

$$\frac{1}{2}(\sigma_1' + \sigma_3') = \frac{1}{2}(\sigma_1 + \sigma_3) - u$$

Stress point P lies on a modified failure envelope defined by the equation

$$\frac{1}{2}(\sigma_1' - \sigma_3') = a' + \frac{1}{2}(\sigma_1' + \sigma_3')\tan\alpha' \qquad (4.7)$$

where a' and α' are the modified shear strength parameters. The parameters c' and ϕ' are then given by

$$\phi' = \sin^{-1}(\tan\alpha') \qquad (4.8)$$

$$c' = \frac{a'}{\cos\phi'} \qquad (4.9)$$

4.2   SHEAR STRENGTH TESTS

The shear strength parameters for a particular soil can be determined by means of laboratory tests on specimens taken from *representative* samples of the *in-situ* soil. Great care and judgement are required in the sampling operation and in the storage and handling of samples prior to testing, especially in the case of undisturbed samples where the object is to preserve the *in-situ* structure and water content of the soil. In the case of clays, test specimens may be obtained from tube or block samples, the latter normally being subjected to the least disturbance. Swelling of a clay specimen will occur due to the release of the *in-situ* total stresses.

Shear strength test procedure is detailed in BS 1377 (Parts 7 and 8) [7].

The direct shear test

The specimen is confined in a metal box (known as the shearbox) of square or circular cross-section split horizontally at mid-height, a small clearance being maintained between the two halves of the box. Porous plates are placed below and on top of the specimen if it is fully or partially saturated to allow free drainage: if the specimen is

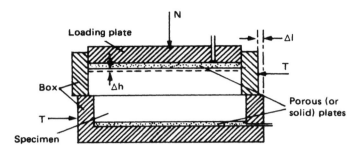

Figure 4.3 Direct shear apparatus.

dry, solid metal plates may be used. The essential features of the apparatus are shown diagrammatically in Figure 4.3. A vertical force (N) is applied to the specimen through a loading plate and shear stress is gradually applied on a horizontal plane by causing the two halves of the box to move relative to each other, the shear force (T) being measured together with the corresponding shear displacement (Δl). Normally, the change in thickness (Δh) of the specimen is also measured. If the initial thickness of the specimen is h_0 then the shear strain (γ) can be represented by $\Delta l / h_0$ and the volumetric strain (ε_v) by $\Delta h / h_0$. A number of specimens of the soil are tested, each under a different vertical force, and the value of shear stress at failure is plotted against the normal stress for each test. The shear strength parameters are then obtained from the best line fitting the plotted points.

The test suffers from several disadvantages, the main one being that drainage conditions cannot be controlled. As pore water pressure cannot be measured, only the total normal stress can be determined, although this is equal to the effective normal stress if the pore water pressure is zero. Only an approximation to the state of pure shear is produced in the specimen and shear stress on the failure plane is not uniform, failure occurring progressively from the edges towards the centre of the specimen. The area under the shear and vertical loads does not remain constant throughout the test. The advantages of the test are its simplicity and, in the case of sands, the ease of specimen preparation.

The triaxial test

This is the most widely used shear strength test and is suitable for all types of soil. The test has the advantages that drainage conditions can be controlled, enabling saturated soils of low permeability to be consolidated, if required, as part of the test procedure, and pore water pressure measurements can be made. A cylindrical specimen, generally having a length/diameter ratio of 2, is used in the test and is stressed under conditions of axial symmetry in the manner shown in Figure 4.4. Typical specimen diameters are 38 and 100 mm. The main features of the apparatus are shown in Figure 4.5. The circular base has a central pedestal on which the specimen is placed, there being access through the pedestal for drainage and for the measurement of pore water pressure. A Perspex cylinder, sealed between a ring and the circular cell top, forms the body of the cell. The cell top has a central bush through which the loading ram passes. The cylinder and cell top clamp onto the base, a seal being made by means of an O-ring.

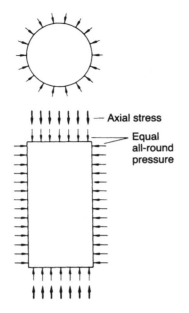

Axial stress

Equal
all-round
pressure

Figure 4.4 Stress system in triaxial test.

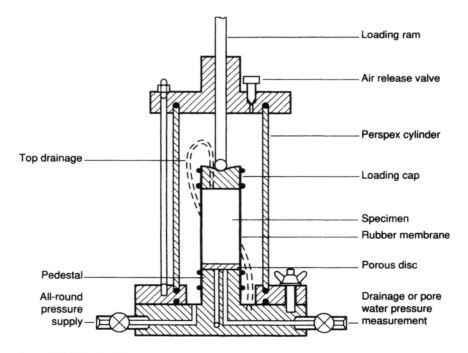

Loading ram

Air release valve

Perspex cylinder

Top drainage

Loading cap

Specimen

Rubber membrane

Porous disc

Pedestal

All-round
pressure
supply

Drainage or pore
water pressure
measurement

Figure 4.5 The triaxial apparatus.

The specimen is placed on either a porous or a solid disc on the pedestal of the apparatus. A loading cap is placed on top of the specimen and the specimen is then sealed in a rubber membrane, O-rings under tension being used to seal the membrane to the pedestal and the loading cap. In the case of sands, the specimen must be prepared in a rubber membrane inside a rigid former which fits around the pedestal. A small negative pressure is applied to the pore water to maintain the stability of the specimen while the former is removed prior to the application of the all-round pressure. A connection may also be made through the loading cap to the top of the specimen, a flexible plastic tube leading from the loading cap to the base of the cell; this connection is normally used for the application of back pressure (as described later in this section). Both the top of the loading cap and the lower end of the loading ram have coned seatings, the load being transmitted through a steel ball. The specimen is subjected to an all-round fluid pressure in the cell, consolidation is allowed to take place, if appropriate, and then the axial stress is gradually increased by the application of compressive load through the ram until failure of the specimen takes place, usually on a diagonal plane. The load is measured by means of a load ring or by a load transducer fitted either inside or outside the cell. The system for applying the all-round pressure must be capable of compensating for pressure changes due to cell leakage or specimen volume change.

In the triaxial test, consolidation takes place under equal increments of total stress normal to the end and circumferential surfaces of the specimen. Lateral strain in the specimen is *not* equal to zero during consolidation under these conditions (unlike in the oedometer test, as described in Section 7.2). Dissipation of excess pore water pressure takes place due to drainage through the porous disc at the bottom (or top) of the specimen. The drainage connection leads to an external burette, enabling the volume of water expelled from the specimen to be measured. The datum for excess pore water pressure is therefore atmospheric pressure, assuming that the water level in the burette is at the same height as the centre of the specimen. Filter paper drains, in contact with the end porous disc, are sometimes placed around the circumference of the specimen; both vertical and radial drainage then take place and the rate of dissipation of excess pore water pressure is increased.

The all-round pressure is taken to be the minor principal stress and the sum of the all-round pressure and the applied axial stress as the major principal stress, on the basis that there are no shear stresses on the surfaces of the specimen. The applied axial stress is thus referred to as the *principal stress difference* (also known as the *deviator stress*). The intermediate principal stress is equal to the minor principal stress; therefore, the stress conditions at failure can be represented by a Mohr circle. If a number of specimens are tested, each under a different value of all-round pressure, the failure envelope can be drawn and the shear strength parameters for the soil determined. In calculating the principal stress difference, the fact that the average cross-sectional area (A) of the specimen does not remain constant throughout the test must be taken into account. If the original cross-sectional area of the specimen is A_0 and the original volume is V_0 then, if the volume of the specimen decreases during the test,

$$A = A_0 \frac{1 - \varepsilon_v}{1 - \varepsilon_a} \tag{4.10}$$

where ε_v is the volumetric strain ($\Delta V / V_0$) and ε_a is the axial strain ($\Delta l / l_0$). If the volume of the specimen increases during the test the sign of ΔV will change and the

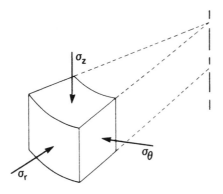

Figure 4.6 Axial, radial and circumferential stresses.

numerator in Equation 4.10 becomes $(1 + \varepsilon_v)$. If required, the radial strain (ε_r) could be obtained from the equation

$$\varepsilon_v = \varepsilon_a + 2\varepsilon_r \tag{4.11}$$

In the case of saturated soils the volume change ΔV is usually determined by measuring the volume of pore water draining from the specimen. The change in axial length Δl corresponds to the movement of the loading ram, which can be measured by a dial gauge.

The above interpretation of the stress conditions in the triaxial test is approximate only. The principal stresses in a cylindrical specimen are in fact the axial, radial and circumferential stresses, σ_z, σ_r and σ_θ, respectively, as shown in Figure 4.6, and the state of stress throughout the specimen is statically indeterminate. If it is assumed that $\sigma_r = \sigma_\theta$ the indeterminacy is overcome and σ_r then becomes constant, equal to the radial stress on the boundary of the specimen. In addition, the strain conditions in the specimen are not uniform due to frictional restraint produced by the loading cap and pedestal disc; this results in dead zones at each end of the specimen, which becomes barrel-shaped as the test proceeds. Non-uniform deformation of the specimen can be largely eliminated by lubrication of the end surfaces. It has been shown, however, that non-uniform deformation has no significant effect on the measured strength of the soil, provided the length/diameter ratio of the specimen is not less than 2.

A special case of the triaxial test is the *unconfined compression test* in which axial stress is applied to a specimen under zero (atmospheric) all-round pressure, no rubber membrane being required. The unconfined test, however, is applicable only for testing intact, fully saturated clays.

A triaxial *extension* test can also be carried out in which an upward load is applied to a ram connected to the loading cap on the specimen. The all-round pressure then becomes the major principal stress and the net vertical stress the minor principal stress.

Pore water pressure measurement

The pore water pressure in a triaxial specimen can be measured, enabling the results to be expressed in terms of effective stress; conditions of *no flow* either out of or into the specimen must be maintained, otherwise the correct pressure will be modified. Pore water pressure is normally measured by means of an electronic pressure transducer. A change in pressure produces a small deflection of the transducer diaphragm, the corresponding strain being calibrated against pressure. The connection between the specimen and the transducer must be filled with de-aired water (produced by boiling water in a near vacuum) and the system should undergo negligible volume change under pressure.

If the specimen is partially saturated a fine porous ceramic disc must be sealed into the pedestal of the cell if the correct pore water pressure is to be measured. Depending on the pore size of the ceramic, only pore water can flow through the disc, provided the difference between the pore air and pore water pressures is below a certain value known as the *air entry value* of the disc. Under undrained conditions the ceramic disc will remain fully saturated with water, provided the air entry value is high enough, enabling the correct pore water pressure to be measured. The use of a coarse porous disc, as normally used for a fully saturated soil, would result in the measurement of the pore air pressure in a partially saturated soil.

Testing under back pressure

Testing under back pressure involves raising the pore water pressure artificially by connecting a source of constant pressure through a porous disc to one end of a triaxial specimen. In a drained test this connection remains open throughout the test, drainage taking place against the back pressure; the back pressure is the datum for excess pore water pressure. In a consolidated–undrained test the connection to the back pressure source is closed at the end of the consolidation stage, before the application of the principal stress difference is commenced.

The object of applying a back pressure is to ensure full saturation of the specimen or to simulate *in-situ* pore water pressure conditions. During sampling the degree of saturation of a clay may fall below 100% owing to swelling on the release of *in-situ* stresses. Compacted specimens will have a degree of saturation below 100%. In both cases a back pressure is applied which is high enough to drive the pore air into solution in the pore water.

It is essential to ensure that the back pressure does not by itself change the effective stresses in the specimen. It is necessary, therefore, to raise the all-round pressure simultaneously with the application of the back pressure and by an equal increment. A specimen is normally considered to be saturated if the pore pressure coefficient B (Section 4.7) has a value of at least 0.95.

The use of back pressure is specified in BS 1377 as a means of ensuring that the test specimen is fully saturated.

Types of test

Many variations of test procedure are possible with the triaxial apparatus but the three principal types of test are as follows:

1 *Unconsolidated–Undrained.* The specimen is subjected to a specified all-round pressure and then the principal stress difference is applied immediately, with no drainage being permitted at any stage of the test.

2   *Consolidated–Undrained.*   Drainage of the specimen is permitted under a speci-fied all-round pressure until consolidation is complete; the principal stress dif-ference is then applied with no drainage being permitted. Pore water pressure measurements may be made during the undrained part of the test.

3   *Drained.*   Drainage of the specimen is permitted under a specified allround pressure until consolidation is complete; with drainage still being permitted, the principal stress difference is then applied at a rate slow enough to ensure that the excess pore water pressure is maintained at zero.

Shear strength parameters determined by means of the above test procedures are relevant only in situations where the field drainage conditions correspond to the test conditions. The shear strength of a soil under undrained conditions is different from that under drained conditions. The undrained strength can be expressed in terms of total stress in the case of fully saturated soils of low permeability, the shear strength parameters being denoted by c_u and ϕ_u. The drained strength is expressed in terms of the effective stress parameters c' and ϕ'.

The vital consideration in practice is the rate at which the changes in total stress (due to construction operations) are applied in relation to the rate of dissipation of excess pore water pressure, which in turn is related to the permeability of the soil. Undrained conditions apply if there has been no significant dissipation during the period of total stress change; this would be the case in soils of low permeability such as clays immediately after the completion of construction. Drained conditions apply in situa-tions where the excess pore water pressure is zero; this would be the case in soils of low permeability after consolidation is complete and would represent the situation a long time, perhaps many years, after the completion of construction. The drained condition would also be relevant if the rate of dissipation were to keep pace with the rate of change of total stress; this would be the case in soils of high permeability such as sands. The drained condition is therefore relevant for sands both immediately after construc-tion and in the long term. Only if there were extremely rapid changes in total stress (e.g. as the result of an explosion or an earthquake) would the undrained condition be relevant for a sand. In some situations, partially drained conditions may apply at the end of construction, perhaps due to a very long construction period or to the soil in question being of intermediate permeability. In such cases the excess pore water pressure would have to be estimated and the shear strength would then be calculated in terms of effective stress.

The vane shear test

This test is used for the *in-situ* determination of the undrained strength of intact, fully saturated clays; the test is not suitable for other types of soil. In particular, this test is very suitable for soft clays, the shear strength of which may be significantly altered by the sampling process and subsequent handling. Generally, this test is only used in clays having undrained strengths less than $100 \, kN/m^2$. This test may not give reliable results if the clay contains sand or silt laminations.

Details of the test are given in BS 1377 (Part 9). The equipment consists of a stainless steel vane (Figure 4.7) of four thin rectangular blades, carried on the end of

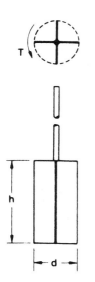

Figure 4.7 The vane test.

a high-tensile steel rod; the rod is enclosed by a sleeve packed with grease. The length of the vane is equal to twice its overall width, typical dimensions being 150 mm by 75 mm and 100 mm by 50 mm. Preferably the diameter of the rod should not exceed 12.5 mm.

The vane and rod are pushed into the clay below the bottom of a borehole to a depth of at least three times the borehole diameter; if care is taken this can be done without appreciable disturbance of the clay. Steady bearings are used to keep the rod and sleeve central in the borehole casing. The test can also be carried out in soft clays, without a borehole, by direct penetration of the vane from ground level; in this case a shoe is required to protect the vane during penetration.

Torque is applied gradually to the upper end of the rod by means of suitable equipment until the clay fails in shear due to rotation of the vane. Shear failure takes place over the surface and ends of a cylinder having a diameter equal to the overall width of the vane. The rate of rotation of the vane should be within the range of 6–12° per minute. The shear strength is calculated from the expression

$$T = \pi c_u \left(\frac{d^2 h}{2} + \frac{d^3}{6} \right) \tag{4.12}$$

where T is the torque at failure, d the overall vane width and h the vane length. However, the shear strength over the cylindrical vertical surface may be different from that over the two horizontal end surfaces, as a result of anisotropy. The shear strength is normally determined at intervals over the depth of interest. If, after the initial test, the vane is rotated rapidly through several revolutions the clay will become remoulded and the shear strength in this condition could then be determined if required.

Small, hand-operated vane testers are also available for use in exposed clay strata.

Special tests

In practice, there are very few problems in which a state of axial symmetry exists as in the triaxial test. In practical states of stress the intermediate principal stress is not usually equal to the minor principal stress and the principal stress directions can undergo rotation as the failure condition is approached. A common condition is that of plane strain in which the strain in the direction of the intermediate principal stress is zero due to restraint imposed by virtue of the length of the structure in question. In the triaxial test, consolidation proceeds under equal all-round pressure (i.e. isotropic consolidation), whereas *in-situ* consolidation takes place under anisotropic stress conditions.

Tests of a more complex nature, generally employing adaptions of triaxial equipment, have been devised to simulate the more complex states of stress encountered in practice but these are used principally in research. The plane strain test uses a prismatic specimen in which strain in one direction (that of the intermediate principal stress) is maintained at zero throughout the test by means of two rigid side plates tied together. The all-round pressure is the minor principal stress and the sum of the applied axial stress and the all-round pressure the major principal stress. A more sophisticated test, also using a prismatic specimen, enables the values of all three principal stresses to be controlled independently, two side pressure bags or jacks being used to apply the intermediate principal stress. Independent control of the three principal stresses can also be achieved by means of tests on soil specimens in the form of hollow cylinders in which different values of external and internal fluid pressure can be applied in addition to axial stress. Torsion applied to the hollow cylinders results in the rotation of the principal stress directions.

Because of its relative simplicity it seems likely that the triaxial test will continue to be the main test for the determination of shear strength characteristics. If considered necessary, corrections can be applied to the results of triaxial tests to obtain the characteristics under more complex states of stress.

4.3   SHEAR STRENGTH OF SANDS

The shear strength characteristics of a sand can be determined from the results of either direct shear tests or drained triaxial tests, only the drained strength of a sand normally being relevant in practice. The characteristics of dry and saturated sands are the same, provided there is zero excess pore water pressure in the case of saturated sands. Typical curves relating shear stress and shear strain for initially dense and loose sand specimens in direct shear tests are shown in Figure 4.8(a). Similar curves are obtained relating principal stress difference and axial strain in drained triaxial compression tests.

In a dense sand there is a considerable degree of interlocking between particles. Before shear failure can take place, this interlocking must be overcome in addition to the frictional resistance at the points of contact. In general, the degree of interlocking is greatest in the case of very dense, well-graded sands consisting of angular particles. The characteristic stress–strain curve for an initially dense sand shows a peak stress at a relatively low strain and thereafter, as interlocking is

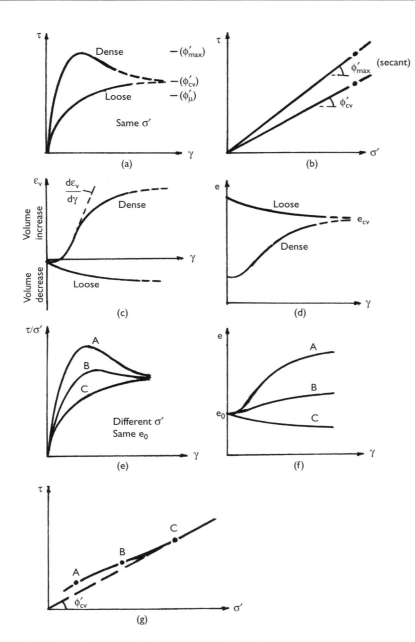

Figure 4.8 Shear strength characteristics of sand.

progressively overcome, the stress decreases with increasing strain. The reduction in the degree of interlocking produces an increase in the volume of the specimen during shearing as characterized by the relationship, shown in Figure 4.8(c), between volumetric strain and shear strain in the direct shear test. In the drained triaxial test a similar relationship would be obtained between volumetric strain and axial strain. The change in volume is also shown in terms of void ratio (e) in Figure 4.8(d). Eventually the specimen would become loose enough to allow particles to move over and around their neighbours without any further net volume change and the shear stress would reach an ultimate value. However, in the triaxial test non-uniform deformation of the specimen becomes excessive as strain is progressively increased and it is unlikely that the ultimate value of principal stress difference can be reached.

The term *dilatancy* is used to describe the increase in volume of a dense sand during shearing and the rate of dilation can be represented by the gradient $d\varepsilon_v/d\gamma$, the maximum rate corresponding to the peak stress. The angle of dilation (ψ) is $\tan^{-1}$ ($d\varepsilon_v/d\gamma$). The concept of dilatancy can be illustrated in the context of the direct shear test. During shearing of a dense sand the macroscopic shear plane is horizontal but sliding between individual particles takes place on numerous microscopic planes inclined at various angles above the horizontal, as the particles move up and over their neighbours. The angle of dilation represents an average value of this angle for the specimen as a whole. The loading plate of the apparatus is thus forced upwards, work being done against the normal stress. For a dense sand the maximum angle of shearing resistance (ϕ'_{max}) determined from peak stresses (Figure 4.8(b)) is significantly greater than the true angle of friction (ϕ_μ) between the surfaces of individual particles, the difference representing the work required to overcome interlocking and rearrange the particles.

In the case of initially loose sand there is no significant particle interlocking to be overcome and the shear stress increases gradually to an ultimate value without a prior peak, accompanied by a decrease in volume. The ultimate values of stress and void ratio for dense and loose specimens under the same values of normal stress in the direct shear test are essentially equal as indicated in Figures 4.8(a) and (d). Thus at the ultimate (or critical) state, shearing takes place at constant volume, the corresponding angle of shearing resistance being denoted ϕ'_{cv} (or ϕ'_{crit}). The difference between ϕ'_μ and ϕ'_{cv} represents the work required to rearrange the particles.

It may be difficult to determine the value of the parameter ϕ'_{cv} because of the relatively high strain required to reach the critical state. In general, the critical state is identified by extrapolation of the stress–strain curve to the point of constant stress, which should also correspond to the point of zero rate of dilation on the volumetric strain–shear strain curve. Stresses at the critical state define a straight line failure envelope intersecting the origin, the slope of which is ϕ'_{cv}.

In practice the parameter ϕ'_{max}, which is a transient value, should only be used for situations in which it can be assumed that strain will remain significantly less than that corresponding to peak stress. If, however, strain is likely to exceed that corresponding to peak stress, a situation that may lead to progressive failure, then the critical-state parameter ϕ'_{cv} should be used.

An alternative method of representing the results from direct shear tests is to plot the stress ratio τ/σ' against shear strain. Plots of stress ratio against shear strain representing tests on three specimens of sand, each having the same initial void ratio, are shown in Figure 4.8(e), the values of effective normal stress (σ') being different in each test. The plots are labelled A, B and C, the effective normal stress being lowest in test A and highest in test C. Corresponding plots of void ratio against shear strain are shown in Figure 4.8(f). Such results indicate that both the maximum stress *ratio* and the ultimate (or critical) void ratio decrease with increasing effective normal stress. The ultimate values of stress ratio, however, are the same. From Figure 4.8(e) it is apparent that the difference between maximum and ultimate stress decreases with increasing effective normal stress; therefore, if the maximum shear stress is plotted against effective normal stress for each individual test, the plotted points will lie on an envelope which is slightly curved, as shown in Figure 4.8(g). The value of ϕ'_{max} for each test can then be represented by a secant parameter, the value decreasing with increasing effective normal stress until it becomes equal to ϕ'_{cv}. The reduction in the difference between maximum and ultimate shear stress with increasing normal stress is mainly due to the corresponding decrease in ultimate void ratio. The lower the ultimate void ratio the less scope there is for dilation. In addition, at high stress levels some fracturing or crushing of particles may occur with the consequence that there will be less particle interlocking to be overcome. Crushing thus causes the suppression of dilatancy and contributes to the reduced value of ϕ'_{max}.

In practice, the routine laboratory testing of sands is not feasible because of the problem of obtaining undisturbed specimens and setting them up, still undisturbed, in the test apparatus. If required, tests can be undertaken on specimens reconstituted in the apparatus at appropriate densities but the *in-situ* structure is then unlikely to be reproduced. Guidance on appropriate values of the parameters ϕ'_{max} and ϕ'_{cv} is given in certain codes of practice. In the case of dense sands it has been shown that the value of ϕ'_{max} under conditions of plane strain can be 4° or 5° higher than the corresponding value obtained by conventional triaxial tests. The increase in the case of loose sands is negligible.

Liquefaction

Liquefaction is a phenomenon in which loose saturated sand loses a large percentage of its shear strength and develops characteristics similar to those of a liquid. It is usually induced by cyclic loading of relatively high frequency, resulting in undrained conditions in the sand. Cyclic loading may be caused, for example, by vibrations from machinery and, more seriously, by earth tremors.

Loose sand tends to compact under cyclic loading. The decrease in volume causes an increase in pore water pressure which cannot dissipate under undrained conditions. Indeed, there may be a cumulative increase in pore water pressure under successive cycles of loading. If the pore water pressure becomes equal to the maximum total stress component, normally the overburden pressure, the value of effective stress will be zero, i.e. interparticle forces will be zero, and the sand will exist in a liquid state with negligible shear strength. Even if the effective stress does not fall to zero the reduction in shear strength may be sufficient to cause failure.

Liquefaction may develop at any depth in a sand deposit where a critical combination of *in-situ* density and cyclic deformation occurs. The higher the void ratio of the sand and the lower the confining pressure the more readily liquefaction will occur. The larger the strains produced by the cyclic loading the lower the number of cycles required for liquefaction.

4.4   SHEAR STRENGTH OF SATURATED CLAYS

Isotropic consolidation

If a saturated clay specimen is allowed to consolidate in the triaxial apparatus under a sequence of equal all-round pressures, sufficient time being allowed between successive increments to ensure that consolidation is complete, the relationship between void ratio (e) and effective stress (σ'_3) can be obtained. Consolidation in the triaxial apparatus under equal all-round pressure is referred to as isotropic consolidation.

The relationship between void ratio and effective stress depends on the stress history of the clay. If the present effective stress is the maximum to which the clay has ever been subjected, the clay is said to be *normally consolidated*. If, on the other hand, the effective stress at some time in the past has been greater than the present value, the clay is said to be *overconsolidated*. The maximum value of effective stress in the past divided by the present value is defined as the *overconsolidation ratio* (OCR). A normally consolidated clay thus has an overconsolidation ratio of unity; an overconsolidated clay has an overconsolidation ratio greater than unity. Overconsolidation is usually the result of geological factors, for example, the erosion of overburden, the melting of ice sheets after glaciation and the permanent rise of the water table. Overconsolidation can also be due to higher stresses previously applied to a specimen in the triaxial apparatus.

The characteristic relationship between e and σ'_3 is shown in Figure 4.9. AB is the curve for a clay in the normally consolidated condition. If after consolidation to point B the effective stress is reduced, the clay will swell or expand and the relationship will be represented by the curve BC. During consolidation from A to B, changes in soil structure continuously take place but the clay does not revert to its original structure during swelling. A clay existing at a state represented by point C is now in the overconsolidated condition, the overconsolidation ratio being the effective stress

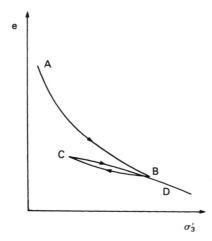

Figure 4.9 Isotropic consolidation.

at point B divided by that at point C. If the effective stress is again increased the consolidation curve is CD, known as the recompression curve, eventually becoming the continuation of the normal consolidation curve AB. It should be realized that a state represented by a point to the right of the normal consolidation curve is impossible.

Undrained strength

In principle, the *unconsolidated–undrained* triaxial test enables the undrained strength of the clay in its *in-situ* condition to be determined, the void ratio of the specimen at the start of the test being unchanged from the *in-situ* value at the depth of sampling. In practice, however, the effects of sampling and preparation result in a small increase in void ratio. Experimental evidence (e.g. Duncan and Seed [10]) has shown that the *in-situ* undrained strength of saturated clays is significantly anisotropic, the strength depending on the direction of the major principal stress relative to the *in-situ* orientation of the specimen. Thus, undrained strength is not a unique parameter.

When a specimen of saturated clay is placed on the pedestal of the triaxial cell the initial pore water pressure is negative due to capillary tension, total stresses being zero and effective stresses positive. After the application of all-round pressure the effective stresses in the specimen remain unchanged because, for a fully saturated soil under undrained conditions, any increase in all-round pressure results in an equal increase in pore water pressure (see Section 4.7). Assuming all specimens to have the same void ratio and composition, a number of unconsolidated–undrained tests, each at a different value of all-round pressure, should result, therefore, in equal values of principal stress difference at failure. The results are expressed in terms of total stress as shown in Figure 4.10, the failure envelope being horizontal, i.e. $\phi_u = 0$, and the shear strength is given by $\tau_f = c_u$. It should be noted that if the values of pore water pressure at failure were measured in a series of tests then in principle only one effective stress circle, shown dotted in Figure 4.10, would be obtained. The circle representing an unconfined compression test would lie to the left of the effective stress circle in Figure 4.10 because of the negative pore water pressure in the specimen. The unconfined strength of a clay is due to a combination of friction and pore water suction.

If the best common tangent to the Mohr circles obtained from a series of tests is not horizontal then the inference is that there has been a reduction in void ratio during

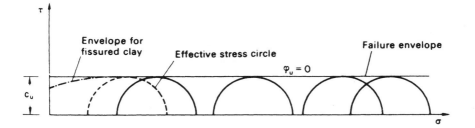

Figure 4.10 Unconsolidated–undrained triaxial test results for saturated clay.

each test due to the presence of air in the voids, i.e. the specimen has not been fully saturated at the outset. It should not be inferred that $\phi_u > 0$. It could also be that an initially saturated specimen has partially dried prior to testing or has been repaired. Another reason could be the entrapment of air between the specimen and the membrane.

In the case of fissured clays the failure envelope at low values of all-round pressure is curved, as shown in Figure 4.10. This is due to the fact that the fissures open to some extent on sampling, resulting in a lower strength and only when the all-round pressure becomes high enough to close the fissures again does the strength become constant. Therefore, the unconfined compression test is not appropriate in the case of fissured clays. The size of a fissured clay specimen should be large enough to represent the mass structure, otherwise the measured strength will be greater than the *in-situ* strength. Large specimens are also required for clays exhibiting other features of macro-fabric. Curvature of the undrained failure envelope at low values of all-round pressure may also be exhibited in heavily overconsolidated clays due to relatively high negative pore water pressure at failure causing cavitation, i.e. pore air comes out of solution.

The results of unconsolidated–undrained tests are usually presented as a plot of c_u against the corresponding depth from which the specimen originated. Considerable scatter can be expected on such a plot as the result of sampling disturbance and macro-fabric features if present. For normally consolidated clays the undrained strength is generally taken to increase linearly with increase in effective vertical stress σ'_v (i.e. with depth if the water table is at the surface); this is comparable to the variation of c_u with σ'_3 (Figure 4.11) in consolidated–undrained triaxial tests. If the water table is below the surface of the clay the undrained strength between the surface and the water table will be significantly higher than that immediately below the water table due to drying of the clay.

The following correlation between the ratio c_u/σ'_v and plasticity index (I_p) for normally consolidated clays was proposed by Skempton:

$$\frac{c_u}{\sigma'_v} = 0.11 + 0.0037I_p \tag{4.13}$$

The *consolidated–undrained* triaxial test enables the undrained strength of the clay to be determined after the void ratio has been changed from the initial value by consolidation. The undrained strength is thus a function of this void ratio or of the corresponding

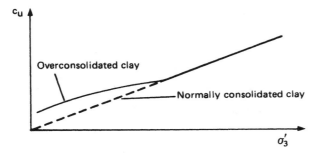

Figure 4.11 Consolidated–undrained triaxial test: variation of undrained strength with consolidation pressure.

all-round pressure (σ'_3) under which consolidation took place. The all-round pressure during the undrained part of the test (i.e. when the principal stress difference is applied) has no influence on the strength of the clay, although it is normally the same pressure as that under which consolidation took place. The results of a series of tests can be represented by plotting the value of c_u (ϕ_u being zero) against the corresponding consolidation pressure σ'_3, as shown in Figure 4.11. For clays in the normally consolidated state the relationship between c_u and σ'_3 is linear, passing through the origin. For clays in the overconsolidated state the relationship is non-linear, as shown in Figure 4.11.

The unconsolidated–undrained test and the undrained part of the consolidated–undrained test can be carried out rapidly (provided no pore water pressure measurements are to be made), failure normally being produced within a period of 10–15 min. However, a slight decrease in strength can be expected if the time to failure is significantly increased and there is evidence that this decrease is more pronounced the greater the plasticity index of the clay. Each test should be continued until the maximum value of principal stress difference has been passed or until an axial strain of 20% has been attained.

It should be realized that clays *in situ* have been consolidated under conditions of zero lateral strain, the effective vertical and horizontal stresses being unequal, i.e. the clay has been consolidated anisotropically. A stress release then occurs on sampling. In the consolidated–undrained triaxial test the specimen is consolidated again under equal all-round pressure, normally equal to the value of the effective vertical stress *in situ*, i.e. the specimen is consolidated isotropically. Isotropic consolidation in the triaxial test under a pressure equal to the *in-situ* effective vertical stress results in a void ratio lower than the *in-situ* value and therefore an undrained strength higher than the *in-situ* value.

The undrained strength of intact soft and firm clays can be measured *in situ* by means of the *vane test*. However, Bjerrum [5] has presented evidence that undrained strength as measured by the vane test is generally greater than the average strength mobilized along a failure surface in a field situation. The discrepancy was found to be greater the higher the plasticity index of the clay and is attributed primarily to the rate effect mentioned earlier in this section. In the vane test shear failure occurs within a few minutes, whereas in a field situation the stresses are usually applied over a period of few weeks or months. A secondary factor may be anisotropy. Bjerrum presented a correction factor (μ), correlated empirically with plasticity index, as shown in Figure 4.12, the vane strength being multiplied by the factor to give the probable field strength.

Clays may be classified on the basis of undrained shear strength as in Table 4.1.

Sensitivity of clays

Some clays are very sensitive to remoulding, suffering considerable loss of strength due to their natural structure being damaged or destroyed. The *sensitivity* of a clay is defined as the ratio of the undrained strength in the undisturbed state to the undrained strength, at the same water content, in the remoulded state. Remoulding for test purposes is normally brought about by the process of kneading. The sensitivity of most clays is between 1 and 4. Clays with sensitivities between 4 and 8 are referred to as *sensitive* and those with sensitivities between 8 and 16 as *extrasensitive*. *Quick* clays are those having sensitivities greater than 16; the sensitivities of some quick clays may be of the order of 100.

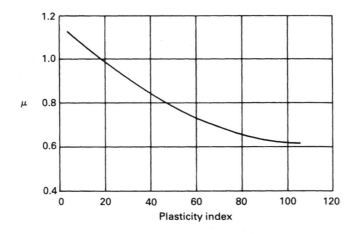

Figure 4.12 Correction factor for undrained strength measured by the vane test. (After Bjerrum [5].)

Table 4.1 Undrained strength classification

Stiffness state	Undrained strength (kN/m²)
Hard	> 300
Very stiff	150–300
Stiff	75–150
Firm	40–75
Soft	20–40
Very soft	< 20

Strength in terms of effective stress

The strength of a clay in terms of effective stress can be determined by either the consolidated–undrained triaxial test with pore water pressure measurement, or the drained triaxial test. The undrained part of the consolidated–undrained test must be run at a rate of strain slow enough to allow equalization of pore water pressure throughout the specimen, this rate being a function of the permeability of the clay. If the pore water pressure at failure is known, the effective principal stresses σ_1' and σ_3' can be calculated and the corresponding Mohr circle drawn.

Failure envelopes for normally consolidated and overconsolidated clays are of the forms shown in Figure 4.13. For a normally consolidated or lightly overconsolidated clay the envelope should pass through the origin and the parameter $c' = 0$. The envelope for a heavily overconsolidated clay is likely to exhibit curvature over the stress range up to preconsolidation pressure and can be represented by either secant or tangent parameters. It should be recalled that a secant parameter ϕ' only applies to a particular stress level and its value will decrease with increasing effective normal stress until it becomes equal to the critical-state parameter ϕ_{cv}' (as illustrated in Example 4.3).

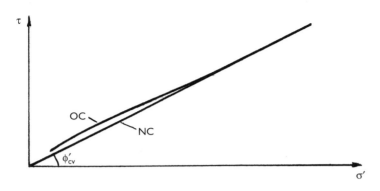

Figure 4.13 Failure envelopes for normally consolidated (NC) and overconsolidated (OC) clays.

Tangent parameters c' and ϕ' apply only to a relatively small stress range. The tangent value of ϕ' is likely to be lower than the critical-state value. Caution is required in the use of the c' parameter, especially at low stress levels where it would represent a significant proportion of shear strength. If the critical-state value of ϕ'_{cv} is required for a heavily overconsolidated clay then, if possible, tests should be performed at stress levels that are high enough to define the critical-state envelope, i.e. specimens should be consolidated at all-round pressures in excess of the preconsolidation value. Alternatively, an estimated value of ϕ'_{cv} can be obtained from tests on normally consolidated specimens reconsolidated from a slurry.

Effective stress parameters can also be obtained by means of drained triaxial tests (or direct shear tests). Clays under drained conditions behave as frictional materials. The rate of strain must be slow enough to ensure full dissipation of excess pore water pressure at any time during application of the principal stress difference. Total and effective stresses will thus be equal throughout the test. The rate of strain must again be related to the permeability of the clay. The volume change taking place during the application of the principal stress difference must be measured in the drained test so that the corrected cross-sectional area of the specimen can be calculated.

Typical test results for specimens of normally consolidated and overconsolidated clays are shown in Figure 4.14. In consolidated–undrained tests, axial stress and pore water pressure are plotted against axial strain. For normally consolidated clays, axial stress reaches an ultimate value at relatively large strain, accompanied by increase in pore water pressure to a steady value. For overconsolidated clays, axial stress increases to a peak value and then decreases with subsequent increase in strain. However, it is not usually possible to reach the ultimate stress due to excessive specimen deformation. Pore water pressure increases initially and then decreases, the higher the overconsolidation ratio the greater the decrease. Pore water pressure may become negative in the case of heavily overconsolidated clays as shown by the dotted line in Figure 4.14(b). In drained tests, axial stress and volume change are plotted against axial strain. For normally consolidated clays an ultimate value of stress is again reached at relatively high strain. A decrease in volume takes place during shearing and the clay hardens. For overconsolidated clays a peak value of axial stress is reached at relatively low strain.

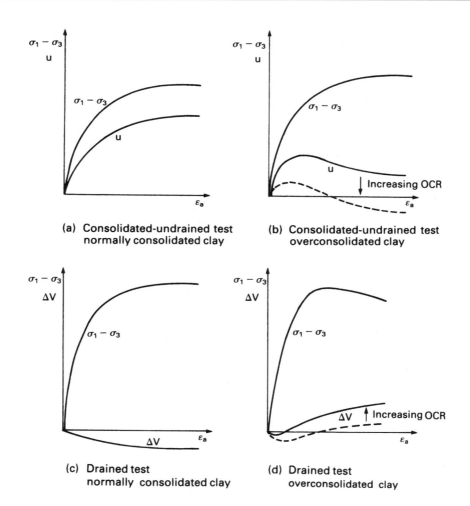

(a) Consolidated-undrained test normally consolidated clay

(b) Consolidated-undrained test overconsolidated clay

(c) Drained test normally consolidated clay

(d) Drained test overconsolidated clay

Figure 4.14 Typical results from consolidated–undrained and drained triaxial tests.

Subsequently, axial stress decreases with increasing strain but, again, it is not usually possible to reach the ultimate stress in the triaxial apparatus. After an initial decrease, the volume of an overconsolidated clay increases prior to and after peak stress and the clay softens. For overconsolidated clays the decrease from peak stress towards the ultimate value becomes less pronounced as the overconsolidation ratio decreases.

In practical situations, if the stress in a particular soil element becomes equal to the peak shear strength, any further increase in strain will result in a reduction in strength. Consequently, additional stress will be transferred to adjacent elements, perhaps resulting in peak strength being also reached in these elements. A sequence of progressive failure could thus be set in train within a soil mass. Therefore, unless it is certain that strains throughout the soil mass will remain less than that corresponding to peak strength, it is necessary to use the critical-state strength in design.

Stress paths

The successive states of stress in a test specimen or an *in-situ* element of soil can be represented by a series of Mohr circles or, in a less confusing way, by a series of stress points. The curve or straight line connecting the relevant stress points is called the *stress path*, giving a clear representation of the successive states of stress. Stress paths may be drawn in terms of either effective or total stresses. The horizontal distance between the effective and total stress paths is the value of pore water pressure at the stresses in question. In general, the horizontal distance between the two stress paths is the sum of the pore water pressure due to the change in total stress and the static pore water pressure. In the normal triaxial test procedure the static pore water pressure (u_s) is zero. However, if a triaxial test is performed under back pressure, the static pore water pressure is equal to the back pressure. The static pore water pressure of an *in-situ* element is the pressure governed by the water table level.

The effective and total stress paths (denoted by ESP and TSP, respectively) for the triaxial tests represented in Figure 4.14 are shown in Figure 4.15, the coordinates being $\frac{1}{2}(\sigma_1' - \sigma_3')$ and $\frac{1}{2}(\sigma_1' + \sigma_3')$ or the total stress equivalents. The effective stress paths terminate on the modified failure envelope. All the total stress paths and the effective stress paths for the drained tests are straight lines at a slope of 45°. The detailed shape of the effective stress paths for the consolidated–undrained tests depends on the clay in question. The effective and total stress paths for the drained tests coincide, provided no back pressure has been applied. The dotted line in Figure 4.15(c) is the effective stress path for a heavily overconsolidated clay in which the pore water pressure at failure (u_f) is negative.

The hydraulic triaxial apparatus

This apparatus, developed by Bishop and Wesley [4], is shown diagrammatically in Figure 4.16. The chamber containing the soil specimen is similar to a conventional triaxial cell; however, the pedestal is connected by a ram to a piston in a lower pressure chamber, the vertical movement of the ram being guided by a linear bearing. Rolling seals are used to accommodate the movement of the ram in both chambers. Axial load is applied to the specimen by increasing the pressure in the lower chamber. Standard constant pressure systems can be used but preferably with a motorized drive arrangement to control the rate of pressure increase. Although the axial load on the specimen can be calculated from a knowledge of the pressures in the two chambers, the mass of the ram and apparatus dimensions, it is preferable to measure the load directly by means of a load cell above the specimen. A cross-arm mounted on the ram passes through the slots in the side of the apparatus (the section of the apparatus containing the linear bearing being open to atmosphere); vertical rods attached to the ends of the cross-arm operate against dial gauges (or transducers), enabling the change in length of the specimen to be measured. The slots in the side of the apparatus also accommodate the pore pressure and drainage connections from the pedestal. The potential of the apparatus can be fully exploited by means of an automatic system controlled by a computer. The apparatus enables a wide range of stress paths, reproducing *in-situ* stress changes, to be imposed on the specimen.

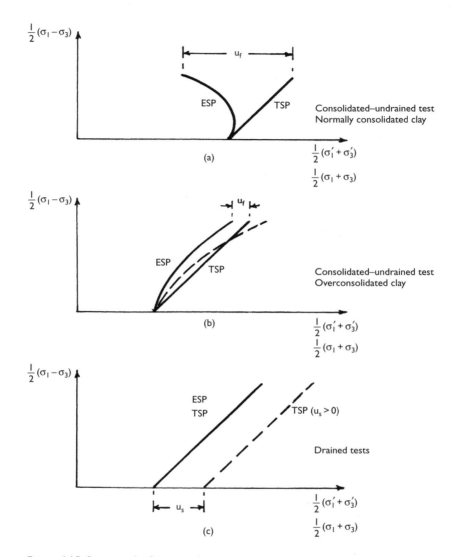

Figure 4.15 Stress paths for triaxial tests.

Example 4.1

The following results were obtained from direct shear tests on specimens of a sand compacted to the *in-situ* density. Determine the value of the shear strength parameter ϕ'.

Normal stress (kN/m²)	50	100	200	300
Shear stress at failure (kN/m²)	36	80	154	235

Would failure occur on a plane within a mass of this sand at a point where the shear stress is $122\,\text{kN/m}^2$ and the effective normal stress $246\,\text{kN/m}^2$?

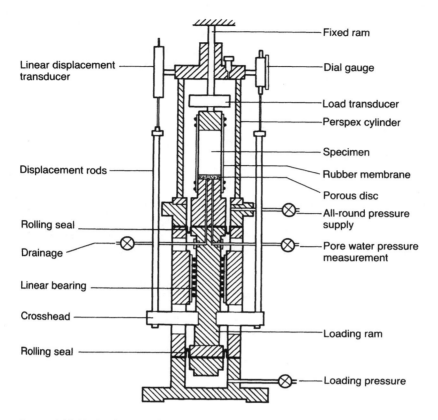

Figure 4.16 Hydraulic triaxial apparatus.

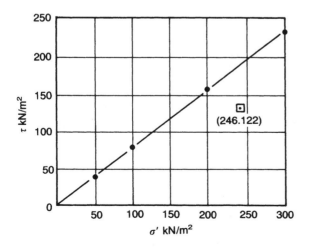

Figure 4.17 Example 4.1.

The values of shear stress at failure are plotted against the corresponding values of normal stress, as shown in Figure 4.17. The failure envelope is the line having the best fit to the plotted points; in this case a straight line through the origin. If the stress scales are the same, the value of ϕ' can be measured directly and is 38°.

The stress state $\tau = 122\,\text{kN/m}^2$, $\sigma' = 246\,\text{kN/m}^2$ plots below the failure envelope, and therefore would not produce failure.

Example 4.2

The results shown in Table 4.2 were obtained at failure in a series of triaxial tests on specimens of a saturated clay initially 38 mm in diameter by 76 mm long. Determine the values of the shear strength parameters with respect to (a) total stress and (b) effective stress.

The principal stress difference at failure in each test is obtained by dividing the axial load by the cross-sectional area of the specimen at failure (Table 4.3). The corrected cross-sectional area is calculated from Equation 4.10. There is, of course, no volume change during an undrained test on a saturated clay. The initial values of length, area and volume for each specimen are:

$$l_0 = 76\,\text{mm}, \qquad A_0 = 1135\,\text{mm}^2, \qquad V_0 = 86 \times 10^3\,\text{mm}^3$$

The Mohr circles at failure and the corresponding failure envelopes for both series of tests are shown in Figure 4.18. In both cases the failure envelope is the line nearest to a common tangent to the Mohr circles. The total stress parameters, representing the undrained strength of the clay, are

$$c_u = 85\,\text{kN/m}^2, \qquad \phi_u = 0$$

Table 4.2

	Type of test	All-round pressure (kN/m²)	Axial load (N)	Axial deformation (mm)	Volume change (ml)
(a)	Undrained	200	222	9.83	–
		400	215	10.06	–
		600	226	10.28	–
(b)	Drained	200	403	10.81	6.6
		400	848	12.26	8.2
		600	1265	14.17	9.5

Table 4.3

	σ_3 (kN/m²)	$\Delta l/l_0$	$\Delta V/V_0$	Area (mm²)	$\sigma_1 - \sigma_3$ (kN/m²)	σ_1 (kN/m²)
(a)	200	0.129	–	1304	170	370
	400	0.132	–	1309	164	564
	600	0.135	–	1312	172	772
(b)	200	0.142	0.077	1222	330	530
	400	0.161	0.095	1225	691	1091
	600	0.186	0.110	1240	1020	1620

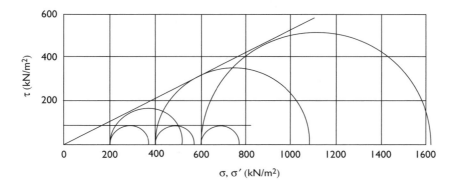

Figure 4.18 Example 4.2.

The effective stress parameters, representing the drained strength of the clay, are

$$c' = 0, \qquad \phi' = 27°$$

Example 4.3

The results shown in Table 4.4 were obtained for peak failure in a series of consolidated–undrained triaxial tests, with pore water pressure measurement, on specimens of a saturated clay. Determine the values of the effective stress parameters.

Values of effective principal stresses σ'_3 and σ'_1 at failure are calculated by subtracting pore water pressure at failure from the total principal stresses as shown in Table 4.5 (all stresses in kN/m^2). The Mohr circles in terms of effective stress are drawn in Figure 4.19. In this case the failure envelope is slightly curved and a different value of the secant parameter ϕ' applies to each circle. For circle (a) the value of ϕ' is the slope of the line OA, i.e. 35°. For circles (b) and (c) the values are 33° and 31°, respectively.

Table 4.4

All-round pressure (kN/m^2)	Principal stress difference (kN/m^2)	Pore water pressure (kN/m^2)
150	192	80
300	341	154
450	504	222

Table 4.5

σ_3	σ_1	σ'_3	σ'_1
150	342	70	262
300	641	146	487
450	954	228	732

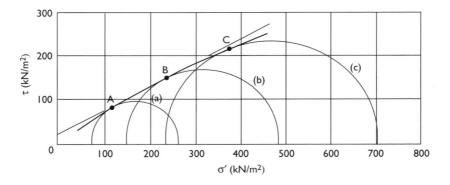

Figure 4.19 Example 4.3.

Tangent parameters can be obtained by approximating the curved envelope to a straight line over the stress range relevant to the problem. In Figure 4.19 a linear approximation has been drawn for the range of effective normal stress 200–300 kN/m², giving parameters $c' = 20\,\text{kN/m}^2$ and $\phi' = 29°$.

4.5 THE CRITICAL-STATE CONCEPT

The critical-state concept, given by Roscoe *et al.* [14], represents an idealization of the observed patterns of behaviour of saturated clays in triaxial compression tests. The concept relates the effective stresses and the corresponding specific volume ($v = 1 + e$) of a clay during shearing under drained or undrained conditions, thus unifying the characteristics of shear strength and deformation. It was demonstrated that a characteristic surface exists which limits all possible states of the clay and that all effective stress paths reach or approach a line on that surface which defines the state at which yielding occurs at constant volume under constant effective stress.

Stress paths are plotted with respect to principal stress difference (or deviator stress) and average effective principal stress, denoted by q' and p', respectively. Thus

$$q' = (\sigma_1' - \sigma_3') \tag{4.14}$$

In the triaxial test the intermediate principal stress (σ_2') is equal to the minor principal stress (σ_3'); therefore, the average principal stress is

$$p' = \frac{1}{3}(\sigma_1' + 2\sigma_3') \tag{4.15}$$

By algebraic manipulation it can be shown that

$$(\sigma_1' + \sigma_3') = \frac{1}{3}(6p' + q') \tag{4.16}$$

Effective, stress paths for a consolidated–undrained test and a drained triaxial test (C′A′ and C′B′, respectively) on specimens of a *normally consolidated* clay are shown in Figure 4.20(a), the coordinate axes being q' and p' (Equations 4.14 and 4.15). Each specimen was allowed to consolidate under the same all-round pressure p'_c and failure occurs at A′ and B′, respectively, these points lying on or close to a straight line OS′ through the origin, i.e. failure occurs if the stress path reaches this line. If a series of consolidated–undrained tests were carried out on specimens each consolidated to a different value of p'_c, the stress paths would all have similar shapes to that shown in Figure 4.20(a). The stress paths for a series of drained tests would be straight lines rising from the points representing p'_c at a slope of 3 vertical to 1 horizontal (because if there is no change in σ'_3, changes in q' and p' are then in the ratio 3:1). In all these tests the state of stress at failure would lie on or close to the straight line OS′.

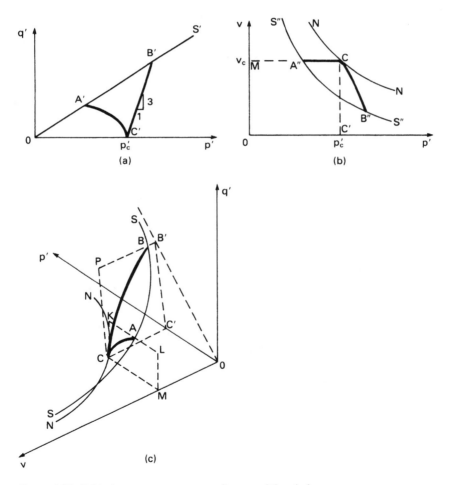

Figure 4.20 Critical-state concept: normally consolidated clays.

The isotropic consolidation curve (NN) for the normally consolidated clay would have the form shown in Figure 4.20(b), the coordinate axes being v and p'. The volume of the specimen during the application of the principal stress difference in a consolidated–undrained test on a saturated clay remains constant, and therefore the relationship between v and p' will be represented by a horizontal line starting from the point C on the consolidation curve corresponding to p'_c and finishing at the point A″ representing the value of p' at failure. During a drained test the volume of the specimen decreases and the relationship between v and p' will be represented by a curve CB″. If a series of consolidated–undrained and drained tests were carried out on specimens each consolidated to a different value of p'_c, the points representing the values of v and p' at failure would lie on or close to a curve S″S″ of similar shape to the consolidation curve.

The data represented in Figures 4.20(a) and (b) can be combined in a three-dimensional plot with coordinates q', p' and v, as shown in Figure 4.20(c). On this plot the line OS′ and the curve S″S″ combine as the single curve SS. The curve SS is known as the *critical-state line*, points on this line representing combinations of q', p' and v at which shear failure and subsequent yielding at constant effective stresses occur. In Figures 4.20(a) and (b), OS′ and S″S″ are the projections of the critical-state line on the q'–p' and v–p' planes, respectively. The stress paths for a consolidated–undrained test (CA) and a drained test (CB), both consolidated to the same pressure p'_c, are also shown in Figure 4.20(c). The stress path for the consolidated–undrained test lies on a plane CKLM parallel to the q'–p' plane, the value of v being constant throughout the undrained part of the test. The stress path for the drained test lies on a plane normal to the q'–p' plane and inclined at a slope of 3:1 to the direction of the q' axis. Both stress paths start at point C on the normal consolidation curve NN which lies on the v–p' plane.

The stress paths for a series of consolidated–undrained and drained tests on specimens each consolidated to different values of p'_c would all lie on a curved surface, spanning between the normal consolidation curve NN and the critical-state line SS, called the *state boundary surface*. It is impossible for a specimen to reach a state represented by a point beyond this surface.

The stress paths for a consolidated–undrained test and a drained triaxial test (D′E′ and D′F′, respectively) on specimens of a *heavily overconsolidated* clay are shown in Figure 4.21(a). The stress paths start from a point D′ on the expansion (or recompression) curve for the clay. The consolidated–undrained specimen reaches failure at point E′ on the line U′H′, above the projection (OS′) of the critical-state line. If the test were continued after failure, the stress path would be expected to continue along U′H′ and to approach point H′ on the critical-state line. However, the higher the overconsolidation ratio the higher the strain required to reach the critical state. The deformation of the consolidated–undrained triaxial specimen would become non-uniform at high strains and it is unlikely that the specimen as a whole would reach the critical state. The drained specimen reaches failure at point F′ also on the line U′H′. After failure, the stresses decrease along the same stress path, approaching the critical-state line at point X′. However, heavily overconsolidated specimens increase in volume (and hence soften) prior to and after failure in a drained test. Narrow zones adjacent to the failure planes become weaker than the remainder of the clay and the specimen as a whole does not reach the critical state. The corresponding relationships between v and p' are

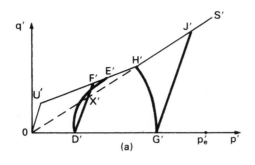

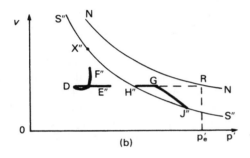

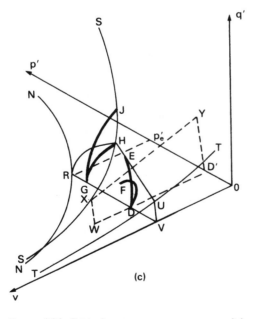

Figure 4.21 Critical-state concept: overconsolidated clays.

represented by the lines DE″ and DF″, respectively, in Figure 4.21(b); these lines approach but do not reach the critical-state line (S″S″) at points H″ and X″, respectively. The volume of the undrained specimen remains constant during shearing but that of the drained specimen, after decreasing initially, increases up to and beyond failure.

The line U′H′ is the projection of the state boundary surface, known as the *Hvorslev surface*, for heavily overconsolidated clays. However, it is assumed that the soil cannot withstand tensile effective stresses, i.e. the effective minor principal stress (σ_3') cannot be less than zero. A line (OU′) through the origin at a slope of 3:1 ($q'/p' = 3$ for $\sigma_3' = 0$ in Equations 4.14 and 4.15) is therefore a limit to the state boundary. On the q'–p'–v plot, shown in Figure 4.21(c), this line becomes a plane lying between the line TT (referred to as the 'no tension' cut-off) and the v axis. Thus, the state boundary surface for heavily overconsolidated clays lies between TT and the critical-state line SS. In Figure 4.21(c) the undrained stress path (DE) lies on a plane RHUV parallel to the q'–p' plane. The drained stress path (DF) lies on a plane WXYD′ normal to the q'–p' plane and inclined at a slope of 3:1 to the direction of the q' axis.

Also shown in Figures 4.21(a) and (c) are the stress paths for consolidated–undrained and drained tests (G′H′ and G′J′, respectively) on *lightly overconsolidated* specimens of the same clay, starting at the same value of specific volume as the heavily over-consolidated specimens. The initial point on the stress paths (G′) is on the expansion (or recompression) curve to the right of the projection S″S″ of the critical-state line in Figure 4.21(b). In both tests, failure is reached at points on or close to the critical-state line. During a drained test, a lightly overconsolidated specimen decreases in volume and hardens; no decrease in stresses therefore occurs after failure. As a result the deformation of the specimen is relatively uniform and the critical state is likely to be reached.

The section of the complete state boundary surface, for normally consolidated and overconsolidated clays, on a plane of constant specific volume is RHU in Figure 4.21(c). The shape of the section will be similar on all planes of constant specific volume. A single section (TSN) can therefore be drawn with respect to coordinate axes q'/p_e' and p'/p_e', as shown in Figure 4.22, where p_e' is the value of p' at the intersection of a given plane of constant specific volume with the normal consolidation curve. In Figure 4.22, point N is on the normal consolidation line, S is on the critical-state line and T is on the 'no tension' cut-off. A specimen whose state is represented by a point

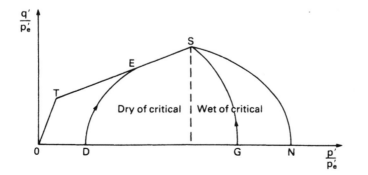

Figure 4.22 Section of the complete state boundary surface.

lying between N and the vertical through S is said to be *wet of critical* (i.e. its water content is higher than that of clay at the critical state, at the same value of p'). A specimen whose state is represented by a point lying between the origin and the vertical through S is said to be *dry of critical*.

To summarize, the state boundary surface joins the lines NN, SS and TT in Figure 4.21(c) and marks the limit to all possible combinations of stresses q' and p' and specific volume v. The plane between TT and the v axis is the boundary for no tension failure. The critical-state line SS defines all possible states of ultimate failure, i.e. of continuing strain at constant volume under constant stresses. In the case of normally consolidated clays the stress paths for both drained and undrained tests lie entirely on the state boundary surface, failure being reached at a point on the critical-state line; the state of the clay remains wet of critical. In the case of overconsolidated clays, the stress paths prior to failure for both drained and undrained tests lie inside the state boundary surface. A distinction must be made between heavily overconsolidated and lightly overconsolidated clays. Heavily overconsolidated clays reach failure at a point on the state boundary surface on the dry side of the critical-state line; subsequently the stress path moves along the state boundary surface but is unlikely to reach the critical-state line. Lightly overconsolidated clays remain wet of critical and reach failure on the critical-state line.

The characteristics of both loose and dense sands during shearing under drained conditions are broadly similar to those for overconsolidated clays, failure occurring on the state boundary surface on the dry side of the critical-state line.

The equation of the projection of the critical-state line (OS$'$ in Figure 4.20(a)) on the q'–p' plane is

$$q' = Mp' \tag{4.17}$$

where M is the slope of OS$'$. Using Equations 4.6(a), 4.14, 4.16 and 4.17 the parameter M can be related to the angle of shearing resistance ϕ':

$$\sin \phi' = \frac{\sigma_1' - \sigma_3'}{\sigma_1' + \sigma_3'}$$

$$= \frac{3M}{6 + M}$$

If the projection of the critical-state line on the v–p' plane is replotted on a v–$\ln p'$ plane it will approximate to a straight line parallel to the corresponding normal consolidation line (of slope $-\lambda$) as shown in Figure 4.23. The equation of the critical-state line, with respect to v and p', can therefore be written as

$$v = \Gamma - \lambda \ln p' \tag{4.18}$$

where Γ is the value of v on the critical-state line at $p' = 1\,\text{kN/m}^2$.

Similarly, the equation of the normal consolidation line (NN) is

$$v = N - \lambda \ln p' \tag{4.19}$$

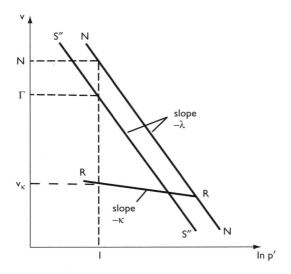

Figure 4.23 Critical-state concept: $e - \ln p'$ relationships.

where N is the value of v at $p' = 1 \, \text{kN/m}^2$. The swelling and recompression relationships can be approximated to a single straight line (RR), of slope $-\kappa$, represented by the equation

$$v = v_\kappa - \kappa \ln p' \qquad (4.20)$$

where v_κ is the value of v at $p' = 1 \, \text{kN/m}^2$.

Example 4.4

The following parameters are known for a saturated normally consolidated clay: $N = 2.48$, $\lambda = 0.12$, $\Gamma = 2.41$ and $M = 1.35$. Estimate the values of principal stress difference and void ratio at failure in undrained and drained triaxial tests on specimens of the clay consolidated under an all-round pressure of $300 \, \text{kN/m}^2$. What would be the expected value of ϕ'_{cv}?

After normal consolidation to $300 \, \text{kN/m}^2$ (p'_c), the specific volume (v_c) is given by

$$v_c = N - \lambda \ln p'_c = 2.48 - 0.12 \ln 300 = 1.80$$

In an undrained test on a saturated clay the volume change is zero, and therefore the specific volume at failure (v_f) will also be 1.80, i.e. the void ratio at failure (e_f) will be 0.80.

Assuming failure to take place on the critical-state line,

$$q'_f = M p'_f$$

and the value of p'_f can be obtained from Equation 4.18. Therefore

$$q'_f = M \exp\left(\frac{\Gamma - v_f}{\lambda}\right)$$
$$= 1.35 \exp\left(\frac{2.41 - 1.80}{0.12}\right)$$
$$= 218\,\text{kN/m}^2 = (\sigma_1 - \sigma_3)_f$$

For a drained test the slope of the stress path on a $q'-p'$ plot is 3, i.e.

$$q'_f = 3(p'_f - p'_c) = 3\left(\frac{q'_f}{M} - p'_c\right)$$

Therefore,

$$q'_f = \frac{3Mp'_c}{3 - M} = \frac{3 \times 1.35 \times 300}{3 - 1.35} = 736\,\text{kN/m}^2$$
$$= (\sigma_1 - \sigma_3)_f$$

Then,

$$p'_f = \frac{q'_f}{M} = \frac{736}{1.35} = 545\,\text{kN/m}^2$$
$$v_f = \Gamma - \lambda \ln p'_f = 2.41 - 0.12 \ln 545 = 1.65$$

Therefore,

$$e_f = 0.65$$
$$\phi'_{cv} = \sin^{-1}\left(\frac{3M}{6 + M}\right)$$
$$= \sin^{-1}\left(\frac{3 \times 1.35}{7.35}\right)$$
$$= 33°$$

4.6 RESIDUAL STRENGTH

In the drained triaxial test, most clays would eventually show a decrease in shear strength with increasing strain after the peak strength has been reached. However, in the triaxial test there is a limit to the strain which can be applied to the specimen. The most satisfactory method of investigating the shear strength of clays at large strains is by means of the ring shear apparatus [3, 8], an annular direct shear apparatus. The annular specimen (Figure 4.24(a)) is sheared, under a given normal stress, on a horizontal plane by the rotation of one half of the apparatus relative to the other; there is no restriction to the magnitude of shear displacement between the two halves

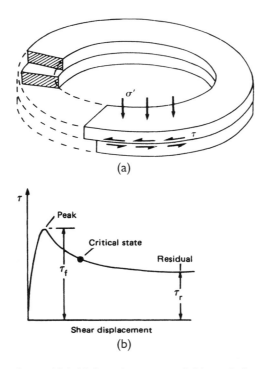

(a)

(b)

Figure 4.24 (a) Ring shear test and (b) residual strength.

of the specimen. The rate of rotation must be slow enough to ensure that the specimen remains in a drained condition. Shear stress, which is calculated from the applied torque, is plotted against shear displacement as shown in Figure 4.24(b).

The shear strength falls below the peak value and the clay in a narrow zone adjacent to the failure plane will soften and reach the critical state. However, because of non-uniform strain in the specimen, the exact point on the curve corresponding to the critical state is uncertain. With continuing shear displacement the shear strength continues to decrease, below the critical-state value, and eventually reaches a residual value at a relatively large displacement. If the clay contains a relatively high proportion of plate-like particles a reorientation of these particles parallel to the failure plane will occur (in the narrow zone adjacent to the failure plane) as the strength decreases towards the residual value. However, reorientation may not occur if the plate-like particles exhibit high interparticle friction. In this case, and in the case of soils containing a relatively high proportion of bulky particles, rolling and translation of particles takes place as the residual strength is approached. It should be appreciated that the critical-state concept envisages continuous deformation of the specimen as a whole, whereas in the residual condition there is preferred orientation or translation of particles in a narrow shear zone. The original soil structure in this narrow shear zone is destroyed as a result of particle reorientation. A remoulded specimen can therefore be used in the ring shear apparatus if only the residual strength (and not the peak strength) is required.

The results from a series of tests, under a range of values of normal stress, enable the failure envelope for both peak and residual strength to be obtained, the residual strength parameters in terms of effective stress being denoted c_r' and ϕ_r'. Residual strength data for a large range of soils have been published [13], which indicate that the value of c_r' can be taken to be zero. Thus, the residual strength can be expressed as

$$\tau_r = \sigma_f' \tan \phi_r' \tag{4.21}$$

4.7 PORE PRESSURE COEFFICIENTS

Pore pressure coefficients are used to express the response of pore water pressure to *changes* in total stress *under undrained conditions* and enable the *initial* value of excess pore water pressure to be determined. Values of the coefficients may be determined in the laboratory and can be used to predict pore water pressures in the field under similar stress conditions.

(1) Increment of isotropic stress

Consider an element of soil, of volume V and porosity n, in equilibrium under total principal stresses σ_1, σ_2 and σ_3, as shown in Figure 4.25, the pore pressure being u_0. The element is subjected to equal increases in total stress $\Delta\sigma_3$ in each direction, resulting in an immediate increase Δu_3 in pore pressure.

> The increase in effective stress in each direction $= \Delta\sigma_3 - \Delta u_3$
> Reduction in volume of the soil skeleton $= C_s V(\Delta\sigma_3 - \Delta u_3)$

where C_s is the compressibility of the soil skeleton under an isotropic effective stress increment.

> Reduction in volume of the pore space $= C_v n V \Delta u_3$

where C_v is the compressibility of pore fluid under an isotropic pressure increment.

If the soil *particles* are assumed to be incompressible and if no drainage of pore fluid takes place then the reduction in volume of the soil skeleton must equal the reduction in volume of the pore space, i.e.

$$C_s V(\Delta\sigma_3 - \Delta u_3) = C_v n V \Delta u_3$$

Therefore,

$$\Delta u_3 = \Delta\sigma_3 \left(\frac{1}{1 + n(C_v/C_s)} \right)$$

Writing $1/[1 + n(C_v/C_s)] = B$, defined as a pore pressure coefficient,

$$\Delta u_3 = B\Delta\sigma_3 \tag{4.22}$$

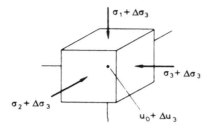

Figure 4.25 Soil element under isotropic stress increment.

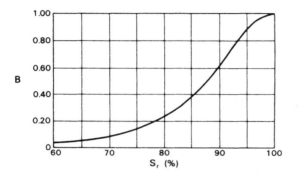

Figure 4.26 Typical relationship between B and degree of saturation.

In fully saturated soils the compressibility of the pore fluid (water only) is considered negligible compared with that of the soil skeleton, and therefore $C_v/C_s \to 0$ and $B \to 1$. Equation 4.22 with $B = 1$ has already been assumed in the discussion on undrained strength earlier in the present chapter. In partially saturated soils the compressibility of the pore fluid is high due to the presence of pore air, and therefore $C_v/C_s > 0$ and $B < 1$. The variation of B with degree of saturation for a particular soil is shown in Figure 4.26.

The value of B can be measured in the triaxial apparatus. A specimen is set up under any value of all-round pressure and the pore water pressure measured (after consolidation if desired). Under undrained conditions the all-round pressure is then increased (or reduced) by an amount $\Delta\sigma_3$ and the *change* in pore water pressure (Δu) from the initial value is measured, enabling the value of B to be calculated from Equation 4.22. This procedure is used as a check for full saturation in triaxial testing.

(2) Major principal stress increment

Consider now an increase $\Delta\sigma_1$ in the total major principal stress only, as shown in Figure 4.27, resulting in an immediate increase Δu_1 in pore pressure.

The increases in effective stress are

$$\Delta\sigma_1' = \Delta\sigma_1 - \Delta u_1$$
$$\Delta\sigma_3' = \Delta\sigma_2' = -\Delta u_1$$

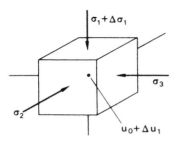

Figure 4.27 Soil element under major principal stress increment.

If the soil behaved as an elastic material then the reduction in volume of the soil skeleton would be

$$\frac{1}{3}C_s V(\Delta\sigma_1 - 3\Delta u_1)$$

The reduction in volume of the pore space is

$$C_v n V \Delta u_1$$

Again, these two volume changes will be equal for undrained conditions, i.e.

$$\frac{1}{3}C_s V(\Delta\sigma_1 - 3\Delta u_1) = C_v n V \Delta u_1$$

Therefore,

$$\Delta u_1 = \frac{1}{3}\left(\frac{1}{1 + n(C_v/C_s)}\right)\Delta\sigma_1$$

$$= \frac{1}{3}B\Delta\sigma_1$$

Soils, however, are not elastic and the above equation is rewritten in the general form:

$$\Delta u_1 = AB\Delta\sigma_1 \tag{4.23}$$

where A is a pore pressure coefficient to be determined experimentally. AB may also be written as $\overline{A}$. In the case of fully saturated soils ($B = 1$)

$$\Delta u_1 = A\Delta\sigma_1 \tag{4.24}$$

The value of A for a fully saturated soil can be determined from measurements of pore water pressure during the application of principal stress difference under undrained conditions in a triaxial test. The *change* in total major principal stress is equal to the value of the principal stress difference applied and if the corresponding *change* in pore water pressure is measured the value of A can be calculated from Equation 4.24. The value of the coefficient at any stage of the test can be obtained but the value at failure is of more interest.

For highly compressible soils such as normally consolidated clays the value of A is found to lie within the range 0.5–1.0. In the case of clays of high sensitivity the increase in major principal stress may cause collapse of the soil structure, resulting in very high pore water pressures and values of A greater than 1. For soils of lower compressibility such as lightly overconsolidated clays the value of A lies within the range 0–0.5. If the clay is heavily overconsolidated there is a tendency for the soil to dilate as the major principal stress is increased but under undrained conditions no water can be drawn into the element and a negative pore water pressure may result. The value of A for heavily overconsolidated soils may lie between -0.5 and 0. A typical relationship between the value of A at failure (A_f) and OCR for a fully saturated clay is shown in Figure 4.28.

For the condition of zero lateral strain in the soil element, reduction in volume is possible in the direction of the major principal stress only. If C_{s0} is the uniaxial compressibility of the soil skeleton then under undrained conditions

$$C_{s0}V(\Delta\sigma_1 - \Delta u_1) = C_v nV\Delta u_1$$

Therefore,

$$\begin{aligned}\Delta u_1 &= \Delta\sigma_1\left(\frac{1}{1 + n(C_v/C_{s0})}\right)\\ &= A\Delta\sigma_1\end{aligned}$$

where $A = 1/[1 + n(C_v/C_{s0})]$. For a fully saturated soil, $C_v/C_{s0} \to 0$ and $A \to 1$, for the condition of zero lateral strain only. This was assumed in the discussion on consolidation in Chapter 3.

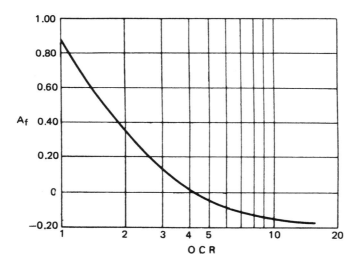

Figure 4.28 Typical relationship between A at failure and overconsolidation ratio.

(3)   Combination of increments

Cases 1 and 2 above may be combined to give the equation for the pore pressure response Δu to an isotropic stress increase $\Delta\sigma_3$ together with an axial stress increase $(\Delta\sigma_1 - \Delta\sigma_3)$ as occurs in the triaxial test. Combining Equations 4.22 and 4.23

$$\Delta u = \Delta u_3 + \Delta u_1$$
$$= B[\Delta\sigma_3 + A(\Delta\sigma_1 - \Delta\sigma_3)] \tag{4.25}$$

An overall coefficient $\bar{B}$ can be obtained by dividing Equation 4.25 by $\Delta\sigma_1$:

$$\frac{\Delta u}{\Delta\sigma_1} = B\left[\frac{\Delta\sigma_3}{\Delta\sigma_1} + A\left(1 - \frac{\Delta\sigma_3}{\Delta\sigma_1}\right)\right]$$

Therefore,

$$\frac{\Delta u}{\Delta\sigma_1} = B\left[1 - (1 - A)\left(1 - \frac{\Delta\sigma_3}{\Delta\sigma_1}\right)\right]$$

or

$$\frac{\Delta u}{\Delta\sigma_1} = \bar{B} \tag{4.26}$$

Under static conditions, if there is no change in the level of the water table during subsequent consolidation, Δu is equal to the initial value of excess pore water pressure in fully saturated soils.

Since soils are not elastic materials the pore pressure coefficients are not constants, their values depending on the stress levels over which they are determined.

Example 4.5

The following results refer to a consolidated–undrained triaxial test on a saturated clay specimen under an all-round pressure of $300\,kN/m^2$:

$\Delta l/l_0$	0	0.01	0.02	0.04	0.08	0.12
$\sigma_1 - \sigma_3$ (kN/m^2)	0	138	240	312	368	410
u (kN/m^2)	0	108	158	178	182	172

Draw the total and effective stress paths and plot the variation of the pore pressure coefficient A during the test.

From the data, the values in Table 4.6 are calculated. For example, when the strain is 0.01, $A = 108/138 = 0.78$. The stress paths and the variation of A are plotted in Figure 4.29 in terms of (a) $\frac{1}{2}(\sigma_1 - \sigma_3)$ and $\frac{1}{2}(\sigma_1 + \sigma_3)$, (b) q and p, or the effective stress equivalents. From the shape of the effective stress path and the value of A at failure it can be concluded that the clay is overconsolidated.

Table 4.6

$\Delta l/l_0$	0	0.01	0.02	0.04	0.08	0.12
$\frac{1}{2}(\sigma_1 - \sigma_3)$	0	69	120	156	184	205
$\frac{1}{2}(\sigma_1 + \sigma_3)$	300	369	420	456	484	505
$\frac{1}{2}(\sigma_1' + \sigma_3')$	300	261	262	278	302	333
q	0	138	240	312	368	410
p	300	346	380	404	423	437
p'	300	238	222	226	241	265
A	–	0.78	0.66	0.57	0.50	0.42

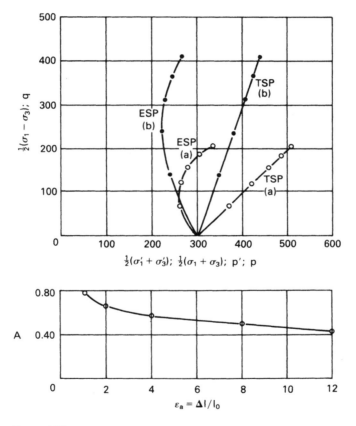

Figure 4.29

Example 4.6

In a triaxial test a soil specimen was consolidated under an all-round pressure of $800 \, \text{kN/m}^2$ and a back pressure of $400 \, \text{kN/m}^2$. Thereafter, under undrained conditions, the all-round pressure was raised to $900 \, \text{kN/m}^2$, resulting in a pore water pressure reading of $495 \, \text{kN/m}^2$; then (with the all-round pressure remaining at $900 \, \text{kN/m}^2$) axial load was applied to give a principal stress difference of $585 \, \text{kN/m}^2$ and a pore water pressure reading of 660 kN/m^2. Calculate the values of the pore pressure coefficients B, $\overline{A}$ and $\overline{B}$.

Corresponding to an increase in all-round pressure from 800 to 900 kN/m^2 the pore pressure increases from the value of the back pressure, 400 kN/m^2, to 495 kN/m^2. Therefore,

$$B = \frac{\Delta u_3}{\Delta \sigma_3} = \frac{495 - 400}{900 - 800} = \frac{95}{100} = 0.95$$

The total major principal stress increases from 900 to $(900 + 585)$ kN/m^2; the corresponding increase in pore pressure is from 495 to 660 kN/m^2. Therefore,

$$\overline{A} = \frac{\Delta u_1}{\Delta \sigma_1} = \frac{660 - 495}{585} = \frac{165}{585} = 0.28$$

The overall increase in pore pressure is from 400 to 660 kN/m^2, corresponding to an increase in total major principal stress from 800 to $(800 + 100 + 585)$ kN/m^2. Therefore,

$$\overline{B} = \frac{\Delta u}{\Delta \sigma_1} = \frac{660 - 400}{100 + 585} = \frac{260}{685} = 0.38$$

PROBLEMS

4.1 What is the shear strength in terms of effective stress on a plane within a saturated soil mass at a point where the total normal stress is 295 kN/m^2 and the pore water pressure 120 kN/m^2? The effective stress parameters of the soil for the appropriate stress range are $c' = 12$ kN/m^2 and $\phi' = 30°$.

4.2 A series of drained triaxial tests were carried out on specimens of a sand prepared at the same porosity and the following results were obtained at failure. Determine the value of the angle of shearing resistance ϕ'.

All-round pressure (kN/m^2)	100	200	400	800
Principal stress difference (kN/m^2)	452	908	1810	3624

4.3 In a series of unconsolidated–undrained triaxial tests on specimens of a fully saturated clay the following results were obtained at failure. Determine the values of the shear strength parameters c_u and ϕ_u.

All-round pressure (kN/m^2)	200	400	600
Principal stress difference (kN/m^2)	222	218	220

4.4 The effective stress parameters for a fully saturated clay are known to be $c' = 15$ kN/m^2 and $\phi' = 29°$. In an unconsolidated–undrained triaxial test on a specimen of the same clay the all-round pressure was 100 kN/m^2 and the principal stress difference at failure 170 kN/m^2. Assuming that the above parameters are appropriate to the failure stress state of the test, what would be the expected value of pore water pressure in the specimen at failure?

4.5   The results below were obtained at failure in a series of consolidated–undrained triaxial tests, with pore water pressure measurement, on specimens of a fully saturated clay. Determine the values of the shear strength parameters c' and ϕ'. If a specimen of the same soil were consolidated under an all-round pressure of $250 \, kN/m^2$ and the principal stress difference applied with the all-round pressure changed to $350 \, kN/m^2$, what would be the expected value of principal stress difference at failure?

σ_3 (kN/m²)	150	300	450	600
$\sigma_1 - \sigma_3$ (kN/m²)	103	202	305	410
u (kN/m²)	82	169	252	331

4.6   The following results were obtained at failure in a series of drained triaxial tests on fully saturated clay specimens originally 38 mm in diameter by 76 mm long. Determine the secant parameter ϕ' for each test and the values of tangent parameters c' and ϕ' for the stress range 300–500 kN/m².

All-round pressure (kN/m²)	200	400	600
Axial compression (mm)	7.22	8.36	9.41
Axial load (N)	565	1015	1321
Volume change (ml)	5.25	7.40	9.30

4.7   Derive Equation 4.12.
       In an *in-situ* vane test on a saturated clay a torque of 35 N m is required to shear the soil. The vane is 50 mm wide by 100 mm long. What is the undrained strength of the clay?

4.8   A consolidated–undrained triaxial test on a specimen of saturated clay was carried out under an all-round pressure of $600 \, kN/m^2$. Consolidation took place against a back pressure of $200 \, kN/m^2$. The following results were recorded during the test.

$\sigma_1 - \sigma_3$ (kN/m²)	0	80	158	214	279	319
u (kN/m²)	200	229	277	318	388	433

Draw the stress paths and give the value of the pore pressure coefficient A at failure.

4.9   In a triaxial test a soil specimen is allowed to consolidate fully under an all-round pressure of $200 \, kN/m^2$. Under undrained conditions the all-round pressure is increased to $350 \, kN/m^2$, the pore water pressure then being measured as $144 \, kN/m^2$. Axial load is then applied under undrained conditions until failure takes place, the following results being obtained.

Axial strain (%)	0	2	4	6	8	10
Principal stress difference (kN/m²)	0	201	252	275	282	283
Pore water pressure (kN/m²)	144	244	240	222	212	209

Determine the value of the pore pressure coefficient B and plot the variation of coefficient $\overline{A}$ with axial strain, stating the value at failure.

REFERENCES

1   Bishop, A.W. (1966) The strength of soils as engineering materials, *Geotechnique*, **16**, 91–128.

2   Bishop, A.W., Alpan, I., Blight, G.E. and Donald, I.B. (1960) Factors controlling the strength of partly saturated cohesive soils, in *Proceedings of the ASCE Conference on Shear Strength of Cohesive Soils*, Boulder, CO, USA, ASCE, New York, pp. 503–32.

3   Bishop, A.W., Green, G.E., Garga, V.K., Andresen, A. and Brown, J.D. (1971) A new ring shear apparatus and its application to the measurement of residual strength, *Geotechnique*, **21**, 273–328.

4   Bishop, A.W. and Wesley, L.D. (1975) A hydraulic triaxial apparatus for controlled stress path testing, *Geotechnique*, **25**, 657–70.

5   Bjerrum, L. (1973) Problems of soil mechanics and construction on soft clays, in *Proceedings of the 8th International Conference of SMFE*, Moscow, Vol. 3, pp. 111–59.

6   Bolton, M.D. (1986) The strength and dilatancy of sands, *Geotechnique*, **36**, 65–78.

7   British Standard 1377 (1990) *Methods of Test for Soils for Civil Engineering Purposes*, British Standards Institution, London.

8   Bromhead, E.N. (1979) A simple ring shear apparatus, *Ground Engineering*, **12** (5), 40–4.

9   Cornforth, D.H. (1964) Some experiments on the influence of strain conditions on the strength of sand, *Geotechnique*, **14**, 143–67.

10 Duncan, J.M. and Seed, H.B. (1966) Strength variation along failure surfaces in clay, *Journal of the ASCE*, **92** (SM6), 81–104.

11 Head, K.H. (1986) *Manual of Soil Laboratory Testing*, three volumes, Pentech, London.

12 Hight, D.W., Gens, A. and Symes, M.J. (1983) The development of a new hollow cylinder apparatus for investigating the effects of principal stress rotation in soils, *Geotechnique*, **33**, 355–83.

13 Lupini, J.F., Skinner, A.E. and Vaughan, P.R. (1981) The drained residual strength of cohesive soils, *Geotechnique*, **31**, 181–213.

14 Roscoe, K.H., Schofield, A.N. and Wroth, C.P. (1958) On the yielding of soils, *Geotechnique*, **8**, 22–53.

15 Rowe, P.W. (1962) The stress–dilatancy relation for static equilibrium of an assembly of particles in contact, *Proceedings of the Royal Society A*, **269**, 500–27.

16 Skempton, A.W. (1954) The pore pressure coefficients A and B, *Geotechnique*, **4**, 143–7.

17 Skempton, A.W. (1985) Residual strength of clays in landslides, folded strata and the laboratory, *Geotechnique*, **35**, 1–18.

18 Skempton, A.W. and Sowa, V.A. (1963) The behaviour of saturated clays during sampling and testing, *Geotechnique*, **13**, 269–90.

Chapter 5

Stresses and displacements

5.1 ELASTICITY AND PLASTICITY

The stresses and displacements in a soil mass due to applied loading are considered in this chapter. Many problems can be treated by analysis in two dimensions, i.e. only the stresses and displacements in a single plane need to be considered. The total normal stresses and shear stresses in the x and z directions on an element of soil are shown in Figure 5.1, the stresses being positive as shown; the stresses vary across the element. The rates of change of the normal stresses in the respective directions are $\partial\sigma_x/\partial x$ and $\partial\sigma_z/\partial z$; the rates of change of the shear stresses are $\partial\tau_{xz}/\partial x$ and $\partial\tau_{zx}/\partial z$. Every such element in a soil mass must be in static equilibrium. By equating moments about the centre point of the element, and neglecting higher-order differentials, it is apparent that $\tau_{xz} = \tau_{zx}$. By equating forces in the x and z directions the following equations are obtained:

$$\frac{\partial\sigma_x}{\partial x} + \frac{\partial\tau_{zx}}{\partial z} - X = 0 \tag{5.1a}$$

$$\frac{\partial\tau_{xz}}{\partial x} + \frac{\partial\sigma_z}{\partial z} - Z = 0 \tag{5.1b}$$

where X and Z are the respective body forces per unit volume. These are the equations of equilibrium in two dimensions; they can also be written in terms of effective stress. In terms of total stress the body forces are $X = 0$ and $Z = \gamma$ (or γ_{sat}). In terms of effective stress the body forces are $X' = 0$ and $Z' = \gamma'$; however, if seepage is taking place these become $X' = i_x\gamma_w$ and $Z' = \gamma' + i_z\gamma_w$ where i_x and i_z are the hydraulic gradients in the x and z directions, respectively.

Due to the applied loading, points within the soil mass will be displaced relative to the axes and to one another. If the components of displacement in the x and z directions are denoted by u and w, respectively, then the normal strains are given by

$$\varepsilon_x = \frac{\partial u}{\partial x}, \qquad \varepsilon_z = \frac{\partial w}{\partial z}$$

and the shear strain by

$$\gamma_{xz} = \frac{\partial u}{\partial z} + \frac{\partial w}{\partial x}$$

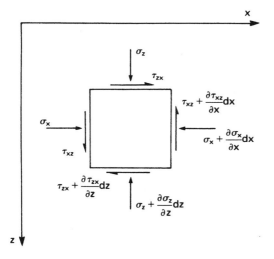

Figure 5.1 Two-dimensional state of stress in an element.

However, these strains are not independent; they must be compatible with each other if the soil mass as a whole is to remain continuous. This requirement leads to the following relationship, known as the equation of compatibility in two dimensions:

$$\frac{\partial^2 \varepsilon_x}{\partial z^2} + \frac{\partial^2 \varepsilon_z}{\partial x^2} - \frac{\partial \gamma_{xz}}{\partial x \partial z} = 0 \qquad (5.2)$$

Equations 5.1 and 5.2, being independent of material properties, can be applied to both elastic and plastic behaviour.

The rigorous solution of a particular problem requires that the equations of equilibrium and compatibility are satisfied for the given boundary conditions; an appropriate stress–strain relationship is also required. In the theory of elasticity [18] a linear stress–strain relationship is combined with the above equations. In general, however, soils are non-homogeneous, exhibit anisotropy and have non-linear stress–strain relationships which are dependent on stress history and the particular stress path followed.

In analysis an appropriate idealization of the stress–strain relationship is employed. One such idealization is shown by the dotted lines in Figure 5.2(a), linearly elastic behaviour being assumed between O and Y′ (the assumed yield point) followed by unrestricted plastic strain (or flow) Y′P at constant stress. This idealization, which is shown separately in Figure 5.2(b), is known as the *elastic–perfectly plastic* model of material behaviour. If only the collapse condition in a practical problem is of interest then the elastic phase can be omitted and the *rigid–perfectly plastic* model, shown in Figure 5.2(c), may be used. A third idealization is the *elastic–strain hardening plastic* model, shown in Figure 5.2(d), in which plastic strain beyond the yield point necessitates further stress increase. If unloading and reloading were to take place subsequent to yielding in the strain hardening model,

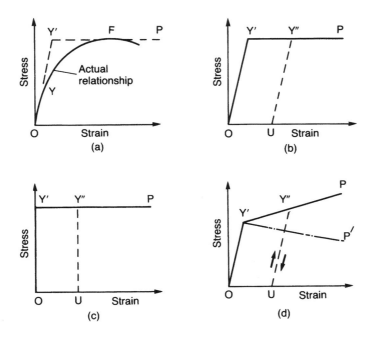

Figure 5.2 (a) Typical stress–strain relationship, (b) elastic–perfectly plastic model, (c) rigid–perfectly plastic model, and (d) elastic–strain hardening and softening plastic models.

as shown by the dotted line Y″U in Figure 5.2(d), there would be a new yield point Y″ at a stress level higher than that at Y′. An increase in yield stress is a characteristic of strain hardening. No such increase takes place in the case of perfectly plastic (i.e. non-hardening) behaviour, the stress at Y″ being equal to that at Y′ as shown in Figures 5.2(b) and (c). A further idealization is the *elastic–strain softening plastic* model, represented by OY′P′ in Figure 5.2(d), in which the plastic strain beyond the yield point is accompanied by stress decrease.

In plasticity theory [8] the characteristics of yielding, hardening and flow are considered; these are described by a yield function, a hardening law and a flow rule, respectively. The yield function is written in terms of stress components or principal stresses. The Mohr–Coulomb criterion (Equation 4.6) is one possible yield function if perfectly plastic behaviour is assumed. Alternatively, the yield function could be expressed in terms of critical-state parameters (Section 4.5). The hardening law represents the relationship between the increase in yield stress and the corresponding plastic strain components. The flow rule specifies the relative (i.e. not absolute) magnitudes of the plastic strain components during yielding under a particular state of stress. The hardening law and the flow rule can also be expressed in terms of critical-state parameters.

In practice the most widely used solutions are those for the vertical stress at a point below a loaded area on the surface of a soil mass. It has been shown [2] that the vertical stress increment at a given point below the surface due to foundation loading is insensitive to a relatively wide range of soil characteristics such as heterogeneity, anisotropy and non-linearity of the stress–strain relationship. Accordingly, solutions from linear elastic theory, in which the soil is assumed to be homogeneous and isotropic, are sufficiently accurate for use in most cases. The main exceptions are loose sands and soft clays, particularly where they are overlain by a relatively dense or stiff stratum. It should be noted, however, that increments of horizontal stress and of shear stress are relatively sensitive to soil characteristics. The insensitivity of the vertical stress increment depends on the assumption of a uniform pressure distribution, as would be the case for a flexible foundation. In the case of a stiff foundation the contact pressure is non-uniform, the exact distribution depending on the soil characteristics. A comprehensive collection of solutions from the theory of elasticity has been published by Poulos and Davis [16].

The finite element method is now widely used for the calculation of stresses and displacements. For example, the finite element program CRISP [1] enables the soil to be treated as either a non-homogeneous or an anisotropic elastic material; alternatively, elastoplastic behaviour can be modelled by means of critical-state parameters, enabling soil displacements up to failure to be determined. A comprehensive treatment of finite element analysis in geotechnical engineering has been published by Potts and Zdravkovic [14, 15].

Displacement solutions from elastic theory can be used at relatively low stress levels. These solutions require a knowledge of the values of Young's modulus (E) and Poisson's ratio (ν) for the soil, either for undrained conditions or in terms of effective stress. Poisson's ratio is required for certain stress solutions. It should be noted that the shear modulus (G), where

$$G = \frac{E}{2(1 + \nu)} \tag{5.3}$$

is independent of the drainage conditions, assuming that the soil is isotropic.

The volumetric strain of an element of linearly elastic material under three principal stresses is given by

$$\frac{\Delta V}{V} = \frac{1 - 2\nu}{E} (\Delta \sigma_1 + \Delta \sigma_2 + \Delta \sigma_3)$$

If this expression is applied to soils over the initial part of the stress–strain curve, then for undrained conditions $\Delta V/V = 0$, hence $\nu = 0.5$. The undrained value of Young's modulus is then related to the shear modulus by the expression $E_u = 3G$. If consolidation takes place then $\Delta V/V > 0$ and $\nu < 0.5$ for drained or partially drained conditions.

In principle, the value of E can be estimated from the curve relating principal stress difference and axial strain in an appropriate triaxial test. The value is usually determined as the secant modulus between the origin and one-third of the peak stress, or over the actual stress range in the particular problem. However, because of the effects of sampling disturbance, it is preferable to determine E (or G) from the results of in-situ tests. One such method is to apply load increments to a test plate, either in a shallow pit or at the bottom of a large-diameter borehole, and to measure the resulting vertical displacements. The value of E is then calculated using the relevant displacement solution, an appropriate value of ν being assumed.

The pressuremeter

The shear modulus (G) can be determined *in situ* by means of the pressuremeter. The original pressuremeter was developed in the 1950s by Menard in an attempt to overcome the problem of sampling disturbance and to ensure that the macro-fabric of the soil is adequately represented. Menard's original design, illustrated in Figure 5.3(a), consists of three cylindrical rubber cells of equal diameter arranged coaxially. The device is lowered into a (slightly oversize) borehole to the required depth and the central measuring cell is expanded against the borehole wall by means of water pressure, measurements of the applied pressure and the corresponding increase in volume of the cell being recorded. Pressure is applied to the water by compressed gas (usually nitrogen) in a control cylinder at the surface. The increase in volume of the measuring cell is determined from the movement of the gas–water interface in the control cylinder, readings normally being taken at times of 15, 30, 60 and 120 s after a pressure increment has been applied. The pressure is corrected for (a) the head difference between the water level in the cylinder and the test level in the borehole, (b) the pressure required to stretch the rubber cell and (c) the expansion of the control cylinder and tubing under pressure. The two outer guard cells are expanded under the same pressure as in the measuring cell but using compressed gas; the increase in volume of the guard cells is not measured. The function of the guard cells is to eliminate end effects, ensuring a state of plane strain adjacent to the measuring cell.

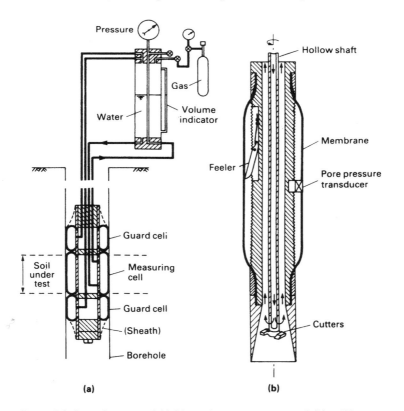

Figure 5.3 Basic features of (a) Menard pressuremeter and (b) self-boring pressuremeter.

The results of a test using the Menard pressuremeter are represented by a plot of corrected pressure (p) against volume (V) as shown in Figure 5.4(a). On this plot a linear section occurs between pressures p_i and p_f. The value p_i is the pressure necessary to achieve initial contact between the cell and the borehole wall and to recompress soil disturbed or softened as a result of boring. The value p_f is the pressure corresponding to the onset of plastic strain in the soil. Eventually a limit pressure (p_l) is approached at which continuous expansion of the borehole cavity would occur. A 'creep' curve, obtained by plotting the volume change between the 30 and 120 s readings against the corresponding pressure, may be a useful aid in fixing the values p_i and p_f, significant breaks occurring at these pressures.

The datum or reference pressure for the interpretation of pressuremeter results is a value (p_0) equal to the *in-situ* total horizontal stress in the soil before boring. Originally this value was assumed to be equal to p_i but the use of a pre-formed borehole means that the soil is being stressed from an unloaded condition, not from the initial undisturbed state; consequently the value of p_0 should be greater than p_i. (It should be appreciated that it is normally very difficult to obtain an independent value of *in-situ* total horizontal stress.) The reference volume V_0 (corresponding to the pressure p_0) is taken to be the initial volume of the borehole cavity over the test length. At any stage during a test the volume V, corresponding to the pressure p, is referred to as the current volume.

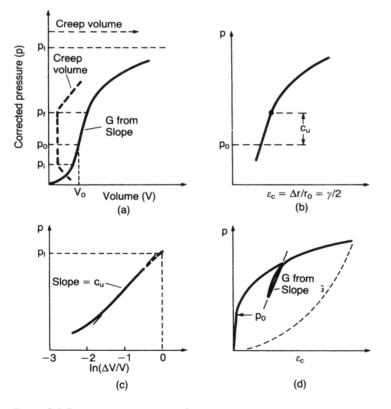

Figure 5.4 Pressuremeter test results.

Alternatively the results of a pressuremeter test can be represented by plotting corrected pressure against the circumferential strain (ε_c) at the borehole wall. The circumferential strain is given by the ratio of the increase in radius of the borehole cavity (Δr) to the radius at the reference state (r_0). The relationship between current volumetric strain and circumferential strain is

$$\frac{\Delta V}{V} = 1 - (1 + \varepsilon_c)^{-2}$$

(Shear strain (γ) is equal to twice circumferential strain.)

Marsland and Randolph [11] proposed a procedure, using the p–ε_c plot, for the determination of p_0, applicable to soils such as stiff clays which exhibit essentially linear stress–strain behaviour up to peak strength. The linear section of the p–ε_c plot should terminate when the shear stress at the borehole wall is equal to the (peak) undrained strength of the clay, i.e. when the pressure becomes equal to ($p_0 + c_u$). The value of c_u is determined using Equation 5.7, for which a value of the reference volume V_0 is required. The method involves an iterative process in which estimates of p_0, and hence V_0, are made and the corresponding value of c_u determined until the point representing ($p_0 + c_u$) corresponds with the point on the plot at which significant curvature begins, as shown in Figure 5.4(b).

The value of the limit pressure (p_1) can be determined by plotting pressure against the logarithm of current volumetric strain and extrapolating to a strain of unity, representing continuous expansion, as shown in Figure 5.4(c).

An analysis of the expansion of the borehole cavity during a pressuremeter test was presented by Gibson and Anderson [6], the soil being considered as an elastic–perfectly plastic material. Within the linear section of the p–V plot the shear modulus is given by

$$G = V \frac{\mathrm{d}p}{\mathrm{d}V} \tag{5.4}$$

where $\mathrm{d}p/\mathrm{d}V$ is the slope of the linear section and V is the current volume of the borehole cavity. However, it is recommended that the modulus is determined from an unloading–reloading cycle to minimize the effect of soil disturbance.

In the case of saturated clays it is possible to obtain the value of the undrained shear strength (c_u), by iteration, from the following expression:

$$p_1 - p_0 = \left[\ln\left(\frac{G}{c_u}\right) + 1 \right] c_u \tag{5.5}$$

In modern developments of the pressuremeter the measuring cell is expanded directly by gas pressure. This pressure and the radial expansion of the rubber membrane are recorded by means of electrical transducers within the cell. In addition, a pore water pressure transducer is fitted into the cell wall such that it is in contact with the soil during the test. A considerable increase in accuracy is obtained with these pressuremeters compared with the original Menard device. It is also possible to adjust the cell pressure continuously, using electronic control equipment, to achieve a constant rate of increase in circumferential strain (i.e. a strain-controlled test), rather than to apply the pressure in increments (a stress-controlled test).

Some soil disturbance adjacent to a borehole is inevitable and the results of pressuremeter tests in pre-formed holes can be sensitive to the method of boring. The self-boring pressuremeter was developed to overcome this problem and is suitable for use in most types of soil; however, special insertion techniques are required in the case of sands. This device, illustrated in Figure 5.3(b), is jacked slowly into the ground and the soil is broken up by a rotating cutter fitted inside a cutting head at the lower end, the optimum position of the cutter being a function of the shear strength of the soil. Water or drilling fluid is pumped down the hollow shaft to which the cutter is attached and the resulting slurry is carried to the surface through the annular space adjacent to the shaft; the device is thus inserted with minimal disturbance of the soil. The only correction required is for the pressure required to stretch the membrane.

A 'push-in' penetrometer has also been developed for insertion below the bottom of a borehole, for use particularly in off-shore work. This pressuremeter is fitted with a cutting shoe, a soil core passing upwards inside the device.

The membrane of a pressuremeter may be protected against possible damage (particularly in coarse soils) by a thin stainless steel sheath with longitudinal cuts, designed to cause only negligible resistance to the expansion of the cell.

Results from a strain-controlled test in clay using the self-boring pressuremeter are of the form shown in Figure 5.4(d), the pressure (p) being plotted against the circumferential strain (ε_c). Use of the self-boring pressuremeter overcomes the difficulty in determining the initial *in-situ* total horizontal stress; because soil disturbance is minimal the pressure at which the membrane starts to expand (referred to as the 'lift-off' pressure) should be equal to p_0, as shown in Figure 5.4(d).

The value of the shear modulus is given by the following equation, derived in a later analysis by Palmer [13], in which no assumption is made regarding the stress–strain characteristics of the soil. For expansion of a borehole cavity at small strains it was shown that

$$G = \frac{1}{2}\frac{dp}{d\varepsilon_c} \tag{5.6}$$

The modulus should be obtained from the slope of an unloading–reloading cycle as shown in Figure 5.4(d), ensuring that the soil remains in the 'elastic' state during unloading. Wroth [20] has shown that, in the case of a clay, this requirement will be satisfied if the reduction in pressure during the unloading stage is less than $2c_u$.

For a saturated clay the undrained shear strength (c_u) can also be obtained from the following equation derived from the analysis of Gibson and Anderson:

$$p = p_1 + c_u \ln\left(\frac{\Delta V}{V}\right) \tag{5.7}$$

where $\Delta V/V$ is the current volumetric strain.

It should be noted that Equation 5.7 is relevant only after the plastic state has been reached in the soil (i.e. when $p_f < p < p_1$). The plot of p against $\ln(\Delta V/V)$ should become essentially linear for the final stage of the test as shown in Figure 5.4(c), and the value of c_u is given by the slope of the line.

In Palmer's analysis it was shown that at small strains the shear stress in the soil at the wall of the expanding cavity is given by

$$\tau = \varepsilon_c \frac{\mathrm{d}p}{\mathrm{d}\varepsilon_c} \tag{5.8}$$

and at larger strains by

$$\tau = \frac{\mathrm{d}p}{\mathrm{d}[\ln(\Delta V / V)]} \tag{5.9}$$

both the circumferential and volumetric strains being defined with respect to the reference state. Equations 5.8 and 5.9 can be used to derive the entire stress–strain curve for the soil.

An analysis for the interpretation of pressuremeter tests in sands has been given by Hughes *et al.* [9]. The analysis enables values for the angle of shearing resistance (ϕ') and the angle of dilation (ψ) to be determined. A comprehensive review of the use of pressuremeters, including examples of test results and their application in design, has been given by Mair and Wood [10].

5.2 STRESSES FROM ELASTIC THEORY

The stresses within a semi-infinite, homogeneous, isotropic mass, with a linear stress–strain relationship, due to a point load on the surface, were determined by Boussinesq in 1885. The vertical, radial, circumferential and shear stresses at a depth z and a horizontal distance r from the point of application of the load were given. The stresses due to surface loads distributed over a particular area can be obtained by integration from the point load solutions. The stresses at a point due to more than one surface load are obtained by superposition. In practice, loads are not usually applied directly on the surface but the results for surface loading can be applied conservatively in problems concerning loads at a shallow depth.

A range of solutions, suitable for determining the stresses below foundations, is given in the following sections. Negative values of loading can be used if the stresses due to excavation are required or in problems in which the principle of superposition is used. The stresses due to surface loading act in addition to the *in-situ* stresses due to the self-weight of the soil.

Point load

Referring to Figure 5.5(a), the stresses at X due to a point load Q on the surface are as follows:

$$\sigma_z = \frac{3Q}{2\pi z^2} \left\{ \frac{1}{1 + (r/z)^2} \right\}^{5/2} \tag{5.10}$$

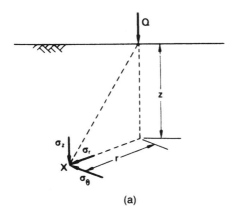

(a)

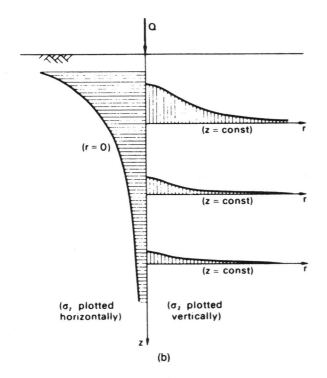

(b)

Figure 5.5 (a) Stresses due to point load and (b) variation of vertical stress due to point load.

$$\sigma_r = \frac{Q}{2\pi}\left\{\frac{3r^2z}{(r^2+z^2)^{5/2}} - \frac{1-2v}{r^2+z^2+z(r^2+z^2)^{1/2}}\right\} \tag{5.11}$$

$$\sigma_\theta = -\frac{Q}{2\pi}(1-2v)\left\{\frac{z}{(r^2+z^2)^{3/2}} - \frac{1}{r^2+z^2+z(r^2+z^2)^{1/2}}\right\} \tag{5.12}$$

$$\tau_{rz} = \frac{3Q}{2\pi}\left\{\frac{rz^2}{(r^2+z^2)^{5/2}}\right\} \tag{5.13}$$

It should be noted that when $v = 0.5$ the second term in Equation 5.11 vanishes and Equation 5.12 gives $\sigma_\theta = 0$.

Equation 5.10 is used most frequently in practice and can be written in terms of an influence factor I_p, where

$$I_p = \frac{3}{2\pi}\left\{\frac{1}{1+(r/z)^2}\right\}^{5/2}$$

Then,

$$\sigma_z = \frac{Q}{z^2}I_p$$

Values of I_p in terms of r/z are given in Table 5.1. The form of the variation of σ_z with z and r is illustrated in Figure 5.5(b). The left-hand side of the figure shows the variation of σ_z with z on the vertical through the point of application of the load Q (i.e. for $r = 0$); the right-hand side of the figure shows the variation of σ_z with r for three different values of z.

It should be noted that the expression for σ_z (Equation 5.10) is independent of elastic modulus (E) and Poisson's ratio (ν).

Table 5.1 Influence factors for vertical stress due to point load

r/z	I_p	r/z	I_p	r/z	I_p
0.00	0.478	0.80	0.139	1.60	0.020
0.10	0.466	0.90	0.108	1.70	0.016
0.20	0.433	1.00	0.084	1.80	0.013
0.30	0.385	1.10	0.066	1.90	0.011
0.40	0.329	1.20	0.051	2.00	0.009
0.50	0.273	1.30	0.040	2.20	0.006
0.60	0.221	1.40	0.032	2.40	0.004
0.70	0.176	1.50	0.025	2.60	0.003

Line load

Referring to Figure 5.6(a), the stresses at point X due to a line load of Q per unit length on the surface are as follows:

$$\sigma_z = \frac{2Q}{\pi} \frac{z^3}{(x^2 + z^2)^2} \tag{5.14}$$

$$\sigma_x = \frac{2Q}{\pi} \frac{x^2 z}{(x^2 + z^2)^2} \tag{5.15}$$

$$\tau_{xz} = \frac{2Q}{\pi} \frac{xz^2}{(x^2 + z^2)^2} \tag{5.16}$$

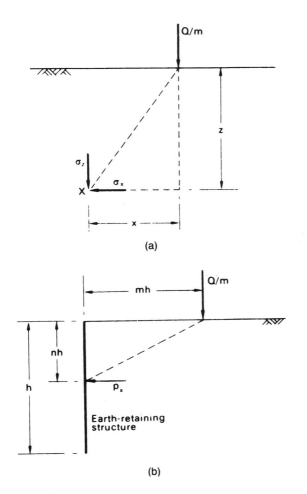

Figure 5.6 (a) Stresses due to line load and (b) lateral pressure due to line load.

Equation 5.15 can be used to estimate the lateral pressure on an earth-retaining structure due to a line load on the surface of the backfill. In terms of the dimensions given in Figure 5.6(b), Equation 5.15 becomes

$$\sigma_x = \frac{2Q}{\pi h} \frac{m^2 n}{(m^2 + n^2)^2}$$

However, the structure will tend to interfere with the lateral strain due to the load Q and to obtain the lateral pressure on a relatively rigid structure a second load Q must be imagined at an equal distance on the other side of the structure. Then, the lateral pressure is given by

$$p_x = \frac{4Q}{\pi h} \frac{m^2 n}{(m^2 + n^2)^2} \qquad (5.17)$$

The total thrust on the structure is given by

$$P_x = \int_0^1 p_x h \, dn = \frac{2Q}{\pi} \frac{1}{m^2 + 1} \qquad (5.18)$$

Strip area carrying uniform pressure

The stresses at point X due to a uniform pressure q on a strip area of width B and infinite length are given in terms of the angles α and β defined in Figure 5.7(a).

$$\sigma_z = \frac{q}{\pi}\{\alpha + \sin\alpha\cos(\alpha + 2\beta)\} \qquad (5.19)$$

$$\sigma_x = \frac{q}{\pi}\{\alpha - \sin\alpha\cos(\alpha + 2\beta)\} \qquad (5.20)$$

$$\tau_{xz} = \frac{q}{\pi}\{\sin\alpha\sin(\alpha + 2\beta)\} \qquad (5.21)$$

Contours of equal vertical stress in the vicinity of a strip area carrying a uniform pressure are plotted in Figure 5.8(a). The zone lying inside the vertical stress contour of value $0.2q$ is described as the *bulb of pressure*.

Strip area carrying linearly increasing pressure

The stresses at point X due to pressure increasing linearly from zero to q on a strip area of width B are given in terms of the angles α and β and the lengths R_1 and R_2, as defined in Figure 5.7(b).

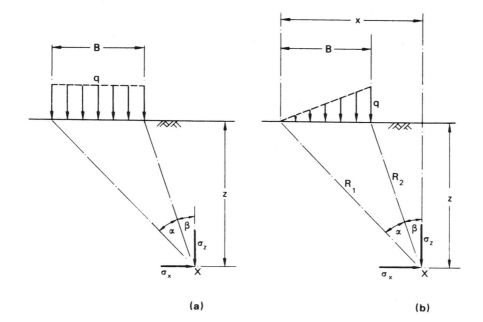

Figure 5.7 Stresses due to (a) uniform pressure and (b) linearly increasing pressure, on strip area.

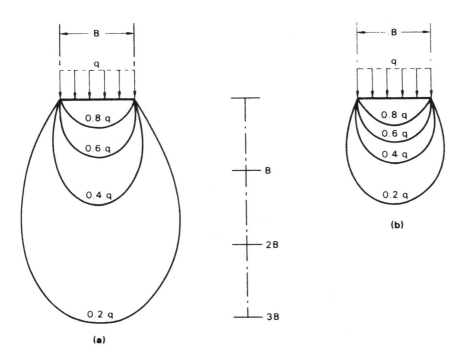

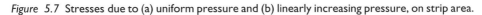

Figure 5.8 Contours of equal vertical stress: (a) under strip area and (b) under square area.

$$\sigma_z = \frac{q}{\pi}\left(\frac{x}{B}\alpha - \frac{1}{2}\sin 2\beta\right) \tag{5.22}$$

$$\sigma_x = \frac{q}{\pi}\left(\frac{x}{B}\alpha - \frac{z}{B}\ln\frac{R_1^2}{R_2^2} + \frac{1}{2}\sin 2\beta\right) \tag{5.23}$$

$$\tau_{xz} = \frac{q}{2\pi}\left(1 + \cos 2\beta - 2\frac{z}{B}\alpha\right) \tag{5.24}$$

Circular area carrying uniform pressure

The vertical stress at depth z under the *centre* of a circular area of diameter $D = 2R$ carrying a uniform pressure q is given by

$$\sigma_z = q\left[1 - \left\{\frac{1}{1 + (R/z)^2}\right\}^{3/2}\right] = qI_c \tag{5.25}$$

Values of the influence factor I_c in terms of D/z are given in Figure 5.9.

The radial and circumferential stresses under the centre are equal and are given by

$$\sigma_r = \sigma_\theta = \frac{q}{2}\left[(1 + 2v) - \frac{2(1 + v)}{\{1 + (R/z)^2\}^{1/2}} + \frac{1}{\{1 + (R/z)^2\}^{3/2}}\right] \tag{5.26}$$

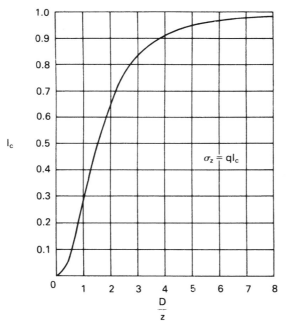

Figure 5.9 Vertical stress under the centre of a circular area carrying a uniform pressure.

Rectangular area carrying uniform pressure

A solution has been obtained for the vertical stress at depth z under a *corner* of a rectangular area of dimensions mz and nz (Figure 5.10) carrying a uniform pressure q. The solution can be written in the form

$$\sigma_z = qI_r$$

Values of the influence factor I_r in terms of m and n are given in the chart, due to Fadum [5], shown in Figure 5.10. The factors m and n are interchangeable. The chart can also be used for a strip area, considered as a rectangular area of infinite length. Superposition enables any area based on rectangles to be dealt with and enables the vertical stress under any point within or outside the area to be obtained.

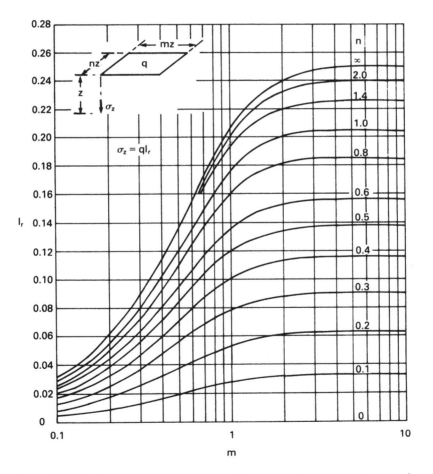

Figure 5.10 Vertical stress under a corner of a rectangular area carrying a uniform pressure. (Reproduced from R.E. Fadum (1948) *Proceedings of the 2nd International Conference of SMFE*, Rotterdam, Vol. 3, by permission of Professor Fadum.)

Contours of equal vertical stress in the vicinity of a square area carrying a uniform pressure are plotted in Figure 5.8(b). Influence factors for σ_x and σ_y (which depend on v) are given in Ref. [16].

Influence chart for vertical stress

Newmark [12] constructed an influence chart, based on the Boussinesq solution, enabling the vertical stress to be determined at any point below an area of any shape carrying a uniform pressure q. The chart (Figure 5.11) consists of influence areas, the boundaries of which are two radial lines and two circular arcs. The loaded area is drawn on tracing paper to a scale such that the length of the scale line on the chart represents the depth z at which the vertical stress is required. The position of the loaded area on the chart is such that the point at which the vertical stress is required is at the centre of the chart. For the chart shown in Figure 5.11 the influence value is 0.005, i.e. each influence area represents a vertical stress of $0.005q$. Hence, if the

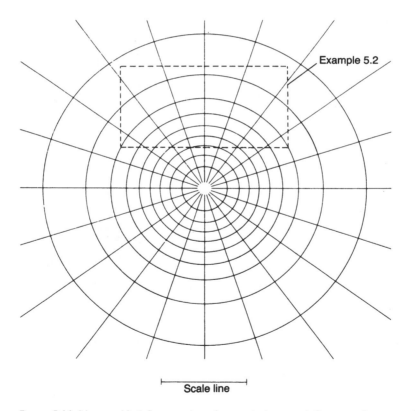

Scale line

Figure 5.11 Newmark's influence chart for vertical stress. Influence value per unit pressure = 0.005. (Reproduced from N.M. Newmark (1942) *Influence Charts for Computation of Stresses in Elastic Foundations*, University of Illinois, Bulletin No. 338, by permission of Professor Newmark.)

number of influence areas covered by the scale drawing of the loaded area is N, the required vertical stress is given by

$$\sigma_z = 0.005\,Nq$$

Example 5.1

A load of 1500 kN is carried on a foundation 2 m square at a shallow depth in a soil mass. Determine the vertical stress at a point 5 m below the centre of the foundation (a) assuming that the load is uniformly distributed over the foundation and (b) assuming that the load acts as a point load at the centre of the foundation.

(a) Uniform pressure,

$$q = \frac{1500}{2^2} = 375\,\text{kN/m}^2$$

The area must be considered as four quarters to enable Figure 5.10 to be used. In this case

$$mz = nz = 1\,\text{m}$$

Then, for $z = 5$ m

$$m = n = 0.2$$

From Figure 5.10,

$$I_r = 0.018$$

Hence,

$$\sigma_z = 4qI_r = 4 \times 375 \times 0.018 = 27\,\text{kN/m}^2$$

(b) From Table 5.1, $I_p = 0.478$ since $r/z = 0$ vertically below a point load.

Hence,

$$\sigma_z = \frac{Q}{z^2}I_p = \frac{1500}{5^2} \times 0.478 = 29\,\text{kN/m}^2$$

The point load assumption should not be used if the depth to the point X (Figure 5.5(a)) is less than three times the larger dimension of the foundation.

Example 5.2

A rectangular foundation 6×3 m carries a uniform pressure of 300 kN/m² near the surface of a soil mass. Determine the vertical stress at a depth of 3 m below a point (A)

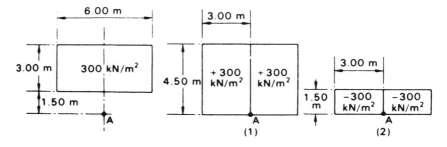

Figure 5.12 Example 5.2.

on the centre line 1.5 m outside a long edge of the foundation (a) using influence factors and (b) using Newmark's influence chart.

(a) Using the principle of superposition the problem is dealt with in the manner shown in Figure 5.12. For the two rectangles (1) carrying a *positive* pressure of 300 kN/m^2, $m = 1.00$ and $n = 1.50$, and therefore

$$I_r = 0.193$$

For the two rectangles (2) carrying a *negative* pressure of 300 kN/m^2, $m = 1.00$ and $n = 0.50$, and therefore

$$I_r = 0.120$$

Hence,

$$\sigma_z = (2 \times 300 \times 0.193) - (2 \times 300 \times 0.120)$$
$$= 44 \, \text{kN/m}^2$$

(b) Using Newmark's influence chart (Figure 5.11) the scale line represents 3 m, fixing the scale to which the rectangular area must be drawn. The area is positioned such that the point A is at the centre of the chart. The number of influence areas covered by the rectangle is approximately 30 (i.e. $N = 30$), hence

$$\sigma_z = 0.005 \times 30 \times 300$$
$$= 45 \, \text{kN/m}^2$$

Example 5.3

A strip footing 2 m wide carries a uniform pressure of 250 kN/m^2 on the surface of a deposit of sand. The water table is at the surface. The saturated unit weight of the sand is 20 kN/m^3 and $K_0 = 0.40$. Determine the effective vertical and horizontal stresses at a point 3 m below the centre of the footing before and after the application of the pressure.

Before loading:

$$\sigma_z' = 3\gamma' = 3 \times 10.2 = 30.6\,\text{kN/m}^2$$

$$\sigma_x' = K_0\sigma_z' = 0.40 \times 30.6 = 12.2\,\text{kN/m}^2$$

After loading: Referring to Figure 5.7(a), for a point 3 m below the centre of the footing,

$$\alpha = 2\tan^{-1}\left(\frac{1}{3}\right) = 36°52' = 0.643 \text{ radians}$$

$$\sin\alpha = 0.600$$

$$\beta = \frac{-\alpha}{2}$$

$$\therefore \ \cos(\alpha + 2\beta) = 1$$

The increases in total stress due to the applied pressure are:

$$\Delta\sigma_z = \frac{q}{\pi}(\alpha + \sin\alpha) = \frac{250}{\pi}(0.643 + 0.600) = 99.0\,\text{kN/m}^2$$

$$\Delta\sigma_x = \frac{q}{\pi}(\alpha - \sin\alpha) = \frac{250}{\pi}(0.643 - 0.600) = 3.4\,\text{kN/m}^2$$

Hence,

$$\sigma_z' = 30.6 + 99.0 = 129.6\,\text{kN/m}^2$$

$$\sigma_x' = 12.2 + 3.4 = 15.6\,\text{kN/m}^2$$

5.3   DISPLACEMENTS FROM ELASTIC THEORY

The vertical displacement (s_i) under an area carrying a uniform pressure q on the surface of a semi-infinite, homogeneous, isotropic mass, with a linear stress–strain relationship, can be expressed as

$$s_i = \frac{qB}{E}(1 - v^2)I_s \tag{5.27}$$

where I_s is an influence factor depending on the shape of the loaded area. In the case of a rectangular area, B is the lesser dimension (the greater dimension being L) and in the case of a circular area, B is the diameter. The loaded area is assumed to be flexible. Values of influence factors are given in Table 5.2 for displacements under the centre and a corner (the edge in the case of a circle) of the area and for the average displacement under the area as a whole. According to Equation 5.27,

Table 5.2 Influence factors for vertical displacement under flexible area carrying uniform pressure

Shape of area	I_s		
	Centre	Corner	Average
Square	1.12	0.56	0.95
Rectangle, $L/B = 2$	1.52	0.76	1.30
Rectangle, $L/B = 5$	2.10	1.05	1.83
Circle	1.00	0.64	0.85

vertical displacement increases in direct proportion to both the pressure and the width of the loaded area. The distribution of vertical displacement is of the form shown in Figure 5.13(a), extending beyond the edges of the area. The contact pressure between the loaded area and the supporting mass is uniform. It should be noted that, unlike the expressions for vertical stress (σ_v) given in Section 5.2, the expression for vertical displacement is dependent on the values of elastic modulus (E) and Poisson's ratio (ν) for the soil in question. Because of the uncertainties involved in obtaining elastic parameters, values of vertical displacement calculated from elastic theory, therefore, are less reliable than values of vertical stress.

In the case of an extensive, homogeneous deposit of saturated clay, it is a reasonable approximation to assume that E is constant throughout the deposit and the distribution of Figure 5.13(a) applies. In the case of sands, however, the value of E varies with confining pressure and, therefore, will increase with depth and vary across the width of the loaded area, being greater under the centre of the area than at the edges. As a result, the distribution of vertical displacement will be of the form shown in Figure 5.13(b); the contact pressure will again be uniform if the area is flexible. Due to the variation of E, and to heterogeneity, elastic theory is little used in practice in the case of sands.

If the loaded area is *rigid* the vertical displacement will be uniform across the width of the area and its magnitude will be only slightly less than the *average* displacement under a corresponding flexible area. For example, the value of I_s for a rigid circular area is $\pi/4$, this value being used in the calculation of E from the results of *in-situ* plate loading tests. The contact pressure under a rigid area is not uniform; for a circular area the forms of the distributions of contact pressure on clay and sand, respectively, are shown in Figures 5.14(a) and (b).

In most cases in practice the soil deposit will be of limited thickness and will be underlain by a hard stratum. Christian and Carrier [3] proposed the use of results by

(a)   (b)

Figure 5.13 Distributions of vertical displacement: (a) clay and (b) sand.

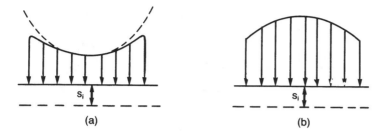

Figure 5.14 Contact pressure under rigid area: (a) clay and (b) sand.

Giroud [7] and by Burland [2] in such cases. The *average* vertical displacement under a flexible area carrying a uniform pressure q is given by

$$s_i = \mu_0 \mu_1 \frac{qB}{E} \tag{5.28}$$

where μ_0 depends on the depth of embedment and μ_1 depends on the layer thickness and the shape of the loaded area. Values of the coefficients μ_0 and μ_1 for Poisson's ratio equal to 0.5 are given in Figure 5.15. The principle of superposition can be used in cases of a number of soil layers each having a different value of E (see Example 5.4).

The above solutions for vertical displacement are used mainly to estimate the immediate settlement of foundations on saturated clays; such settlement occurs under undrained conditions, the appropriate value of Poisson's ratio being 0.5. The value of the undrained modulus E_u is therefore required and the main difficulty in predicting immediate settlement is in the determination of this parameter. A value of E_u could be determined by means of the undrained triaxial test. However, such a value would be very sensitive to sampling disturbance and would be too low if the unconsolidated–undrained test were used. If the specimen were initially reconsolidated then a more realistic value of E_u would be obtained. Consolidation may be either isotropic under $\frac{1}{2}$ to $\frac{2}{3}$ of the *in-situ* effective overburden pressure, or under K_0 conditions to simulate the actual *in-situ* effective stresses. If possible, however, the value of E_u should be determined from the results of *in-situ* load tests or pressuremeter tests. It should be recognized, however, that the value obtained from load tests is sensitive to the time interval between excavation and testing, because there will be a gradual change from the undrained condition with time; the greater the time interval between excavation and testing the lower the value of E_u. The value of E_u can be obtained directly if settlement observations are taken during the initial loading of full-scale foundations. For particular clays, correlations can be established between E_u and the undrained shear strength parameter c_u.

It has been demonstrated that for certain soils, such as normally consolidated clays, there is a significant departure from linear stress–strain behaviour within the range of working stress, i.e. local yielding will occur within this range, and the immediate settlement will be underestimated. A method of correction for local yield has been given by D'Appolonia *et al.* [4].

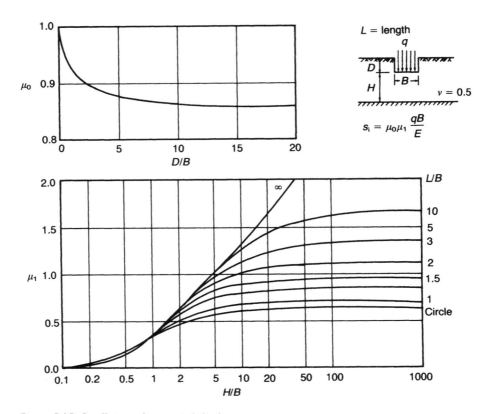

Figure 5.15 Coefficients for vertical displacement.

In principle the vertical displacement under fully drained conditions could be estimated using elastic theory if the value of the modulus for this condition (E') and the value of Poisson's ratio for the soil skeleton (ν') could be determined.

Example 5.4

A foundation $4 \times 2\,\mathrm{m}$, carrying a uniform pressure of $150\,\mathrm{kN/m^2}$, is located at a depth of $1\,\mathrm{m}$ in a layer of clay $5\,\mathrm{m}$ thick for which the value of E_u is $40\,\mathrm{MN/m^2}$. The layer is underlain by a second clay layer $8\,\mathrm{m}$ thick for which the value of E_u is $75\,\mathrm{MN/m^2}$. A hard stratum lies below the second layer. Determine the average immediate settlement under the foundation.

Now, $D/B = 0.5$, and therefore from Figure 5.15, $\mu_0 = 0.94$.

1   Considering the upper clay layer, with $E_u = 40\,\mathrm{MN/m^2}$:

$$H/B = 4/2 = 2, \qquad L/B = 2$$

$$\therefore \mu_1 = 0.60$$

Hence, from Equation 5.28

$$s_{i_1} = 0.94 \times 0.60 \times \frac{150 \times 2}{40} = 4.2\,\text{mm}$$

2 Considering the two layers combined, with $E_u = 75\,\text{MN/m}^2$:

$$H/B = 12/2 = 6, \qquad L/B = 2$$
$$\therefore \mu_1 = 0.85$$
$$s_{i_2} = 0.94 \times 0.85 \times \frac{150 \times 2}{75} = 3.2\,\text{mm}$$

3 Considering the upper layer, with $E_u = 75\,\text{MN/m}^2$:

$$H/B = 2, \qquad L/B = 2$$
$$\therefore \mu_1 = 0.60$$
$$s_{i_3} = 0.94 \times 0.60 \times \frac{150 \times 2}{75} = 2.3\,\text{mm}$$

Hence, using the principle of superposition, the settlement of the foundation is given by

$$s_i = s_{i_1} + s_{i_2} - s_{i_3}$$
$$= 4.2 + 3.2 - 2.3 = 5\,\text{mm}$$

PROBLEMS

5.1 Calculate the vertical stress in a soil mass at a depth of 5 m vertically below a point load of 5000 kN acting near the surface. Plot the variation of vertical stress with radial distance (up to 10 m) at a depth of 5 m.

5.2 Three point loads, 10 000, 7500 and 9000 kN, act in line 5 m apart near the surface of a soil mass. Calculate the vertical stress at a depth of 4 m vertically below the centre (7500 kN) load.

5.3 Determine the vertical stress at a depth of 3 m below the centre of a shallow foundation 2×2 m carrying a uniform pressure of 250 kN/m^2. Plot the variation of vertical stress with depth (up to 10 m) below the centre of the foundation.

5.4 A shallow foundation 25×18 m carries a uniform pressure of 175 kN/m^2. Determine the vertical stress at a point 12 m below the mid-point of one of the longer sides (a) using influence factors, (b) by means of Newmark's chart.

5.5 A line load of 150 kN/m acts 2 m behind the back surface of an earth-retaining structure 4 m high. Calculate the total thrust, and plot the distribution of pressure, on the structure due to the line load.

5.6 A foundation 4×2 m carries a uniform pressure of 200 kN/m^2 at a depth of 1 m in a layer of saturated clay 11 m deep and underlain by a hard stratum. If E_u for the clay is 45 MN/m^2, determine the average value of immediate settlement under the foundation.

REFERENCES

1   Britto, A.M. and Gunn, M.J. (1987) *Critical State Soil Mechanics Via Finite Elements*, Ellis Horwood, Chichester.

2   Burland, J.B. (1970) Discussion, in *Proceedings of Conference on In-Situ Investigations in Soils and Rocks*, British Geotechnical Society, London, p. 61.

3   Christian, J.T. and Carrier III, W.D. (1978) Janbu, Bjerrum and Kjaernsli's chart reinterpreted, *Canadian Geotechnical Journal*, **15**, 123, 436.

4   D'Appolonia, D.J., Poulos, H.G. and Ladd, C.C. (1971) Initial settlement of structures on clay, *Journal of the ASCE*, **97**, No. SM10, 1359–77.

5   Fadum, R.E. (1948) Influence values for estimating stresses in elastic foundations, in *Proceedings of the 2nd International Conference of SMFE*, Rotterdam, Vol. 3, pp. 77–84.

6   Gibson, R.E. and Anderson, W.F. (1961) *In-situ* measurement of soil properties with the pressuremeter, *Civil Engineering and Public Works Review*, **56**, 615–18.

7   Giroud, J.P. (1972) Settlement of rectangular foundation on soil layer, *Journal of the ASCE*, **98**, No. SM1, 149–54.

8   Hill, R. (1950) *Mathematical Theory of Plasticity*, Oxford University Press, New York.

9   Hughes, J.M.O., Wroth, C.P. and Windle, D. (1977) Pressuremeter tests in sands, *Geotechnique*, **27**, 455–77.

10 Mair, R.J. and Wood, D.M. (1987) *Pressuremeter Testing: Methods and interpretation*, CIRIA/Butterworths, London.

11 Marsland, A. and Randolph, M.F. (1977) Comparisons of the results from pressuremeter tests and large *in-situ* plate tests in London clay, *Geotechnique*, **27**, 217–43.

12 Newmark, N.M. (1942) *Influence Charts for Computation of Stresses in Elastic Foundations*, University of Illinois, Bulletin No. 338.

13 Palmer, A.C. (1972) Undrained plane strain expansion of a cylindrical cavity in clay: a simple interpretation of the pressuremeter test, *Geotechnique*, **22**, 451–7.

14 Potts, D.M. and Zdravkovic, L. (1999) *Finite Element Analysis in Geotechnical Engineering: Theory*, Thomas Telford, London.

15 Potts, D.M. and Zdravkovic, L. (2001) *Finite Element Analysis in Geotechnical Engineering: Application*, Thomas Telford, London.

16 Poulos, H.G. and Davis, E.H. (1974) *Elastic Solutions for Soil and Rock Mechanics*, John Wiley & Sons, New York.

17 Scott, R.F. (1963) *Principles of Soil Mechanics*, Addison-Wesley, Reading, MA.

18 Timoshenko, S. and Goodier, J.N. (1970) *Theory of Elasticity*, 3rd edn, McGraw-Hill, New York.

19 Windle, D. and Wroth, C.P. (1977) The use of a self-boring pressuremeter to determine the undrained properties of clay, *Ground Engineering*, **10** (6), 37–46.

20 Wroth, C.P. (1984) The interpretation of *in-situ* soil tests, *Geotechnique*, **34**, 447–89.

Chapter 6

Lateral earth pressure

6.1 INTRODUCTION

This chapter deals with the magnitude and distribution of lateral pressure between a soil mass and an adjoining retaining structure. Conditions of plane strain are assumed, i.e. strains in the longitudinal direction of the structure are assumed to be zero. The rigorous treatment of this type of problem, with both stresses and displacements being considered, would involve a knowledge of appropriate equations defining the stress–strain relationship for the soil and the solution of the equations of equilibrium and compatibility for the given boundary conditions. It is possible to determine displacements by means of the finite element method using suitable computer software, provided realistic values of the relevant deformation parameters are available. However, it is the failure condition of the retained soil mass which is of primary interest and in this context, provided a consideration of displacements is not required, it is possible to use the concept of plastic collapse. Earth pressure problems can thus be considered as problems in plasticity.

It is assumed that the stress–strain behaviour of the soil can be represented by the rigid–perfectly plastic idealization, shown in Figure 6.1, in which both yielding and shear failure occur at the same state of stress: unrestricted plastic flow takes place at this stress level. A soil mass is said to be in a state of plastic equilibrium if the shear stress at every point within the mass reaches the value represented by point Y′.

Plastic collapse occurs after the state of plastic equilibrium has been reached in *part* of a soil mass, resulting in the formation of an unstable mechanism: that part of the soil

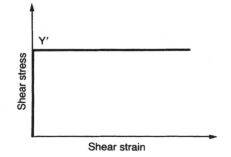

Figure 6.1 Idealized stress–strain relationship.

mass slips relative to the rest of the mass. The applied load system, including body forces, for this condition is referred to as the collapse load. Determination of the collapse load using plasticity theory is complex and would require that the equilibrium equations, the yield criterion and the flow rule were satisfied within the plastic zone. The compatibility condition would not be involved unless specific deformation conditions were imposed. However, plasticity theory also provides the means of avoiding complex analyses. The limit theorems of plasticity can be used to calculate lower and upper bounds to the true collapse load. In certain cases, the theorems produce the same result which would then be the exact value of the collapse load. The limit theorems can be stated as follows.

Lower bound theorem

If a state of stress can be found, which at no point exceeds the failure criterion for the soil and is in equilibrium with a system of external loads (which includes the self-weight of the soil), then collapse cannot occur; the external load system thus constitutes a lower bound to the true collapse load (because a more efficient stress distribution may exist, which would be in equilibrium with higher external loads).

Upper bound theorem

If a mechanism of plastic collapse is postulated and if, in an increment of displacement, the work done by a system of external loads is equal to the dissipation of energy by the internal stresses, then collapse must occur; the external load system thus constitutes an upper bound to the true collapse load (because a more efficient mechanism may exist resulting in collapse under lower external loads).

In the lower bound approach, the conditions of equilibrium and yield are satisfied without consideration of the mode of deformation. The Mohr–Coulomb failure criterion is also taken to be the yield criterion. In the upper bound approach, a mechanism of plastic collapse is formed by choosing a slip surface and the work done by the external forces is equated to the loss of energy by the stresses acting along the slip surface, without consideration of equilibrium. The chosen collapse mechanism is not necessarily the true mechanism but it must be kinematically admissible, i.e. the motion of the sliding soil mass must be compatible with its continuity and with any boundary restrictions. It can be shown that for undrained conditions the slip surface, in section, should consist of a straight line or a circular arc (or a combination of the two); for drained conditions the slip surface should consist of a straight line or a logarithmic spiral (or a combination of the two). Examples of lower and upper bound plasticity solutions have been given by Atkinson [1] and Parry [16].

Lateral pressure calculations are normally based on the classical theories of Rankine or Coulomb, described in Sections 6.2 and 6.3, and these theories can be related to the concepts of plasticity.

6.2   RANKINE'S THEORY OF EARTH PRESSURE

Rankine's theory (1857) considers the state of stress in a soil mass when the condition of plastic equilibrium has been reached, i.e. when shear failure is on the point of occurring

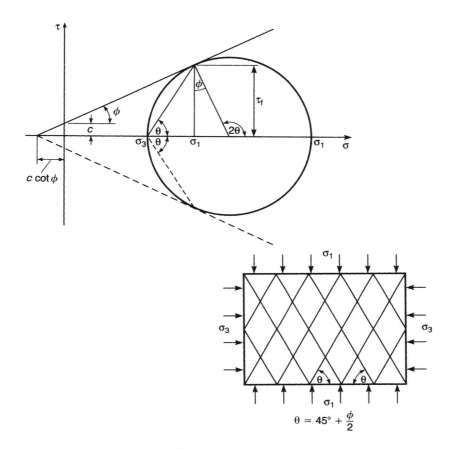

Figure 6.2 State of plastic equilibrium.

throughout the mass. The theory satisfies the conditions of a lower bound plasticity solution. The Mohr circle representing the state of stress at failure in a two-dimensional element is shown in Figure 6.2, the relevant shear strength parameters being denoted by c and ϕ. Shear failure occurs along a plane at an angle of $45° + \phi/2$ to the major principal plane. If the soil mass as a whole is stressed such that the principal stresses at every point are in the same directions then, theoretically, there will be a network of failure planes (known as a slip line field) equally inclined to the principal planes, as shown in Figure 6.2. It should be appreciated that the state of plastic equilibrium can be developed only if sufficient deformation of the soil mass can take place.

Consider now a semi-infinite mass of soil with a horizontal surface and having a vertical boundary formed by a *smooth* wall surface extending to semi-infinite depth, as represented in Figure 6.3(a). The soil is assumed to be homogeneous and isotropic. A soil element at any depth z is subjected to a vertical stress σ_z and a horizontal stress σ_x and, since there can be no lateral transfer of weight if the surface is horizontal, no shear stresses exist on horizontal and vertical planes. The vertical and horizontal stresses, therefore, are principal stresses.

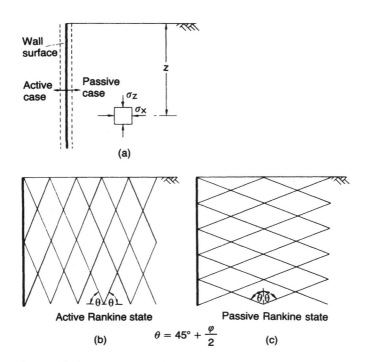

Figure 6.3 Active and passive Rankine states.

If there is a movement of the wall away from the soil, the value of σ_x decreases as the soil dilates or expands outwards, the decrease in σ_x being an unknown function of the lateral strain in the soil. If the expansion is large enough, the value of σ_x decreases to a minimum value such that a state of plastic equilibrium develops. Since this state is developed by a decrease in the horizontal stress σ_x, this must be the minor principal stress (σ_3). The vertical stress σ_z is then the major principal stress (σ_1).

The stress σ_1 ($=\sigma_z$) is the overburden pressure at depth z and is a fixed value for any depth. The value of σ_3 ($=\sigma_x$) is determined when a Mohr circle through the point representing σ_1 touches the failure envelope for the soil. The relationship between σ_1 and σ_3 when the soil reaches a state of plastic equilibrium can be derived from this Mohr circle. Rankine's original derivation assumed a value of zero for the shear strength parameter c but a general derivation with c greater than zero is given below to cover the cases in which undrained parameter c_u or tangent parameter c' is used.

Referring to Figure 6.2,

$$\sin\phi = \frac{\frac{1}{2}(\sigma_1 - \sigma_3)}{\frac{1}{2}(\sigma_1 + \sigma_3 + 2c\cot\phi)}$$

$$\therefore \quad \sigma_3(1 + \sin\phi) = \sigma_1(1 - \sin\phi) - 2c\cos\phi$$

$$\therefore \quad \sigma_3 = \sigma_1 \left(\frac{1 - \sin \phi}{1 + \sin \phi} \right) - 2c \left(\frac{\sqrt{(1 - \sin^2 \phi)}}{1 + \sin \phi} \right)$$

$$\therefore \quad \sigma_3 = \sigma_1 \left(\frac{1 - \sin \phi}{1 + \sin \phi} \right) - 2c \sqrt{\left(\frac{1 - \sin \phi}{1 + \sin \phi} \right)} \tag{6.1}$$

Alternatively, $\tan^2 (45° - \phi/2)$ can be substituted for $(1 - \sin \phi)/(1 + \sin \phi)$.

As stated, σ_1 is the overburden pressure at depth z, i.e.

$$\sigma_1 = \gamma z$$

The horizontal stress for the above condition is defined as the *active pressure* (p_a) being due directly to the self-weight of the soil. If

$$K_a = \frac{1 - \sin \phi}{1 + \sin \phi}$$

is defined as the active pressure coefficient, then Equation 6.1 can be written as

$$p_a = K_a \gamma z - 2c \sqrt{K_a} \tag{6.2}$$

When the horizontal stress becomes equal to the active pressure the soil is said to be in the *active Rankine state*, there being two sets of failure planes each inclined at $45° + \phi/2$ to the *horizontal* (the direction of the major principal plane) as shown in Figure 6.3(b).

In the above derivation, a movement of the wall away from the soil was considered. On the other hand, if the wall is moved against the soil mass, there will be lateral compression of the soil and the value of σ_x will increase until a state of plastic equilibrium is reached. For this condition, σ_x becomes a maximum value and is the major principal stress σ_1. The stress σ_z, equal to the overburden pressure, is then the minor principal stress, i.e.

$$\sigma_3 = \gamma z$$

The maximum value σ_1 is reached when the Mohr circle through the point representing the fixed value σ_3 touches the failure envelope for the soil. In this case, the horizontal stress is defined as the *passive pressure* (p_p) representing the maximum inherent resistance of the soil to lateral compression. Rearranging Equation 6.1:

$$\sigma_1 = \sigma_3 \left(\frac{1 + \sin \phi}{1 - \sin \phi} \right) + 2c \sqrt{\left(\frac{1 + \sin \phi}{1 - \sin \phi} \right)} \tag{6.3}$$

If

$$K_p = \frac{1 + \sin \phi}{1 - \sin \phi}$$

is defined as the passive pressure coefficient, then Equation 6.3 can be written as

$$p_{\mathrm{p}} = K_{\mathrm{p}}\gamma z + 2c\sqrt{K_{\mathrm{p}}} \qquad (6.4)$$

When the horizontal stress becomes equal to the passive pressure the soil is said to be in the *passive Rankine state*, there being two sets of failure planes each inclined at $45° + \phi/2$ to the *vertical* (the direction of the major principal plane) as shown in Figure 6.3(c).

Inspection of Equations 6.2 and 6.4 shows that the active and passive pressures increase linearly with depth as represented in Figure 6.4. When $c = 0$, triangular distributions are obtained in each case.

When c is greater than zero, the value of p_{a} is zero at a particular depth z_0. From Equation 6.2, with $p_{\mathrm{a}} = 0$,

$$z_0 = \frac{2c}{\gamma\sqrt{K_{\mathrm{a}}}} \qquad (6.5)$$

This means that in the active case the soil is in a state of tension between the surface and depth z_0. In practice, however, this tension cannot be relied upon to act on the wall, since cracks are likely to develop within the tension zone and the part of the pressure distribution diagram above depth z_0 should be neglected.

The force per unit length of wall due to the active pressure distribution is referred to as the *total active thrust* (P_{a}). For a vertical wall surface of height H:

$$P_{\mathrm{a}} = \int_{z_0}^{H} p_{\mathrm{a}}\, \mathrm{d}z$$

$$= \frac{1}{2} K_{\mathrm{a}}\gamma(H^2 - z_0^2) - 2c(\sqrt{K_{\mathrm{a}}})(H - z_0) \qquad (6.6\mathrm{a})$$

$$= \frac{1}{2} K_{\mathrm{a}}\gamma(H - z_0)^2 \qquad (6.6\mathrm{b})$$

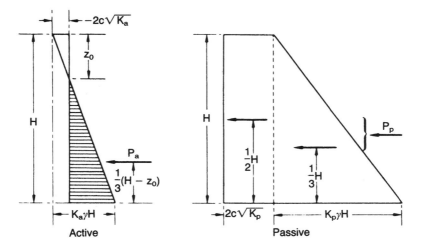

Figure 6.4 Active and passive pressure distributions.

The force P_a acts at a distance of $\frac{1}{3}(H - z_0)$ above the bottom of the wall surface.

The force due to the passive pressure distribution is referred to as the *total passive resistance* (P_p). For a vertical wall surface of height H:

$$P_p = \int_0^H p_p \, \mathrm{d}z$$
$$= \frac{1}{2} K_p \gamma H^2 + 2c(\sqrt{K_p})H \tag{6.7}$$

The two components of P_p act at distances of $\frac{1}{3}H$ and $\frac{1}{2}H$, respectively, above the bottom of the wall surface.

If a uniformly distributed surcharge pressure of q per unit area acts over the entire surface of the soil mass, the vertical stress σ_z at any depth is increased to $\gamma z + q$, resulting in an *additional* pressure of $K_a q$ in the active case or $K_p q$ in the passive case, both distributions being constant with depth as shown in Figure 6.5. The corresponding forces on a vertical wall surface of height H are $K_a q H$ and $K_p q H$, respectively, each acting at mid-height. The surcharge concept can be used to obtain the pressure distribution in stratified soil deposits. In the case of two layers of soil having different shear strengths, the weight of the upper layer can be considered as a surcharge acting on the lower layer. There will be a discontinuity in the pressure diagram at the boundary between the two layers due to the different values of shear strength parameters.

If the soil below the water table is in the fully drained condition, the active and passive pressures must be expressed in terms of the effective weight of the soil and the effective stress parameters c' and ϕ'. For example, if the water table is at the surface and if no seepage is taking place, the active pressure at depth z is given by

$$p_a = K_a \gamma' z - 2c' \sqrt{K_a}$$

where

$$K_a = \frac{1 - \sin \phi'}{1 + \sin \phi'}$$

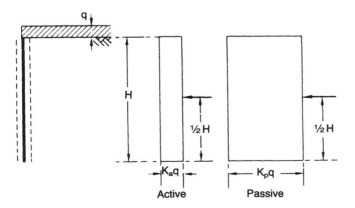

Figure 6.5 Additional pressure due to surcharge.

Corresponding equations apply in the passive case. The hydrostatic pressure $\gamma_w z$ due to the water in the soil pores must be considered *in addition* to the active or passive pressure.

For the undrained condition in a fully saturated clay, the active and passive pressures are calculated using the parameter c_u (ϕ_u being zero) and the total unit weight γ_{sat} (i.e. the water in the soil pores is not considered separately). The effect of the tension zone must be considered for this condition. In theory, a (dry) crack could open to a depth (z_0) of $2c_u/\gamma_{sat}$ (i.e. Equation 6.5 with $K_a = 1$ for $\phi_u = 0$). Cracking is most likely to occur at the clay/wall interface where the resistance to fracture is lower than that within the clay. If a crack at the interface were to fill with water (due to heavy rainfall or another source of inflow), then hydrostatic pressure would act on the wall. Thus the clay would be supported by the water filling the crack to the depth (z_{0w}) at which the active pressure equals the hydrostatic pressure. Thus, assuming no surface surcharge:

$$\gamma_{sat} z_{0w} - 2c_u = \gamma_w z_{0w}$$

$$\therefore z_{0w} = \frac{2c_u}{(\gamma_{sat} - \gamma_w)}$$

In the Rankine theory it is assumed that the wall surface is smooth whereas in practice considerable friction may be developed between the wall and the adjacent soil, depending on the wall material. In principle, the theory results either in an over-estimation of active pressure and an underestimation of passive pressure (i.e. lower bounds to the respective 'collapse loads') or in exact values of active and passive pressures.

Example 6.1

(a) Calculate the total active thrust on a vertical wall 5 m high retaining a sand of unit weight $17\,kN/m^3$ for which $\phi' = 35°$; the surface of the sand is horizontal and the water table is below the bottom of the wall. (b) Determine the thrust on the wall if the water table rises to a level 2 m below the surface of the sand. The saturated unit weight of the sand is $20\,kN/m^3$.

(a) $K_a = \dfrac{1 - \sin 35°}{1 + \sin 35°} = 0.27$

$P_a = \dfrac{1}{2} K_a \gamma H^2 = \dfrac{1}{2} \times 0.27 \times 17 \times 5^2 = 57.5\,kN/m$

(b) The pressure distribution on the wall is now as shown in Figure 6.6, including hydrostatic pressure on the lower 3 m of the wall. The components of the thrust are:

(1) $\dfrac{1}{2} \times 0.27 \times 17 \times 2^2 = 9.2\,kN/m$

(2) $0.27 \times 17 \times 2 \times 3 = 27.6$

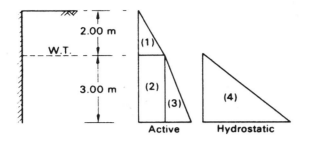

Figure 6.6 Example 6.1.

(3) $\dfrac{1}{2} \times 0.27 \times (20 - 9.8) \times 3^2 = 12.4$

(4) $\dfrac{1}{2} \times 9.8 \times 3^2 \qquad\qquad = 44.1$

Total thrust $\qquad\qquad\qquad = 93.3\,\text{kN/m}$

Example 6.2

The soil conditions adjacent to a sheet pile wall are given in Figure 6.7, a surcharge pressure of $50\,\text{kN/m}^2$ being carried on the surface behind the wall. For soil 1, a sand above the water table, $c' = 0$, $\phi' = 38°$ and $\gamma = 18\,\text{kN/m}^3$. For soil 2, a saturated clay, $c' = 10\,\text{kN/m}^2$, $\phi' = 28°$ and $\gamma_{\text{sat}} = 20\,\text{kN/m}^3$. Plot the distributions of active pressure behind the wall and passive pressure in front of the wall.

For soil 1,

$$K_a = \frac{1 - \sin 38°}{1 + \sin 38°} = 0.24, \qquad K_p = \frac{1}{0.24} = 4.17$$

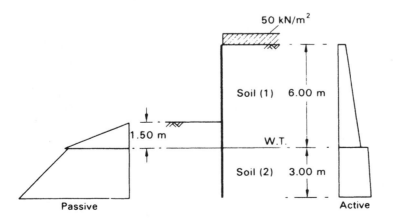

Figure 6.7 Example 6.2.

Table 6.1

Soil	Depth (m)	Pressure (kN/m²)	
Active pressure			
1	0	0.24×50	$= 12.0$
1	6	$(0.24 \times 50) + (0.24 \times 18 \times 6) = 12.0 + 25.9$	$= 37.9$
2	6	$0.36[50 + (18 \times 6)] - (2 \times 10 \times \sqrt{0.36}) = 56.9 - 12.0$	$= 44.9$
2	9	$0.36[50 + (18 \times 6)] - (2 \times 10 \times \sqrt{0.36}) + (0.36 \times 10.2 \times 3)$	
		$= 56.9 - 12.0 + 11.0$	$= 55.9$
Passive pressure			
1	0	0	
1	1.5	$4.17 \times 18 \times 1.5$	$= 112.6$
2	1.5	$(2.78 \times 18 \times 1.5) + (2 \times 10 \times \sqrt{2.78}) = 75.1 + 33.3$	$= 108.4$
2	4.5	$(2.78 \times 18 \times 1.5) + (2 \times 10 \times \sqrt{2.78})$	
		$+ (2.78 \times 10.2 \times 3) = 75.1 + 33.3 + 85.1$	$= 193.5$

For soil 2,

$$K_a = \frac{1 - \sin 28°}{1 + \sin 28°} = 0.36, \qquad K_p = \frac{1}{0.36} = 2.78$$

The pressures in soil 1 are calculated using $K_a = 0.24$, $K_p = 4.17$ and $\gamma = 18\,\mathrm{kN/m^3}$. Soil 1 is then considered as a surcharge of $(18 \times 6)\,\mathrm{kN/m^2}$ on soil 2, in addition to the surface surcharge. The pressures in soil 2 are calculated using $K_a = 0.36$, $K_p = 2.78$ and $\gamma' = (20 - 9.8) = 10.2\,\mathrm{kN/m^3}$ (see Table 6.1). The active and passive pressure distributions are shown in Figure 6.7. In addition, there is equal hydrostatic pressure on each side of the wall below the water table.

Sloping soil surface

The Rankine theory will now be applied to cases in which the soil surface slopes at a constant angle β to the horizontal. It is assumed that the active and passive pressures act in a direction parallel to the sloping surface. Consider a rhombic element of soil, with sides vertical and at angle β to the horizontal, at depth z in a semi-infinite mass. The vertical stress and the active or passive pressure are each inclined at β to the appropriate sides of the element, as shown in Figure 6.8(a). Since these stresses are not normal to their respective planes (i.e. there are shear components), they are not principal stresses.

In the active case, the vertical stress at depth z on a plane inclined at angle β to the horizontal is given by

$$\sigma_z = \gamma z \cos \beta$$

and is represented by the distance OA on the stress diagram (Figure 6.8(b)). If lateral expansion of the soil is sufficient to induce the state of plastic equilibrium, the Mohr circle representing the state of stress in the element must pass through point A (such that the greater part of the circle lies on the side of A towards the origin) and touch the

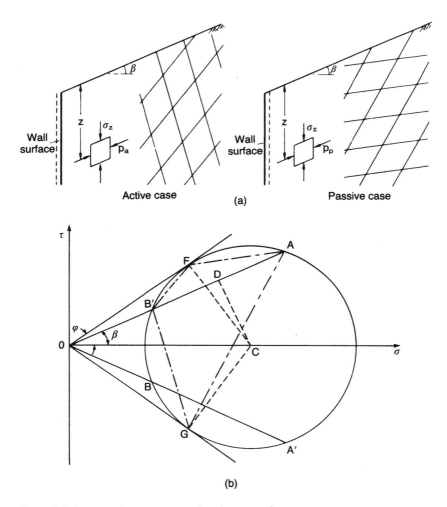

(b)

Figure 6.8 Active and passive states for sloping surface.

failure envelope for the soil. The active pressure p_a is then represented by OB (numerically equal to OB$'$) on the diagram. When $c = 0$ the relationship between p_a and σ_z, giving the active pressure coefficient, can be derived from the diagram:

$$K_a = \frac{p_a}{\sigma_z} = \frac{\text{OB}}{\text{OA}} = \frac{\text{OB}'}{\text{OA}} = \frac{\text{OD} - \text{AD}}{\text{OD} + \text{AD}}$$

Now

$$\text{OD} = \text{OC} \cos \beta$$

$$\text{AD} = \sqrt{(\text{OC}^2 \sin^2 \phi - \text{OC}^2 \sin^2 \beta)}$$

therefore

$$K_a = \frac{\cos\beta - \sqrt{(\cos^2\beta - \cos^2\phi)}}{\cos\beta + \sqrt{(\cos^2\beta - \cos^2\phi)}} \tag{6.8}$$

Thus the active pressure, acting parallel to the slope, is given by

$$p_a = K_a \gamma z \cos\beta \tag{6.9}$$

and the total active thrust on a vertical wall surface of height H is

$$P_a = \frac{1}{2} K_a \gamma H^2 \cos\beta \tag{6.10}$$

In the passive case, the vertical stress σ_z is represented by the distance OB$'$ in Figure 6.8(b). The Mohr circle representing the state of stress in the element, after a state of plastic equilibrium has been induced by lateral compression of the soil, must pass through B$'$ (such that the greater part of the circle lies on the side of B$'$ away from the origin) and touch the failure envelope. The passive pressure p_p is then represented by OA$'$ (numerically equal to OA) and when $c = 0$ the passive pressure coefficient (equal to p_p/σ_z) is given by

$$K_p = \frac{\cos\beta + \sqrt{(\cos^2\beta - \cos^2\phi)}}{\cos\beta - \sqrt{(\cos^2\beta - \cos^2\phi)}} \tag{6.11}$$

Then the passive pressure, acting parallel to the slope, is given by

$$p_p = K_p \gamma z \cos\beta \tag{6.12}$$

and the total passive resistance on a vertical wall surface of height H is

$$P_p = \frac{1}{2} K_p \gamma H^2 \cos\beta \tag{6.13}$$

The active and passive pressures can, of course, be obtained graphically from Figure 6.8(b). The above formulae apply only when the shear strength parameter c is zero; when c is greater than zero the graphical procedure should be used.

The directions of the two sets of failure planes can be obtained from Figure 6.8(b). In the active case, the coordinates of point A represent the state of stress on a plane inclined at angle β to the horizontal, therefore point B$'$ is the origin of planes, also known as the pole. (A line drawn from the origin of planes intersects the circumference of the circle at a point whose coordinates represent the state of stress on a plane parallel to that line.) The state of stress on a vertical plane is represented by the coordinates of point B. Then the failure planes, which are shown in Figure 6.8(a), are parallel to B$'$F and B$'$G (F and G lying on the failure envelope). In the passive case, the coordinates of point B$'$ represent the state of stress on a plane inclined at angle β

to the horizontal, and therefore point A is the origin of planes: the state of stress on a vertical plane is represented by the coordinates of point A'. Then the failure planes in the passive case are parallel to AF and AG.

Referring to Equations 6.8 and 6.11, it is clear that both K_a and K_p become equal to unity when $\beta = \phi$; this is incompatible with real soil behaviour. Use of the theory is inappropriate, therefore, in such circumstances.

Example 6.3

A vertical wall 6 m high, above the water table, retains a 20° soil slope, the retained soil having a unit weight of $18\,\mathrm{kN/m^3}$; the appropriate shear strength parameters are $c' = 0$ and $\phi' = 40°$. Determine the total active thrust on the wall and the directions of the two sets of failure planes relative to the horizontal.

In this case the total active thrust can be obtained by calculation. Using Equation 6.8,

$$K_a = \frac{\cos 20° - \sqrt{(\cos^2 20° - \cos^2 40°)}}{\cos 20° + \sqrt{(\cos^2 20° - \cos^2 40°)}} = 0.265$$

Then

$$
\begin{aligned}
P_a &= \frac{1}{2}K_a\gamma H^2 \cos\beta \\
&= \frac{1}{2} \times 0.265 \times 18 \times 6^2 \times 0.940 = 81\,\mathrm{kN/m}
\end{aligned}
$$

The result can also be determined using a stress diagram (Figure 6.9). Draw the failure envelope on the τ/σ plot and a straight line through the origin at 20° to the horizontal. At a depth of 6 m,

$$\sigma_z = \gamma z \cos\beta = 18 \times 6 \times 0.940 = 102\,\mathrm{kN/m^2}$$

and this stress is set off to scale (distance OA) along the 20° line. The Mohr circle is then drawn as in Figure 6.9 and the active pressure (distance OB or OB') is scaled from the diagram, i.e.

$$p_a = 27\,\mathrm{kN/m^2}$$

Then

$$P_a = \frac{1}{2}p_a H = \frac{1}{2} \times 27 \times 6 = 81\,\mathrm{kN/m}$$

The failure planes are parallel to B'F and B'G in Figure 6.9. The directions of these lines are measured as 59° and 71°, respectively, to the horizontal (adding up to $90° + \phi$).

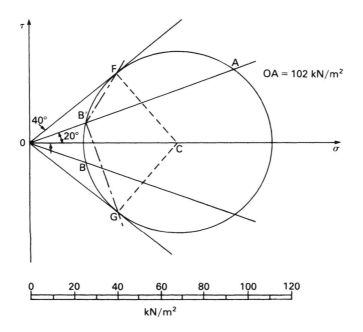

Figure 6.9 Example 6.3.

Earth pressure at-rest

It has been shown that active pressure is associated with lateral expansion of the soil and is a minimum value; passive pressure is associated with lateral compression of the soil and is a maximum value. The active and passive values may thus be referred to as limit pressures. If the lateral strain in the soil is zero, the corresponding lateral pressure is called the *earth pressure at-rest* and is usually expressed in terms of effective stress by the equation

$$p_0 = K_0 \gamma' z \tag{6.14}$$

where K_0 is defined as the coefficient of earth pressure at-rest, in terms of effective stress.

Since the at-rest condition does not involve failure of the soil, the Mohr circle representing the vertical and horizontal stresses does not touch the failure envelope and the horizontal stress cannot be evaluated. The value of K_0, however, can be determined experimentally by means of a triaxial test in which the axial stress and the all-round pressure are increased simultaneously such that the lateral strain in the specimen is maintained at zero: the hydraulic triaxial apparatus is most suitable for this purpose. For soft clays, methods of measuring lateral pressure *in situ* have been developed by Bjerrum and Andersen [3] and, using the pressuremeter, by Wroth and Hughes [28].

Generally, for any condition intermediate to the active and passive states, the value of the lateral stress is unknown. The range of possible conditions can only be determined experimentally and Figure 6.10 shows the form of the relationship between

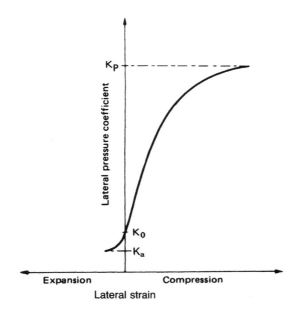

Figure 6.10 Relationship between lateral strain and lateral pressure coefficient.

strain and the lateral pressure coefficient. The exact relationship depends on the initial value of K_0 and on whether excavation or backfilling is involved in construction. The strain required to mobilize the passive pressure is considerably greater than that required to mobilize the active pressure. Experimental evidence indicates, for example, that the mobilization of full passive resistance requires a wall movement of the order of 2–4% of embedded depth in the case of dense sands and of the order of 10–15% in the case of loose sands. The corresponding percentages for the mobilization of active pressure are of the order of 0.25 and 1%, respectively.

For normally consolidated soils, the value of K_0 can be related approximately to the effective stress parameter ϕ' by the following formula proposed by Jaky:

$$K_0 = 1 - \sin \phi' \tag{6.15a}$$

For overconsolidated soils the value of K_0 depends on the stress history and can be greater than unity, a proportion of the at-rest pressure developed during initial consolidation being retained in the soil when the effective vertical stress is subsequently reduced. Mayne and Kulhawy [14] proposed the following correlation for overconsolidated soils during expansion (but not recompression):

$$K_0 = (1 - \sin \phi')(R_{oc})^{\sin \phi'} \tag{6.15b}$$

R_{oc} being the OCR. In Eurocode 7 it is proposed that

$$K_0 = (1 - \sin \phi')(R_{oc})^{0.5} \tag{6.15c}$$

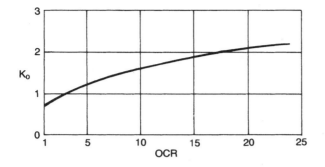

Figure 6.11 Typical relationship between K_0 and overconsolidation ratio for a clay.

Table 6.2 Coefficient of earth pressure at-rest

Soil	K_0
Dense sand	0.35
Loose sand	0.6
Normally consolidated clays (Norway)	0.5–0.6
Clay, OCR = 3.5 (London)	1.0
Clay, OCR = 20 (London)	2.8

A typical relationship between K_0 and OCR for a clay, determined in the triaxial apparatus, is shown in Figure 6.11 and some typical values of K_0 for different soils are given in Table 6.2.

6.3 COULOMB'S THEORY OF EARTH PRESSURE

Coulomb's theory (1776) involves consideration of the stability, as a whole, of the wedge of soil between a retaining wall and a trial failure plane. The *force* between the wedge and the wall surface is determined by considering the equilibrium of forces acting on the wedge when it is on the point of sliding either up or down the failure plane, i.e. when the wedge is in a condition of *limiting equilibrium*. Friction between the wall and the adjacent soil is taken into account. The angle of friction between the soil and the wall material, denoted by δ, can be determined in the laboratory by means of a direct shear test. At any point on the wall surface a shearing resistance per unit area of $p_n \tan \delta$ will be developed, where p_n is the normal pressure on the wall at that point. A constant component of shearing resistance or 'wall adhesion', c_w, can also be assumed if appropriate in the case of clays. Due to wall friction the shape of the failure surface is curved near the bottom of the wall in both the active and passive cases, as indicated in Figure 6.12, but in the Coulomb theory the failure surfaces are assumed to be plane in each case. In the active case, the curvature is slight and the error involved in assuming a plane surface, is relatively small. This is also true in the passive case for values of δ less than $\phi/3$, but for the higher values of δ normally appropriate in practice the error becomes relatively large.

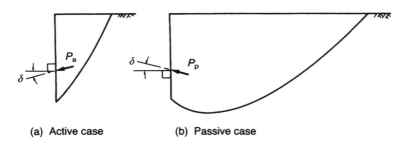

(a) Active case   (b) Passive case

Figure 6.12 Curvature due to wall friction.

The Coulomb theory is now interpreted as an upper bound plasticity solution (although analysis is based on force equilibrium and not on the work–energy balance defined in Section 6.1); collapse of the soil mass above the chosen failure plane occurring as the wall moves away from or into the soil. Thus, in general, the theory underestimates the total active thrust and overestimates the total passive resistance (i.e. upper bounds to the true collapse loads). When $\delta = 0$, the Coulomb theory gives results which are identical to those of the Rankine theory for the case of a vertical wall and a horizontal soil surface, i.e. the solution for this case is exact because the upper and lower bound results coincide.

Active case

Figure 6.13(a) shows the forces acting on the soil wedge between a wall surface AB, inclined at angle α to the horizontal, and a trial failure plane BC, at angle θ to the horizontal. The soil surface AC is inclined at angle β to the horizontal. The shear strength parameter c will be taken as zero, as will be the case for most backfills. For the failure condition, the soil wedge is in equilibrium under its own weight (W), the

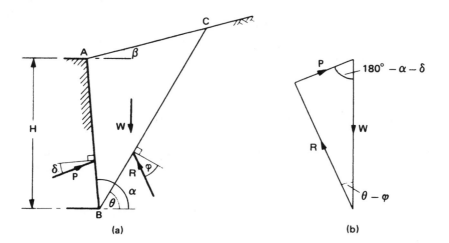

Figure 6.13 Coulomb theory: active case with $c = 0$.

reaction to the force (P) between the soil and the wall, and the reaction (R) on the failure plane. Because the soil wedge tends to move down the plane BC at failure, the reaction P acts at angle δ below the normal to the wall. (If the wall were to settle more than the backfill, the reaction P would act at angle δ above the normal.) At failure, when the shear strength of the soil has been fully mobilized, the direction of R is at angle ϕ below the normal to the failure plane (R being the resultant of the normal and shear forces on the failure plane). The directions of all three forces, and the magnitude of W, are known, and therefore the triangle of forces (Figure 6.13(b)) can be drawn and the magnitude of P determined for the trial in question.

A number of trial failure planes would have to be selected to obtain the maximum value of P, which would be the total active thrust on the wall. However, using the sine rule, P can be expressed in terms of W and the angles in the triangle of forces. Then the maximum value of P, corresponding to a particular value of θ, is given by $\partial P/\partial\theta = 0$. This leads to the following solution for P_a:

$$P_a = \frac{1}{2}K_a\gamma H^2 \qquad (6.16)$$

where

$$K_a = \left(\frac{\dfrac{\sin(\alpha - \phi)}{\sin\alpha}}{\sqrt{[\sin(\alpha + \delta)]} + \sqrt{\left[\dfrac{\sin(\phi + \delta)\sin(\phi - \beta)}{\sin(\alpha - \beta)}\right]}} \right)^2 \qquad (6.17)$$

The point of application of the total active thrust is not given by the Coulomb theory but is assumed to act at a distance of $\frac{1}{3}H$ above the base of the wall.

The Coulomb theory can be extended to cases in which the shear strength parameter c is greater than zero (i.e. c_u in the undrained case or if tangent parameter c' is used in the drained case). A value is then selected for the wall adhesion parameter c_w. It is assumed that tension cracks may extend to a depth z_0, the trial failure plane (at angle θ to the horizontal) extending from the heel of the wall to the bottom of the tension zone, as shown in Figure 6.14. The forces acting on the soil wedge at failure are as follows:

1   The weight of the wedge (W).
2   The reaction (P) between the wall and the soil, acting at angle δ below the normal.
3   The force due to the constant component of shearing resistance on the wall ($C_w = c_w \times$ EB).
4   The reaction (R) on the failure plane, acting at angle ϕ below the normal.
5   The force on the failure plane due to the constant component of shear strength ($C = c \times$ BC).

The directions of all five forces are known together with the magnitudes of W, C_w and C, and therefore the value of P can be determined from the force diagram for the trial failure plane. Again, a number of trial failure planes would be selected to obtain the maximum value of P.

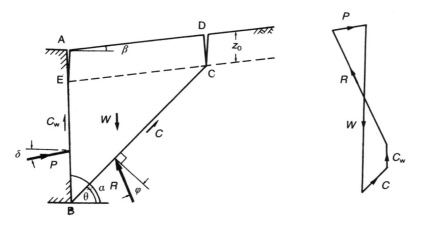

Figure 6.14 Coulomb theory: active case with $c > 0$.

The special case of a vertical wall and a horizontal soil surface will now be considered. For the undrained condition ($\phi_u = 0$), an expression for P can be obtained by resolving forces vertically and horizontally. The total active thrust is given by the maximum value of P, for which $\partial P / \partial \theta = 0$. The resulting value is

$$P_a = \frac{1}{2}\gamma(H^2 - z_0^2) - 2c_u(H - z_0)\sqrt{\left(1 + \frac{c_w}{c_u}\right)}$$

For $\phi_u = 0$ the earth pressure coefficient K_a is unity and it is convenient to introduce a second coefficient K_{ac}, where

$$K_{ac} = 2\sqrt{\left(1 + \frac{c_w}{c_u}\right)}$$

For the fully drained condition in terms of tangent parameters c' and ϕ', it can be assumed that

$$K_{ac} = 2\sqrt{\left[K_a\left(1 + \frac{c_w}{c'}\right)\right]}$$

In general, the active pressure at depth z can be expressed as

$$p_a = K_a\gamma z - K_{ac}c \tag{6.18}$$

where

$$K_{ac} = 2\sqrt{\left[K_a\left(1 + \frac{c_w}{c}\right)\right]} \tag{6.19}$$

the shear strength parameters being those appropriate to the drainage conditions of the problem. The depth of a dry tension crack (at which $p_a = 0$) is given by

$$z_0 = \frac{2c\sqrt{\left(1 + \frac{c_w}{c}\right)}}{\gamma\sqrt{K_a}} \tag{6.20}$$

The depth of a water-filled crack (z_{0w}) is obtained from the condition $p_a = \gamma_w z_{0w}$. Hydrostatic pressure in tension cracks can be eliminated by means of a horizontal filter.

Passive case

In the passive case, the reaction P acts at angle δ above the normal to the wall surface (or δ below the normal if the wall were to settle more than the adjacent soil) and the reaction R at angle ϕ above the normal to the failure plane. In the triangle of forces, the angle between W and P is $180° - \alpha + \delta$ and the angle between W and R is $\theta + \phi$. The total passive resistance, equal to the minimum value of P, is given by

$$P_p = \frac{1}{2}K_p\gamma H^2 \tag{6.21}$$

where

$$K_p = \left(\frac{\dfrac{\sin(\alpha + \phi)}{\sin\alpha}}{\sqrt{[\sin(\alpha - \delta)]} - \sqrt{\left[\dfrac{\sin(\phi + \delta)\sin(\phi + \beta)}{\sin(\alpha - \beta)}\right]}}\right)^2 \tag{6.22}$$

However, in the passive case it is not generally realistic to neglect the curvature of the failure surface and use of Equation 6.22 overestimates passive resistance, seriously so for the higher values of ϕ, representing an error on the unsafe side. It is recommended that passive pressure coefficients derived by Caquot and Kerisel [8] should be used. Caquot and Kerisel derived both active and passive coefficients by integrating the differential equations of equilibrium, the failure surfaces being logarithmic spirals. Coefficients have also been obtained by Sokolovski [24] by numerical integration.

In general, the passive pressure at depth z can be expressed as

$$p_p = K_p\gamma z + K_{pc}c \tag{6.23}$$

where

$$K_{pc} = 2\sqrt{\left[K_p\left(1 + \frac{c_w}{c}\right)\right]} \tag{6.24}$$

Kerisel and Absi [13] published tables of active and passive coefficients for a wide range of values of ϕ, δ, α and β, the active coefficients being very close to those calculated from Equation 6.17. For $\alpha = 90°$ and $\beta = 0°$, values of coefficients (denoted K_{ah} and K_{ph}) for *horizontal components* of pressure (i.e. $K_a \cos \delta$ and $K_p \cos \delta$, respectively) are plotted in Figure 6.15.

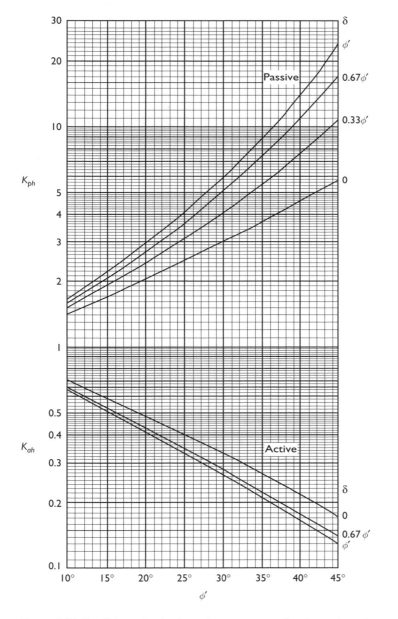

Figure 6.15 Coefficients for horizontal components of active and passive pressure.

Compaction-induced pressure

In the case of backfilled walls, lateral pressure is also influenced by the compaction process, an effect that is not considered in the earth pressure theories. During back-filling, the weight of the compaction plant produces additional lateral pressure on the wall. Pressures significantly in excess of the active value can result near the top of the wall, especially if it is restrained by propping during compaction. As each layer is compacted, the soil adjacent to the wall is pushed downwards against frictional resistance on the wall surface. When the compaction plant is removed the potential rebound of the soil is restricted by wall friction, thus inhibiting reduction of the additional lateral pressure. Also, the lateral strains induced by compaction have a significant plastic component which is irrecoverable. Thus, there is a residual lateral pressure on the wall. A simple analytical method of estimating the residual lateral pressure has been proposed by Ingold [10].

Compaction of backfill behind a retaining wall is normally effected by rolling. The compaction plant can be represented approximately by a line load equal to the weight of the roller. If a vibratory roller is employed, the centrifugal force due to the vibrating mechanism should be added to the static weight. The vertical stress immediately below a line load Q per unit length is derived from Equation 5.14:

$$\sigma_z = \frac{2Q}{\pi z}$$

Then the lateral pressure on the wall at depth z is given by

$$p_c = K_a(\gamma z + \sigma_z)$$

When the stress σ_z is removed, the lateral stress may not revert to the original value ($K_a \gamma z$). At shallow depth the residual lateral pressure could be high enough, relative to the vertical stress γz, to cause passive failure in the soil. Therefore, assuming there is no reduction in lateral stress on removal of the compaction plant, the maximum (or critical) depth (z_c) to which failure could occur is given by

$$p_c = K_p \gamma z_c$$

Thus

$$K_a(\gamma z_c + \sigma_z) = \frac{1}{K_a}\gamma z_c$$

If it is assumed that γz_c is negligible compared to σ_z, then

$$z_c = \frac{K_a^2 \sigma_z}{\gamma}$$
$$= \frac{K_a^2}{\gamma}\frac{2Q}{\pi z_c}$$

Therefore

$$z_c = K_a \sqrt{\frac{2Q}{\pi\gamma}}$$

The maximum value of lateral pressure (p_{max}) occurs at the critical depth, therefore (again neglecting γz_c):

$$p_{max} = \frac{2QK_a}{\pi z_c}$$

$$= \sqrt{\frac{2Q\gamma}{\pi}} \qquad (6.25)$$

The fill is normally placed and compacted in layers. Assuming that the pressure p_{max} is reached, and remains, in each successive layer, a vertical line can be drawn as a pressure envelope below the critical depth. Thus the distribution shown in Figure 6.16 represents a conservative basis for design. However, at a depth z_a the active pressure will exceed the value p_{max}. The depth z_a, being the limiting depth of the vertical envelope, is obtained from the equation

$$K_a \gamma z_a = \sqrt{\frac{2Q\gamma}{\pi}}$$

Thus

$$z_a = \frac{1}{K_a} \sqrt{\frac{2Q}{\pi\gamma}} \qquad (6.26)$$

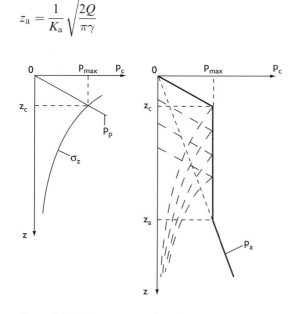

Figure 6.16 Compaction-induced pressure.

6.4   APPLICATION OF EARTH PRESSURE THEORY
TO RETAINING WALLS

In the Rankine theory the state of stress in a semi-infinite soil mass is considered, the entire mass being subjected to lateral expansion or compression. However, the movement of a retaining wall of finite dimensions cannot develop the active or passive state in the soil mass as a whole. The active state, for example, would be developed only within a wedge of soil between the wall and a failure plane passing through the lower end of the wall and at an angle of $45° + \phi/2$ to the horizontal, as shown in Figure 6.17(a); the remainder of the soil mass would not reach a state of plastic equilibrium. A specific (minimum) value of lateral strain would be necessary for the development of the active state within the above wedge. A uniform strain within the wedge would be produced by a rotational movement (A'B) of the wall, away from the soil, about its lower end and a deformation of this type, of sufficient magnitude, constitutes the minimum deformation requirement for the development of the active state. Any deformation configuration enveloping A'B, for example, a uniform translational movement A'B', would also result in the development of the active state. If the deformation of the wall were not to satisfy the minimum deformation requirement, the soil adjacent to the wall would not reach a state of plastic equilibrium and the lateral pressure would be between the active and at-rest values. If the wall were to deform by rotation about its upper end (due, for example, to restraint by a prop), the conditions for the complete development of the active state would not be satisfied because of inadequate strain in the soil near the surface; consequently, the pressure near the top of the wall would approximate to the at-rest value.

In the passive case the minimum deformation requirement is a rotational movement of the wall, about its lower end, into the soil. If this movement were of sufficient magnitude, the passive state would be developed within a wedge of soil between the wall and a failure plane at an angle of $45° + \phi/2$ to the vertical as shown in Figure 6.17(b). In practice, however, only part of the potential passive resistance would normally be mobilized. The relatively large deformation necessary for the full development of passive resistance would be unacceptable, with the result that the pressure under working conditions would be between the at-rest and passive values, as indicated in Figure 6.10 (and consequently providing a factor of safety against passive failure).

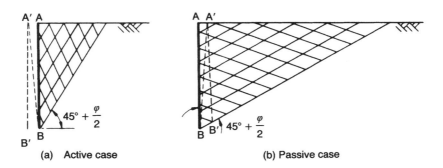

(a) Active case   (b) Passive case

Figure 6.17 Minimum deformation conditions.

The selection of an appropriate value of ϕ' is of prime importance in the passive case. The difficulty is that strains vary significantly throughout the soil mass and in particular along the failure surface. The effect of strain, which is governed by the mode of wall deformation, is neglected both in the failure criterion and in analysis. In the earth pressure theories, a constant value of ϕ' is assumed throughout the soil above the failure surface whereas, in fact, the mobilized value of ϕ' varies. In the case of dense sands the average value of ϕ' along the failure surface, as the passive condition is approached, corresponds to a point beyond the peak on the stress–strain curve (e.g. Figure 4.8(a)); use of the peak value of ϕ' would result, therefore, in an overestimation of passive resistance. It should be noted, however, that peak values of ϕ' obtained from triaxial tests are normally less than the corresponding values in plane strain, the latter being relevant in most retaining wall problems. In the case of loose sands, the wall deformation required to mobilize the ultimate value of ϕ' would be unacceptably large in practice. Guidance on design values of ϕ' is given in codes of practice [5, 9].

The values of lateral strain required to mobilize active and passive pressures in a particular case depend on the value of K_0, representing the initial state of stress, and on the subsequent stress path, which depends on the construction technique and, in particular, on whether backfilling or excavation is involved in construction. In general, the required deformation in the backfilled case is greater than that in the excavated case for a particular soil. It should be noted that for backfilled walls, the lateral strain at a given point is interpreted as that occurring after backfill has been placed and compacted to the level of that point.

6.5 DESIGN OF EARTH-RETAINING STRUCTURES

There are two broad categories of retaining structures: (1) gravity, or freestanding walls, in which stability is due mainly to the weight of the structure; (2) embedded walls, in which stability is due to the passive resistance of the soil over the embedded depth and, in most cases, external support. According to the principles of limit state design, an earth-retaining structure must not (a) collapse or suffer major damage, (b) be subject to unacceptable deformations in relation to its location and function and (c) suffer minor damage which would necessitate excessive maintenance, render it unsightly or reduce its anticipated life. Ultimate limit states are those involving the collapse or instability of the structure as a whole or the failure of one of its components. Serviceability limit states are those involving excessive deformation, leading to damage or loss of function. Both ultimate and serviceability limit states must always be considered. The philosophy of limit states is the basis of Eurocode 7 [9], a standard specifying all the situations which must be considered in design.

The design of retaining structures has traditionally been based on the specification of a factor of safety in terms of moments, i.e. the ratio of the resisting (or restoring) moment to the disturbing (or overturning) moment. This is known as a lumped factor of safety and is given a value high enough to allow for all the uncertainties in the analytical method and in the values of soil parameters. It must be recognized that relatively large deformations are required for the mobilization of available passive resistance and that a structure could be deemed to have failed due to excessive deformation before reaching a condition of collapse. The approach, therefore, is to base design on ultimate limit states with the incorporation of an appropriate factor of safety to satisfy the requirements of serviceability limit states. In general, the higher the factor of safety, the lower will be the deformation required to mobilize the proportion of passive resistance necessary for stability.

The *limit state* approach is based on the application of partial factors to actions and soil properties. Partial factors are denoted by the symbol γ, unfortunately the same symbol as is used for unit weight. Partial load factors are denoted by γ_F, material factors by γ_m and resistance factors by γ_R. In general, actions include loading, soil weight, *in-situ* stresses, pore water pressure, seepage pressure and ground movements. Actions are further classified as being either permanent or variable and as having either favourable or unfavourable effects in relation to limit states. Soil properties relevant to the design of retaining structures are c', $\tan \phi'$ and c_u, as appropriate. Conservative values of the shear strength parameters, deduced from reliable ground investigation and soil test results, are referred to as the *characteristic* values: they may be either upper or lower values, whichever is the more unfavourable. Characteristic values of shear strength parameters are divided by an appropriate partial factor to give the *design* value. The design values of actions, on the other hand, are obtained by multiplying characteristic values by an appropriate factor. The subscripts k and d can be used to denote characteristic and design values, respectively. If the resisting moment in a particular design problem is greater than or equal to the disturbing moment, using design values of actions and soil properties, then the limit state in question will be satisfied. These principles form the basis of Eurocode 7 (EC7): *Geotechnical Design* [9], currently in the form of a prestandard prior to the publication of a final version.

Three design cases (A, B and C) are specified in EC7 and are described in detail in Section 8.1. The geotechnical design of retaining structures is normally governed by Case C which is primarily concerned with uncertainties in soil properties. Partial factors currently recommended for this case are 1.25, 1.60 and 1.40 for $\tan \phi'$, c' and c_u, respectively. The factor for both favourable and unfavourable permanent actions is 1.00 and for variable unfavourable actions 1.30. Variable favourable actions are not considered, i.e. the partial factor is zero. Case A is relevant to the overturning, and Case B to the structural design, of a retaining structure.

In the limit state approach, the design values of active thrust and passive resistance are calculated from the design values of the relevant shear strength parameters. In Case C, the active thrust and passive resistance due to the self-weight of the soil are permanent unfavourable and favourable actions, respectively, for both of which the partial factor is 1.00. The active thrust due to surface surcharge loading is normally a variable unfavourable action and should be multiplied by a partial factor of 1.30.

In the current UK code of practice for earth-retaining structures, BS 8002: 1994 [5] (which will be superseded by EC7), it is simply stated that design loads (including unit weight), derived by factoring or otherwise, are intended to be the most pessimistic or unfavourable values. The factors applied to shear strength are referred to in BS 8002 as mobilization factors, minimum values of 1.20 and 1.50 (with respect to maximum strength) being specified for drained and undrained conditions, respectively, the same factor being applied to both components of shear strength in the drained case. These values are intended to ensure that both ultimate and serviceability limit states are satisfied; in particular, except for loose soils, their use should ensure that wall displacement is unlikely to exceed 0.5% of wall height.

In determining characteristic values of shear strength parameters consideration should be given to the possibility of variations in soil conditions and to the quality of construction likely to be achieved. According to BS 8002, the shear strength used in design should be the lesser of:

1 the value mobilized at a strain which satisfies serviceability limit states, represented by the maximum (peak) strength divided by an appropriate mobilization factor;
2 the value which would be mobilized at collapse after large strain has taken place, represented by the critical-state strength.

The use of the value specified above is intended to ensure that conservative values of active pressure and passive resistance are used in design. Thus, under working conditions (at a deformation governed by the serviceability limit state), the active pressure is greater than, and the passive resistance less than, the respective values at maximum shear strength. If collapse were to occur (an ultimate limit state), the shear strength would approach the critical-state value: consequently, the active pressure would again be greater than, and the passive resistance less than, the corresponding values at maximum shear strength. However, Puller and Lee [20] have questioned the application of a constant mobilization factor because lateral wall movement is not constant with depth, especially in the case of flexible walls. Also, in the case of layered soils, peak strength is attained at different strains in different soil types.

The design value of the angle of wall friction (δ) depends on the type of soil and the wall material. In BS 8002 it is recommended that the design value of δ should be the lesser of the characteristic value determined by test and $\tan^{-1}(0.75 \tan \phi')$, where ϕ' is the design value of angle of shearing resistance. Similarly, in total stress analysis, the design value of wall adhesion (c_w) should be the lesser of the characteristic test value and $0.75c_u$, where c_u is the design value of undrained shear strength. However, for steel sheet piling in clay, the value of c_w may be close to zero immediately after driving but should increase with time.

6.6 GRAVITY WALLS

The stability of gravity (or freestanding) walls is due to the self-weight of the wall, perhaps aided by passive resistance developed in front of the toe. The traditional gravity wall (Figure 6.18(a)), constructed of masonry or mass concrete, is uneconomic because the material is used only for its dead weight. Reinforced concrete cantilever walls (Figure 6.18(b)) are more economic because the backfill itself, acting on the base, is employed to provide most of the required dead weight. Other types of gravity structure include gabion and crib walls (Figures 6.18(c) and (d)). Gabions are cages of steel mesh, rectangular in plan and elevation, filled with particles generally of cobble size, the units being used as the building blocks of a gravity structure. Cribs are open structures assembled from precast concrete or timber members and enclosing coarse-grained fill, the structure and fill acting as a composite unit to form a gravity wall.

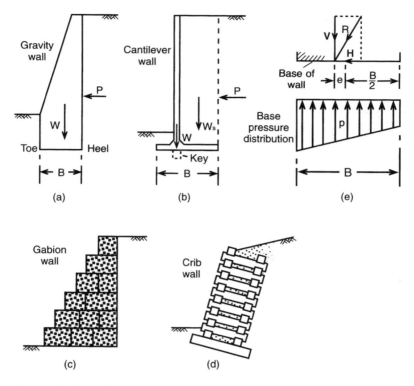

Figure 6.18 Retaining structures.

Limit states which must be considered in wall design are as follows:

1   Overturning of the wall due to instability of the retained soil mass.
2   Base pressure must not exceed the ultimate bearing capacity of the supporting soil (Section 8.2), the maximum base pressure occurring at the toe of the wall because of the eccentricity and inclination of the resultant load.
3   Sliding between the base of the wall and the underlying soil.
4   The development of a deep slip surface which envelops the structure as a whole (analysed using the methods described in Chapter 9).
5   Soil and wall deformations which cause adverse effects on the wall itself or on adjacent structures and services.
6   Adverse seepage effects, internal erosion or leakage through the wall: consideration should be given to the consequences of the failure of drainage systems to operate as intended.
7   Structural failure of any element of the wall or combined soil/structure failure.

The first step in design is to determine all the forces on the wall, from which the horizontal and vertical components, H and V, respectively, of the resultant force (R) acting on the base of the wall are obtained. Soil and water levels should represent the most unfavourable conditions conceivable in practice. Allowance must be made for the possibility of future (planned or unplanned) excavation in front of the wall, a minimum depth of 0.5 m being recommended: accordingly, passive resistance in front of the wall is normally neglected. For design purposes it is recommended that a minimum surcharge pressure of $10 \, \text{kN/m}^2$ should be assumed to act on the soil surface behind the wall. In the case of cantilever walls (Figure 6.18(b)), the vertical plane through the heel is taken to be the virtual wall surface. No shear stresses act on this surface, i.e. $\delta = 0$; therefore the Rankine value of K_a is appropriate. Surcharge pressure in front of the virtual back of a cantilever wall is a variable favourable action and should be neglected. The position of the base resultant (Figure 6.18(e)) is determined by dividing the algebraic sum of the moments of all forces about any point on the base by the vertical component V. To ensure that base pressure remains compressive over the entire base width, the resultant must act within the middle third of the base, i.e. the eccentricity (e) must not exceed $\tfrac{1}{6} B$, where B is the width of the base. If pressure is compressive over the entire width of the base, there will be no possibility of the wall overturning. However, the stability of the wall can be verified by ensuring that the total resisting moment about the toe exceeds the total overturning moment. If a linear distribution of pressure (p) is assumed under the base, the maximum and minimum base pressures can be calculated from the following expression (analogous to that for combined bending and direct stress):

$$ p = \frac{V}{B}\left(1 \pm \frac{6e}{B}\right) \tag{6.27} $$

The sliding resistance between the base and the soil is given by

$$ S = V \tan \delta \tag{6.28} $$

where δ is the angle of friction between the base and the underlying soil. Ignoring passive resistance in front of the wall, the sliding limit state will be satisfied if

$$ S \geq H $$

If necessary, the resistance against sliding can be increased by incorporating a shear key in the base. In the traditional method of design, the ratio S/H represents the lumped factor of safety against sliding, the serviceability limit state being satisfied by an adequate value of this factor.

Example 6.4

Details of a cantilever retaining wall are shown in Figure 6.19, the water table being below the base of the wall. The unit weight of the backfill is $17 \, \text{kN/m}^3$ and a surcharge pressure of $10 \, \text{kN/m}^2$ acts on the surface. Characteristic values of the shear strength parameters for the backfill are $c' = 0$ and $\phi' = 36°$. The angle of friction between the base and the foundation soil is $27°$ (i.e. $0.75\phi'$). Is the design of the wall satisfactory according to (a) the traditional approach and (b) the limit state (EC7) approach?

The position of the base reaction is determined by calculating the moments of all forces about the toe of the wall, the unit weight of concrete being taken as $23.5 \, \text{kN/m}^3$. The active thrust is calculated on the vertical plane through the heel of the wall, thus $\delta = 0$ and the Rankine value of K_a is appropriate.

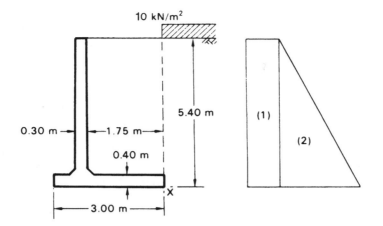

Figure 6.19 Example 6.4.

Table 6.3

(per m)	Force (kN)		Arm (m)	Moment (kN m)
(1)	$0.26 \times 10 \times 5.40$	$= 14.0$	2.70	37.9
(2)	$\dfrac{1}{2} \times 0.26 \times 17 \times 5.40^2$	$= 64.4$	1.80	115.9
		$H = \overline{78.4}$		$M_H = \overline{153.8}$
(Stem)	$5.00 \times 0.30 \times 23.5$	$= 35.3$	1.10	38.8
(Base)	$0.40 \times 3.00 \times 23.5$	$= 28.2$	1.50	42.3
(Soil)	$5.00 \times 1.75 \times 17$	$= \overline{148.8}$	2.125	316.2
		$V = \overline{212.3}$		$M_V = \overline{397.3}$
				$\overline{153.8}$
				$\Sigma M = \overline{243.5}$

(a) For $\phi' = 36°$ and $\delta = 0, K_a = 0.26$.

The calculations are set out in Table 6.3.
    Lever arm of base resultant is given by

$$\frac{\Sigma M}{V} = \frac{243.5}{212.3} = 1.15\,\text{m}$$

i.e. the resultant acts within the middle third of the base.
The factor of safety against overturning is: $397.3/153.8 = 2.58$
Eccentricity of base reaction, $e = 1.50 - 1.15 = 0.35\,\text{m}$
    The maximum and minimum base pressures are given by Equation 6.27, i.e.

$$p = \frac{212.3}{3.0}\left(1 \pm \frac{6 \times 0.35}{3.0}\right)$$

Thus, $p_{max} = 120\,\text{kN/m}^2$ (at the toe of the wall) and $p_{min} = 21\,\text{kN/m}^2$ (at the heel).

The factor of safety against sliding is given by

$$F = \frac{V \tan \delta}{H}$$
$$= \frac{212.3 \tan 27°}{78.4}$$
$$= 1.38$$

The design of the wall is satisfactory.

(b) The design value of $\phi' = \tan^{-1}(\tan 36°/1.25) = 30°$, for which $K_a = 0.33$. Case C is relevant, therefore a partial factor of 1.30 is applied to force (1), the surcharge pressure being a variable unfavourable action. The partial factor for all other forces is 1.00. Then

$$H = (0.33 \times 10 \times 5.40 \times 1.30) + \left(\frac{1}{2} \times 0.33 \times 17 \times 5.40^2\right)$$
$$= 23.2 + 81.8 = 105.0\,\text{kN}$$
$$M_H = (23.2 \times 2.70) + (81.8 \times 1.80) = 209.9\,\text{kN\,m}$$
$$V = 212.3\,\text{kN (as before)}$$
$$M_V = 397.3\,\text{kN\,m (as before)}$$
$$\Sigma M = 187.4\,\text{kN\,m}$$

Lever arm of base resultant $= 187.4/212.3 = 0.88\,\text{m}$
Eccentricity of base reaction, $e = 1.50 - 0.88 = 0.62\,\text{m}$
Then from Equation 6.27, $p_{max} = 159\,\text{kN/m}^2$ and $p_{min} = -17\,\text{kN/m}^2$
The resultant acts outside the middle third of the base, giving a negative value of p_{min}.
The design value of δ is $(0.75 \times 30°) = 22.5°$
The sliding limit state is not satisfied, the resisting force $V \tan \delta$ (88.0 kN) being less than the disturbing force H (105.0 kN).

The width of the base would have to be increased because of the negative base pressure at the heel of the wall and the sliding resistance limit state not being satisfied. The overturning limit state is satisfied, the resisting moment M_V being greater than the disturbing moment M_H. It should be noted that passive resistance in front of the wall has been neglected to allow for unplanned excavation. It is likely that the design would be satisfactory if passive resistance could be relied on throughout the life of the wall.

Case B is likely to govern the structural design of the wall, with partial factors of 1.50 and 1.35 being applied to forces (1) and (2), respectively, and shear strength being unfactored.

Example 6.5

Details of a gravity-retaining wall are shown in Figure 6.20, the unit weight of the wall material being $23.5\,\text{kN/m}^3$. The unit weight of the backfill is $18\,\text{kN/m}^3$ and *design* values of the shear strength parameters are $c' = 0$ and $\phi' = 33°$. The value of δ between

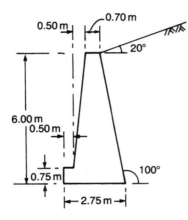

Figure 6.20 Example 6.5.

wall and backfill and between wall and foundation soil is 26°. The pressure on the foundation soil should not exceed 250 kN/m². Is the design of the wall satisfactory?

As the back of the wall and the soil surface are both inclined, the value of K_a will be calculated from Equation 6.17. The values of the angles in this equation are $\alpha = 100°, \beta = 20°, \phi = 33°$ and $\delta = 26°$. Thus,

$$K_a = \left(\frac{\sin 67°/\sin 100°}{\sqrt{\sin 125°} + \sqrt{\sin 58° \sin 13°/\sin 80°}}\right)^2$$

$$= 0.48$$

Then, from Equation 6.16,

$$P_a = \frac{1}{2} \times 0.48 \times 18 \times 6^2 = 155.5\,\text{kN/m}$$

acting at $\frac{1}{3}$ height and at 26° above the normal, or 36° above the horizontal. Moments are considered about the toe of the wall, the calculations being set out in Table 6.4.

Lever arm of base resultant is given by

$$\frac{\Sigma M}{V} = \frac{287.1}{312.5} = 0.92\,\text{m}$$

The overturning limit state is satisfied since the restoring moment (M_V) is greater than the disturbing moment (M_H).

Eccentricity of base reaction, $e = 1.375 - 0.92 = 0.455\,\text{m}$

The maximum and minimum base pressures are given by Equation 6.27, i.e.

$$p = \frac{312.5}{2.75}\left(1 \pm \frac{6 \times 0.455}{2.75}\right)$$

Table 6.4

(per m)	Force (kN)		Arm (m)	Moment (kN m)
$P_a \cos 36°$	$= 125.8$		2.00	251.6
	$H = \overline{125.8}$			$M_H = \overline{251.6}$
$P_a \sin 36°$	$= 91.4$		2.40	219.4
Wall	$\frac{1}{2} \times 1.05 \times 6 \times 23.5 = 74.0$		2.05	151.7
	$0.70 \times 6 \times 23.5 = 98.7$		1.35	133.2
	$\frac{1}{2} \times 0.50 \times 5.25 \times 23.5 = 30.8$		0.83	25.6
	$1.00 \times 0.75 \times 23.5 = \underline{17.6}$		0.50	$\underline{8.8}$
	$V = \overline{312.5}$			$M_V = \overline{538.7}$
				251.6
				$\Sigma M = \overline{287.1}$

Thus $p_{max} = 226\,\text{kN/m}^2$ and $p_{min} = 1\,\text{kN/m}^2$. The maximum base pressure is less than the allowable bearing capacity of the foundation soil, therefore the bearing resistance limit state is satisfied.

With respect to base sliding, the restoring force, $V \tan \delta = 312.5 \tan 26° = 152.4\,\text{kN}$. The disturbing force, H is 125.8 kN. Therefore the sliding limit state is satisfied.

The design of the wall is satisfactory, the relevant limit states being satisfied in terms of design values of the shear strength parameters.

Example 6.6

Details of a retaining structure, with a vertical drain adjacent to the back surface, are shown in Figure 6.21(a), the saturated unit weight of the backfill being $20\,\text{kN/m}^3$. The design parameters for the backfill are $c' = 0$, $\phi' = 38°$ and $\delta = 15°$. Assuming a failure plane at 55° to the horizontal, determine the total horizontal thrust on the wall when the backfill becomes fully saturated due to continuous rainfall, with steady seepage towards the drain. Determine also the thrust on the wall (a) if the vertical drain were replaced by an inclined drain below the failure plane and (b) if there were no drainage system behind the wall.

The flow net for seepage towards the vertical drain is shown in Figure 6.21(a). Since the permeability of the drain must be considerably greater than that of the backfill, the drain remains unsaturated and the pore pressure at every point within the drain is zero (atmospheric). Thus, at every point on the boundary between the drain and the backfill, total head is equal to elevation head. The equipotentials, therefore, must intersect this boundary at points spaced at equal vertical intervals Δh: the boundary itself is neither a flow line nor an equipotential.

The combination of total weight and boundary water force is used. The values of pore water pressure at the points of intersection of the equipotentials with the failure plane are evaluated and plotted normal to the plane. The boundary water force (U), acting normal to the plane, is equal to the area of the pressure diagram, thus

$$U = 55\,\text{kN/m}$$

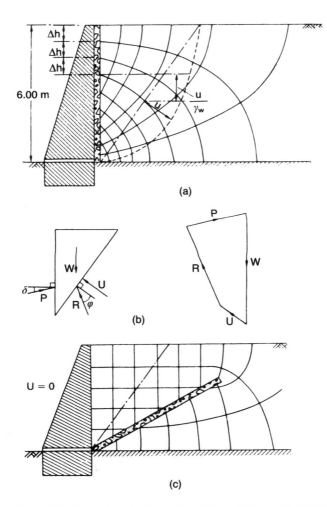

Figure 6.21 Example 6.6. (Reproduced from K. Terzaghi (1943) *Theoretical Soil Mechanics*, John Wiley & Sons Inc., by permission.)

The water forces on the other two boundaries of the soil wedge are zero.

The total weight (W) of the soil wedge is now calculated, i.e.

$$W = 252\,\text{kN/m}$$

The forces acting on the wedge are shown in Figure 6.21(b). Since the directions of the four forces are known, together with the values of W and U, the force polygon can be drawn, from which

$$P_a = 108\,\text{kN/m}$$

The horizontal thrust on the wall is given by

$$P_a \cos \delta = 105 \, \text{kN/m}$$

Other failure surfaces would have to be chosen in order that the maximum value of total active thrust can be determined.

For the inclined drain shown in Figure 6.21(c), the flow lines and equipotentials above the drain are vertical and horizontal, respectively. Thus at every point on the failure plane the pore water pressure is zero. This form of drain is preferable to the vertical drain. In this case

$$P_a = \frac{1}{2} K_a \gamma_{\text{sat}} H^2$$

For $\phi' = 38°$ and $\delta = 15°$, K_{ah} $(=K_a \cos \delta) = 0.21$ (from Figure 6.15). The horizontal thrust is

$$P_a \cos \delta = \frac{1}{2} \times 0.21 \times 20 \times 6^2 = 76 \, \text{kN/m}$$

For the case of no drainage system behind the wall, the pore water is static, and therefore the horizontal thrust

$$= \frac{1}{2} K_a \gamma' H^2 \cos \delta + \frac{1}{2} \gamma_w H^2$$

$$= \left(\frac{1}{2} \times 0.21 \times 10.2 \times 6^2 \right) + \left(\frac{1}{2} \times 9.8 \times 6^2 \right)$$

$$= 39 + 176 = 215 \, \text{kN/m}$$

6.7 EMBEDDED WALLS

Cantilever walls

Walls of this type are mainly of steel sheet piling and are used only when the retained height of soil is relatively low. In sands and gravels these walls may be used as permanent structures but in general they are used only for temporary support. The stability of the wall is due entirely to passive resistance mobilized in front of the wall. The principal limit state is instability of the retained soil mass causing rotation or translation of the wall. Limit states (4) to (7) listed for gravity walls (Section 6.6) should also be considered. The mode of failure is by rotation about a point O near the lower end of the wall as shown in Figure 6.22(a). Consequently, passive resistance acts in front of the wall above O and behind the wall below O, as shown in Figure 6.22(b), thus providing a fixing moment. However this pressure distribution is an idealization as there is unlikely to be a complete change in passive resistance from the front to the back of the wall at point O. To allow for over-excavation it is recommended that the

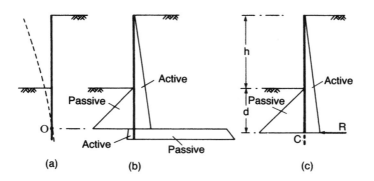

Figure 6.22 Cantilever sheet pile wall.

soil level in front of the wall should be reduced by 10% of the retained height, subject to a maximum of 0.5 m. A minimum surcharge pressure of $10 \, \text{kN/m}^2$ should be assumed to act on the soil surface behind the wall.

Design is generally based on the simplification shown in Figure 6.22(c), it being assumed that the net passive resistance below point O is represented by a concentrated force R acting at a point C, slightly below O, at depth d below the lower soil surface. The traditional method of analysis involves determining the depth d by equating moments about C, a factor of safety F being applied to the restoring moment, i.e. the available passive resistance in front of the wall is divided by F. The value of d is then increased arbitrarily by 20% to allow for the simplification involved in the method, i.e. the required depth of embedment is $1.2d$. However, it is advisable to evaluate R by equating horizontal forces and to check that net passive resistance available over the additional 20% embedded depth is equal to or greater than R. The translation limit state is satisfied if the horizontal resisting force is greater than or equal to the disturbing force. Cantilever walls can also be analysed by applying partial factors.

Anchored or propped walls

Generally, structures of this type are either of steel sheet piling or reinforced concrete diaphragm walls, the construction of which is described in Section 6.9. Additional support to embedded walls is provided by a row of tie-backs or props near the top of the wall, as illustrated in Figure 6.23(a). Tie-backs are normally high-tensile steel cables or rods, anchored in the soil some distance behind the wall. Walls of this type are used extensively in the support of deep excavations and in waterfront construction. In the case of sheet pile walls there are two basic modes of construction. Excavated walls are constructed by driving a row of sheet piling, followed by excavation or dredging to the required depth in front of the wall. Backfilled walls are constructed by partial driving, followed by backfilling to the required height behind the piling. In the case of diaphragm walls, excavation takes place in front of the wall after it has been cast *in situ*. Stability is due to the passive resistance developed in front of the wall together with the supporting forces in the ties or props.

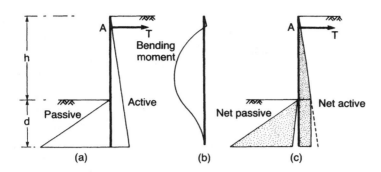

Figure 6.23 Anchored sheet pile wall: free earth support method.

Free earth support analysis

It is assumed that the depth of embedment below excavation level is insufficient to produce fixity at the lower end of the wall. Thus the wall is free to rotate at its lower end, the bending moment diagram being of the form shown in Figure 6.23(b). The limit states to be considered are instability of the retained soil mass causing rotation or translation of the wall, the vertical equilibrium of the wall and states (4) to (7) listed in Section 6.6. To satisfy the rotation limit state, the restoring moment about the anchor or prop must be greater than or equal to the overturning moment. The horizontal forces on the wall are then equated to zero, yielding the minimum value of anchor or prop force required to satisfy the translation limit state. Finally, if appropriate, the vertical forces on the wall are calculated, it being a requirement that the downward force (e.g. the component of the force in an inclined tie back) should not exceed the (upward) frictional resistance available between the wall and the soil on the passive side minus the (downward) frictional force on the active side.

Four methods of introducing safety factors into the calculations have evolved, as described below. These methods also apply to the analysis of embedded cantilever walls.

1   The depth of embedment at which the wall is on the point of collapse is calculated by equating moments to zero, using the fully mobilized values of active and passive pressures. This depth is then multiplied by a factor (F_d), known as the embedment factor. This method is not recommended because of the empirical way in which the factor F_d is introduced but if it were to be used the design should also be checked by one of the other methods.

2   The factor of safety (F_p) is expressed as the ratio of the restoring moment to the overturning moment, i.e. the former must exceed the latter by a specified margin. Fully mobilized values of active and passive pressures are used in calculating the moments. Gross soil pressures are used in the calculations, i.e. the active and passive pressures are not combined in any way. Being the source of the restoring moment, gross passive resistance only is factored.

3   The factor of safety is defined in terms of shear strength. The shear strength parameters are divided by a factor (F_s) before the active and passive pressures are

calculated, the depth of embedment then being determined by equating the overturning and restoring moments. The factor is thus applied to the parameters of greatest uncertainty. The water pressures in an effective stress analysis, of course, should *not* be factored. This method satisfies the requirements of EC7, F_s being the equivalent of the partial material factor γ_m, and of BS 8002, F_s becoming the mobilization factor M.

4   The factor of safety (F_r) is applied to moments, as in method 2, but is defined as the ratio of the moment of net passive resistance to the moment of net active thrust. The concept of net passive resistance is illustrated in Figure 6.23(c), the active thrust over the embedded depth being subtracted from the passive resistance. This method was proposed by Burland *et al.* [7] because of the lack of consistency between values of the factors F_s and F_p over the practical ranges of wall geometry and shear strength parameters, particularly in the case of clays. Burland *et al.* proposed that the factor of safety should be based on net passive resistance on the basis of an analogy with the ultimate bearing capacity of a foundation.

Guidance on the selection of suitable factors of safety for use in the above methods is given by Padfield and Mair [15]. Values of γ_m and M are given in EC7 and BS 8002, respectively. Either gross or net water pressures can be used in design (because, unlike earth pressures, no coefficient is involved), the latter being more convenient. To allow for over-excavation or dredging, the soil level in front of the wall should be reduced by 10% of the depth below the lowest tie or prop, subject to a maximum of 0.5 m. Again, a minimum surcharge pressure of $10\,\mathrm{kN/m^2}$ should be assumed to act on the soil surface behind the wall.

It should be realized that full passive resistance is only developed under conditions of limiting equilibrium, i.e. when the safety or mobilization factor is unity. Under working conditions, with a factor greater than unity, analytical and experimental work has indicated that the distribution of lateral pressure is likely to be of the form shown in Figure 6.24, with passive resistance being fully mobilized close to the lower surface. The extra depth of embedment required to provide adequate safety (whether measured by lumped or partial factors) results in a partial fixing moment at the lower end of the wall and, consequently, a lower maximum bending moment than the value under limiting equilibrium or collapse conditions. In view of the uncertainty regarding the pressure distributions under working conditions it is recommended that bending moments and tie or prop force under limiting equilibrium conditions (i.e. $F = 1$)

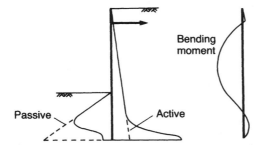

Figure 6.24 Anchored sheet pile wall: pressure distribution under working conditions.

should be used in the structural design of the wall. The tie or prop force thus calculated should be increased by 25% to allow for possible redistribution of pressure due to arching (see below). Bending moments should be calculated on the same basis in the case of cantilever walls. In limit state design, Case B applies to the determination of maximum bending moment.

Effect of flexibility and K_0

The behaviour of an anchored wall is also influenced by its degree of flexibility or stiffness. In the case of *flexible* sheet pile walls, experimental and analytical results indicate that redistributions of lateral pressure take place. The pressures on the most yielding parts of the wall (between the tie and excavation level) are reduced and those on the relatively unyielding parts (in the vicinity of the tie and below excavation level) are increased with respect to the theoretical values, as illustrated in Figure 6.25. These redistributions of lateral pressure are the result of the phenomenon known as *arching*. No such redistributions take place in the case of stiff walls, such as concrete diaphragm walls.

Arching was defined by Terzaghi [25] in the following way. 'If one part of the support of a soil mass yields while the remainder stays in place, the soil adjoining the yielding part moves out of its original position between adjacent stationary soil masses. The relative movement within the soil is opposed by shearing resistance within the zone of contact between the yielding and stationary masses. Since the shearing resistance tends to keep the yielding mass in its original position, the pressure on the yielding part of the support is reduced and the pressure on the stationary parts is increased. This transfer of pressure from a yielding part to adjacent non-yielding parts of a soil mass is called the arching effect. Arching also occurs when one part of a support yields more than the adjacent parts.'

The conditions for arching are present in anchored sheet pile walls when they deflect. If yield of the anchor takes place, arching effects are reduced to an extent depending on the amount of yielding. On the passive side of the wall, the pressure is increased just below excavation level as a result of larger deflections *into* the soil. In the case of backfilled walls, arching is only partly effective until the fill is above tie level.

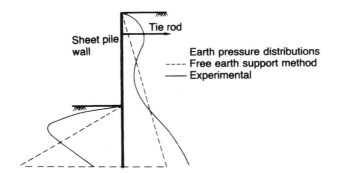

Figure 6.25 Arching effects.

Arching effects are much greater in sands than in silts or clays and are greater in dense sands than in loose sands.

Redistributions of earth pressure result in lower bending moments than those obtained from the free earth support method of analysis; the greater the flexibility of the wall, the greater the moment reduction. Rowe [21, 22] proposed the use of moment reduction coefficients, to be applied to the results of free earth support analyses, based on the flexibility of the wall. Wall flexibility is represented by the parameter $\rho = H^4/EI$ (units m^2/kN per m), where H is the overall height of the wall and EI the flexural rigidity. The tie force is also influenced by earth pressure redistribution and factors are also given for the adjustment of the free earth support value of this force. Details of Rowe's procedure are given by Barden [2].

Rowe's moment reduction factors should only be used if a factored passive resistance $(F > 1)$ has been used for the calculation of bending moments. If bending moments have been calculated for the limiting equilibrium condition $(F = 1)$, Rowe's factors should not be used.

Potts and Fourie [18, 19] analysed a propped cantilever wall in clay by means of the finite element method, incorporating an elastic–perfectly plastic stress–strain relationship. The results indicated that the required depth of embedment was in agreement with the value obtained by the free earth support method. However, the results also showed that in general the behaviour of the wall depended on the wall stiffness (confirming Rowe's earlier findings), the initial value of K_0 (the coefficient of earth pressure at-rest) for the soil and the method of construction (i.e. backfilling or excavation).

In particular, maximum bending moment and prop force increased as wall stiffness increased. For backfilled walls and for excavated walls in soils having a low K_0 value (of the order of 0.5), both maximum bending moment and prop force were lower than those obtained using the free earth support method. However, for stiff walls, such as diaphragm walls, formed by excavation in soils having a high K_0 value (in the range 1–2), such as overconsolidated clays, both maximum bending moment and prop force were significantly higher than those obtained using the free earth support method. For the particular (excavated) wall and material properties considered by Potts and Fourie, the patterns of variation shown in Figure 6.26 were obtained for a factor of

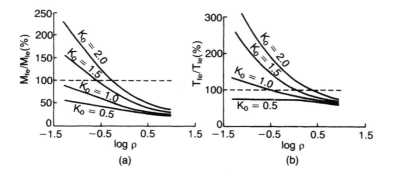

Figure 6.26 Analysis of propped cantilever wall in clay by the finite element method. (Reproduced from D.M. Potts and A.B. Fourie (1985) *Geotechnique*, **35**, No. 3, by permission of Thomas Telford Ltd.)

safety (F_{r}) of 2.0. In this figure, M_{fe} and T_{fe} denote the maximum bending moment and prop force, respectively, obtained from the finite element analysis, and M_{le} and T_{le} denote the corresponding values obtained from a limiting equilibrium (free earth support) analysis.

Pore water pressure distribution

Sheet pile and diaphragm walls are normally analysed in terms of effective stress. Care is therefore required in deciding on the appropriate distribution of pore water pressure. Several different situations are illustrated in Figure 6.27.

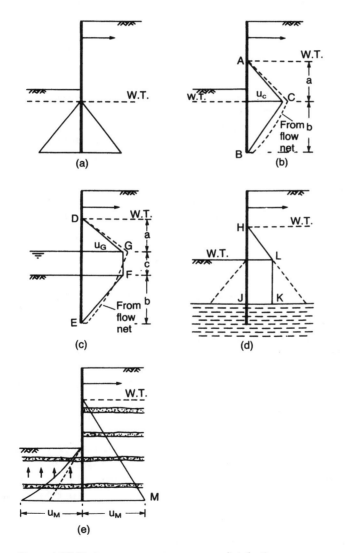

Figure 6.27 Various pore water pressure distributions.

If the water table levels are the same on both sides of the wall, the pore water pressure distributions will be hydrostatic and will balance (Figure 6.27(a)); they can thus be eliminated from the calculations.

If the water table levels are different and if steady seepage conditions have developed and are maintained, the distributions on the two sides of the wall will be unbalanced. The pressure distributions on each side of the wall may be combined because no pressure coefficient is involved. The net distribution on the back of the wall could be determined from the flow net, as illustrated in Example 2.1. However, in most situations an approximate distribution, ABC in Figure 6.27(b), can be obtained by assuming that the total head is dissipated uniformly along the back and front wall surfaces between the two water table levels. The maximum net pressure occurs opposite the lower water table level and, referring to Figure 6.27(b), is given by

$$u_C = \frac{2ba}{2b + a} \gamma_w$$

In general, the approximate method will underestimate net water pressure, especially if the bottom of the wall is relatively close to the lower boundary of the flow region (i.e. if there are large differences in the sizes of curvilinear squares in the flow net). The approximation should not be used in the case of a narrow excavation between two lines of sheet piling where curvilinear squares are relatively small (and seepage pressure relatively high) approaching the base of the excavation.

In Figure 6.27(c), a depth of water is shown in front of the wall, the water level being below that of the water table behind the wall. In this case the approximate distribution DEFG should be used in appropriate cases, the net pressure at G being given by

$$u_G = \frac{(2b + c)a}{2b + c + a} \gamma_w$$

A wall constructed mainly in a soil of relatively high permeability but penetrating a layer of clay of low permeability is shown in Figure 6.27(d). If undrained conditions apply within the clay the pore water pressure in the overlying soil would be hydrostatic and the net pressure distribution would be HJKL as shown.

A wall constructed in a clay which contains thin layers or partings of fine sand or silt is shown in Figure 6.27(e). In this case it should be assumed that the sand or silt allows water at hydrostatic pressure to reach the back surface of the wall. This implies pressure in excess of hydrostatic, and consequent upward seepage, in front of the wall.

For short-term situations for walls in clay (e.g. during and immediately after excavation), there exists the possibility of tension cracks developing or fissures opening. If such cracks or fissures fill with water, hydrostatic pressure should be assumed over the depth in question: the water in the cracks or fissures would also result in softening of the clay. Softening would also occur near the soil surface in front of the wall as a result of stress relief on excavation. An effective stress analysis would ensure a safe design in the event of rapid softening of the clay taking place or if work were delayed during the temporary stage of construction; however, a relatively low factor of safety could be used in such cases. Padfield and Mair [15] give details of a mixed total and effective stress design method as an alternative for short-term situations in clay, i.e. effective stress conditions within the zones liable to soften and total stress conditions below.

Seepage pressure

Under conditions of steady seepage, use of the approximation that total head is dissipated uniformly along the wall has the consequent advantage that the seepage pressure is constant. For the conditions shown in Figure 6.27(b), for example, the seepage pressure at any depth is

$$ j = \frac{a}{2b + a} \gamma_w $$

The effective unit weight of the soil below the water table, therefore, would be increased to $\gamma' + j$ behind the wall, where seepage is downwards, and reduced to $\gamma' - j$ in front of the wall, where seepage is upwards. These values should be used in the calculation of active and passive pressures, respectively, if groundwater conditions are such that steady seepage is maintained. Thus, active pressures are increased and passive pressures are decreased relative to the corresponding static values.

Anchorages

Tie rods are normally anchored in beams, plates or concrete blocks (known as dead-man anchors) some distance behind the wall (Figure 6.28). Ultimate limit states are pullout of the anchor and fracture of the tie. The serviceability limit state is that anchor yield should be minimal. The tie rod force is resisted by the passive resistance developed in front of the anchor, reduced by the active pressure on the back, both calculated either (a) by applying a lumped factor of safety to gross or net passive resistance or (b) by applying partial factors to the shear strength parameters, to ensure that the serviceability limit state is satisfied. The passive resistance should be calculated assuming no surface surcharge and the active pressure calculated assuming that at least a minimum surcharge pressure of $10 \, \text{kN/m}^2$ is imposed. To avoid the possibility

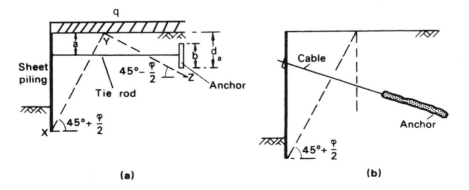

Figure 6.28 (a) Plate anchor and (b) Ground anchor.

of progressive failure of a line of ties, it should be assumed that any single tie could fail either by fracture or by becoming detached and that its load could be redistributed safely to the two adjacent ties. Accordingly, it is recommended that a load factor of at least 2.0 should be applied to the tie rod force. If the height (b) of the anchor is not less than half the depth (d_a) from the surface to the bottom of the anchor, it can be assumed that passive resistance is developed over the depth d_a. The anchor must be located beyond the plane YZ (Figure 6.28(a)) to ensure that the passive wedge of the anchor does not encroach on the active wedge behind the wall.

The equation of equilibrium is

$$\frac{1}{2}(K_p - K_a)\gamma d_a^2 l - K_a q d_a l \geq Ts \qquad (6.29)$$

where

T = tie force per unit length of wall,

s = spacing of ties,

l = length of anchor per tie,

q = surface surcharge pressure.

Ties can also be anchored to the tops of inclined piles. Tensioned cables, attached to the wall and anchored in a mass of cement grout or grouted soil (Figure 6.28(b)), are another means of support. These are known as ground anchors and are described in Section 8.8.

Example 6.7

The sides of an excavation 2.25 m deep in sand are to be supported by a cantilever sheet pile wall, the water table being 1.25 m below the bottom of the excavation. The unit weight of the sand above the water table is $17 \, kN/m^3$ and below the water table the saturated unit weight is $20 \, kN/m^3$. Characteristic parameters are $c' = 0$, $\phi' = 35°$ and $\delta = 0$. Allowing for a surcharge pressure of $10 \, kN/m^2$ on the surface, (a) determine the required depth of embedment of the piling to ensure a factor of safety of 2.0 with respect to gross passive resistance and (b) check that the rotational and translational limit states would be satisfied for the above embedment depth, using appropriate partial factors.

(a) For $\phi' = 35°$ and $\delta = 0$, $K_a = 0.27$ and $K_p = 3.7$.

Below the water table the effective unit weight of the soil is $(20 - 9.8) = 10.2 \, kN/m^3$.

To allow for possible over-excavation the soil level should be reduced by 10% of the retained height of 2.25 m, i.e. by 0.225 m. The depth of the excavation, therefore, becomes 2.475 m, say 2.50 m, and the water table will be 1.00 m below this level.

The design dimensions and the earth pressure diagrams are shown in Figure 6.29. The distributions of hydrostatic pressure on the two sides of the wall balance and can be eliminated from the calculations. The procedure is to equate moments about C, the

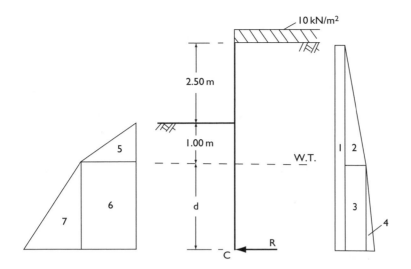

Figure 6.29 Example 6.7.

Table 6.5(a)

(per m) Force (kN)	Arm (m)	Moment (kN m)
(1) $0.27 \times 10 \times (d + 3.5) = 2.70d + 9.45$	$\frac{1}{2}d + \frac{3.5}{2}$	$1.35d^2 + 9.44d + 16.54$
(2) $\frac{1}{2} \times 0.27 \times 17 \times 3.5^2 = 28.11$	$d + \frac{3.5}{3}$	$28.11d + 32.79$
(3) $0.27 \times 17 \times 3.5 \times d = 16.06d$	$\frac{1}{2}d$	$8.03d^2$
(4) $\frac{1}{2} \times 0.27 \times 10.2 \times d^2 = 1.38d^2$	$\frac{1}{3}d$	$0.46d^3$
(5) $-\frac{1}{2} \times 3.7 \times 17 \times 1^2 \times \frac{1}{2} = -15.72$	$d + \frac{1}{3}$	$-15.72d - 5.24$
(6) $-3.7 \times 17 \times 1 \times d \times \frac{1}{2} = -31.45d$	$\frac{1}{2}d$	$-15.72d^2$
(7) $-\frac{1}{2} \times 3.7 \times 10.2 \times d^2 \times \frac{1}{2} = -9.43d^2$	$\frac{1}{3}d$	$-3.14d^3$

point of application of the force representing the net passive resistance below the point of rotation. The forces, lever arms and moments are set out in Table 6.5(a), forces (5), (6) and (7) being divided by the specified factor of safety.

Equating the algebraic sum of the moments about C to zero produces the following equation

$$-2.68d^3 - 6.34d^2 + 21.83d + 44.09 = 0$$
$$\therefore\ d^3 + 2.37d^2 - 8.15d = 16.45$$

By trial, the solution is

$$d = 2.75\,\text{m}$$

The required depth of embedment $= 1.2(2.75 + 1.00) + 0.25 = 4.75\,\text{m}$

The force R should be evaluated and compared with the net passive resistance available over the additional 20% embedment depth. Thus for $d = 2.75\,\text{m}$:

$$R = -(7.42 + 9.45 + 28.11 + 44.16 + 10.44 - 15.72 - 86.49 - 71.31)$$
$$= 74\,\text{kN}$$

Passive pressure acts on the back of the wall between depths of 6.25 and 7.00 m. At a depth of 6.625 m, the net passive pressure is given by

$$p_{\text{p}} - p_{\text{a}} = (3.7 \times 10 \times 6.625) + (3.7 \times 17 \times 3.5) - (0.27 \times 17 \times 1.00)$$
$$+ \{(3.7 - 0.27) \times 10.2 \times 3.125\}$$
$$= 245.1 + 220.1 - 4.6 + 109.3$$
$$= 569.9\,\text{kN/m}^2$$

The net passive resistance available over the additional embedded depth

$$= 569.9(7.00 - 6.25)$$
$$= 427\,\text{kN}(>R \text{ therefore satisfactory}).$$

(b) According to the limit state approach (Case C) the design value of ϕ' is $\tan^{-1}(\tan 35°/1.25)$, i.e. 29°. The corresponding values of K_{a} and K_{p} are 0.35 and 2.9, respectively. Force (1), the active thrust due to the surcharge, is multiplied by a partial factor of 1.30. For $d = 2.75\,\text{m}$, the forces and moments are set out in Table 6.5(b).

Table 6.5(b)

(per m)	Force (kN)		Arm (m)	Moment (kN m)
(1)	$0.35 \times 10 \times 6.25 \times 1.30 =$	28.4	3.125	88.9
(2)	$\frac{1}{2} \times 0.35 \times 17 \times 3.5^2 =$	36.4	3.92	142.9
(3)	$0.35 \times 17 \times 3.5 \times 2.75 =$	57.3	1.375	78.7
(4)	$\frac{1}{2} \times 0.35 \times 10.2 \times 2.75^2 =$	13.5	0.92	12.4
		135.6		322.9
(5)	$\frac{1}{2} \times 2.9 \times 17 \times 1.0^2 =$	24.6	3.08	75.9
(6)	$2.9 \times 17 \times 1.0 \times 2.75 =$	135.6	1.375	186.4
(7)	$\frac{1}{2} \times 2.9 \times 10.2 \times 2.75^2 =$	111.8	0.92	102.9
		272.0		365.2

The resisting moment (365.2 kN m) exceeds the disturbing moment (322.9 kN m), therefore the rotational limit state is satisfied. The resisting force (272.0 kN) exceeds the disturbing force (135.6 kN), therefore the translational limit state is satisfied.

Example 6.8

Details of an anchored sheet pile wall are given in Figure 6.30, the *design* ground and water levels being as shown. The ties are spaced at 2.0 m centres. Above the water table the unit weight of the soil is 17 kN/m^3 and below the water table the saturated unit weight is 20 kN/m^3. Characteristic soil parameters are $c' = 0$, $\phi' = 36°$ and δ is taken to be $\frac{1}{2}\phi'$. Determine the required depth of embedment and the force in each tie (a) for a factor of safety of 2.0 with respect to gross passive resistance and (b) in accordance with the limit state recommendations (Case C). Design a continuous anchor to support the ties using the values obtained in (b).

(a) For $\phi' = 36°$ and $\delta = \frac{1}{2}\phi'$ the coefficients for the horizontal components of lateral pressure are $K_{ah} = 0.23$ and $K_{ph} = 7.2$ (obtained from Figure 6.15). Below the water table the effective unit weight of the soil is 10.2 kN/m^3. The lateral pressure diagrams are shown in Figure 6.30. The water levels on the two sides of the wall are equal, therefore the hydrostatic pressure distributions are in balance and can be eliminated from the calculations. The procedure is to equate the disturbing and resisting moments about the anchor point A. The forces and their lever arms are set out in Table 6.6, the factor of safety (F_p) being applied to the total passive resistance (force (5)).

Equating moments about A yields the following equation:

$$-11.46d^3 - 109.0d^2 + 240.5d + 691.1 = 0$$
$$\therefore d^3 + 9.51d^2 - 20.99d = 60.31$$

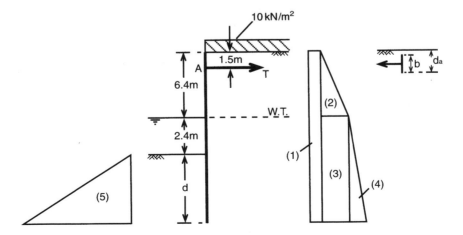

Figure 6.30 Example 6.8.

Table 6.6

(per m) Force (kN)			Arm (m)
(1)	$0.23 \times 10 \times (d + 8.8)$	$= 2.30d + 20.24$	$\frac{1}{2}d + 2.9$
(2)	$\frac{1}{2} \times 0.23 \times 17 \times 6.4^2$	$= 80.1$	2.77
(3)	$0.23 \times 17 \times 6.4 \times (d + 2.4)$	$= 25.0d + 60.1$	$\frac{1}{2}d + 6.1$
(4)	$\frac{1}{2} \times 0.23 \times 10.2 \times (d + 2.4)^2$	$= 1.17d^2 + 5.63d + 6.76$	$\frac{2}{3}d + 6.5$
(5)	$-\frac{1}{2} \times \frac{7.2}{2.0} \times 10.2 \times d^2$	$= -18.36d^2$	$\frac{2}{3}d + 7.3$
Tie		$= -T$	0

By trial, the solution is

$$d = 3.16\,\text{m}$$

The required depth of embedment is 3.16 m.

The algebraic sum of the forces in Table 6.6 must equate to zero. Thus, for $d = 3.16$ m:

$$27.5 + 80.1 + 139.1 + 36.2 - 183.3 - T = 0$$
$$\therefore\ T = 99.6\,\text{kN/m}$$

Hence the force in each tie $= 2 \times 99.6 = 200\,\text{kN}$

(b) The recommended partial factor for shear strength is 1.25, therefore the design value of $\phi' = \tan^{-1}\ (\tan 36°/1.25) = 30°$. Hence (for $\delta = \frac{1}{2}\phi'$) $K_{ah} = 0.29$ and $K_{ph} = 4.6$. Force (1) in Table 6.6 is multiplied by the partial factor 1.30, surcharge being a variable unfavourable action. The partial factor for all other forces, being permanent unfavourable actions, is 1.00. The moment equation then becomes

$$d^3 + 9.51d^2 - 21.14d = 60.97$$

By trial, the required depth of embedment (d) is 3.18 m.

For $d = 3.18$ m, the force equation becomes

$$45.2 + 101.0 + 176.0 + 46.1 - 237.2 - T = 0$$
$$\therefore\ T = 131.1\,\text{kN/m}$$

Hence the force in each tie $= 2 \times 131.1 = 262\,\text{kN}$.

For a continuous anchor, $s = l$ in Equation 6.29. The design load to be resisted by the anchor is 131.1 kN/m. Therefore the minimum value of d_a is given by

$$\frac{1}{2}(4.6-0.29)17d_a^2 - 0.29 \times 10 \times d_a = 131.1$$

$$36.64d_a^2 - 2.90d_a = 131.1$$

$$\therefore d_a = 1.932\,\text{m}$$

Then the vertical dimension (b) of the anchor $= 2(1.932 - 1.5) = 0.864\,\text{m}$.

Vertical equilibrium, neglecting the weight of the wall, can be checked as follows. Downward component of active thrust $= (45.2 + 101.0 + 176.0 + 46.1)\tan 15° = 98.7\,\text{kN/m}$. Upward component of passive resistance $= 237.2\tan 15° = 63.6\,\text{kN/m}$. The imbalance in vertical equilibrium could be improved if δ were taken to be $\tan^{-1}(0.75\tan\phi')$, on the passive side of the wall, recalculating the passive forces accordingly.

Example 6.9

A propped cantilever wall supporting the sides of an excavation in stiff clay is shown in Figure 6.31. The saturated unit weight of the clay (above and below the water table) is $20\,\text{kN/m}^3$. The active and passive coefficients for horizontal components of pressure are 0.30 and 4.2, respectively. Using the Burland–Potts–Walsh method and assuming conditions of steady seepage, determine the required depth of embedment to give a factor of safety of 2.0 with respect to net passive resistance. Determine the force in each prop and draw the shearing force and bending moment diagrams for the wall.

The distributions of earth pressure and net pore water pressure (assuming uniform decrease of total head around the wall) are shown in Figure 6.31; the diagram (7) represents the net available passive resistance. The procedure is to equate moments about A to zero. However, if the forces, lever arms and moments were expressed in terms of the unknown embedment depth d, complex algebraic expressions would result and it is preferable to assume a series of trial values of d and calculate the corresponding values of factor of safety F_r (a computer program could be written). The depth of embedment for $F_r = 2.0$ can then be obtained by interpolation.

Following this procedure, a trial value of $d = 6.0\,\text{m}$ will be selected. Then the maximum net water pressure, at level D is:

$$u_D = \frac{12.0}{16.5} \times 4.5 \times 9.8 = 32.1\,\text{kN/m}^2$$

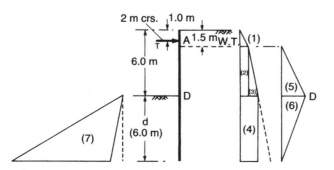

Figure 6.31 Example 6.9.

and the average seepage pressure is

$$j = \frac{4.5}{16.5} \times 9.8 = 2.7 \, \text{kN/m}^3$$

Thus, below the water table, active forces are calculated using

$$(\gamma' + j) = 10.2 + 2.7 = 12.9 \, \text{kN/m}^3$$

and passive forces are calculated using

$$(\gamma' - j) = 10.2 - 2.7 = 7.5 \, \text{kN/m}^3$$

The calculations are set out in Table 6.7.
   Hence the factor of safety is

$$F_r = \frac{4475.7}{2441.1} = 1.83$$

The calculations are repeated for three other values of d, the results being as follows:

d (m)	3.9	5.1	6.0	6.9
F_r	0.87	1.40	1.83	2.33

Referring to Figure 6.32 it is apparent that for $F_r = 2.0$ the required depth of embedment is 6.3 m.
   The prop load, shearing forces and bending moments will be calculated for the condition of limiting equilibrium, i.e. for $F_r = 1.0$: the corresponding value of d is 4.2 m. For this value of d:

Table 6.7

(per m)	Force (kN)			Arm (m)	Moment (kN m)
(1)	$\frac{1}{2} \times 0.30 \times 20 \times 1.5^2$	=	6.8	0	0
(2)	$0.30 \times 20 \times 1.5 \times 4.5$	=	40.5	2.75	111.4
(3)	$\frac{1}{2} \times 0.30 \times 12.9 \times 4.5^2$	=	39.1	3.5	136.8
(4)	$0.30\{(20 \times 1.5) + (12.9 \times 4.5)\}6.0$	=	158.2	8.0	1266.0
(5)	$\frac{1}{2} \times 32.1 \times 4.5$	=	72.2	3.5	252.8
(6)	$\frac{1}{2} \times 32.1 \times 6.0$	=	96.3	7.0	674.1
					2441.1
(7)	$-\frac{1}{2}\{(4.2 \times 7.5) - (0.30 \times 12.9)\}6.0^2$	=	-497.3	9.0	-4475.7

$$u_D = \frac{8.4}{12.9} \times 4.5 \times 9.8 = 28.7\,\text{kN/m}^2$$

$$j = \frac{4.5}{12.9} \times 9.8 = 3.4\,\text{kN/m}^3$$

$$\gamma' + j = 10.2 + 3.4 = 13.6\,\text{kN/m}^3$$

$$\gamma' - j = 10.2 - 3.4 = 6.8\,\text{kN/m}^3$$

The forces on the wall are calculated for $d = 4.2$ m, as shown in Table 6.8.

Multiplying the calculated value by 1.25 to allow for arching, the prop force for 2 m spacing is

$$1.25 \times 2 \times 112.5 = 281\,\text{kN}$$

The shearing forces and bending moments, calculated for $d = 4.2$ m, are given in Table 6.9 and are plotted to scale in Figure 6.32. For the required embedment depth of 6.3 m it is recommended that bending moments given by the dotted line should be used in design.

Table 6.8

(per m) Force (kN)		
(1)	$\frac{1}{2} \times 0.30 \times 20 \times 1.5^2$	$= \quad 6.8$
(2)	$0.30 \times 20 \times 1.5 \times 4.5$	$= \quad 40.5$
(3)	$\frac{1}{2} \times 0.30 \times 13.6 \times 4.5^2$	$= \quad 41.3$
(4)	$0.30\{(20 \times 1.5) + (13.6 \times 4.5)\}4.2$	$= \quad 114.9$
(5)	$\frac{1}{2} \times 28.7 \times 4.5$	$= \quad 64.6$
(6)	$\frac{1}{2} \times 28.7 \times 4.2$	$= \quad 60.3$
		328.4
(7)	$-\frac{1}{2}\{(4.2 \times 6.8) - (0.30 \times 13.6)\}4.2^2 =$	-215.9
$\therefore$ Prop force (T)		$= \quad 112.5$

Table 6.9

Depth (m)	Shearing force (kN)	Bending moment (kN m)
0	0	0
1	$-3/+109.5$	$+1$
2	$+93.6$	-103.3
4	$+34.6$	-248.5
6	-40.7	-278.5
8	-90.2	-126.8
10	-14.3	-1.5

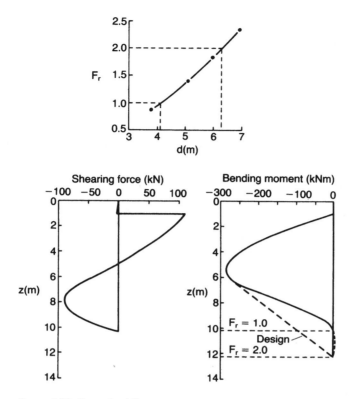

Figure 6.32 Example 6.9.

6.8 BRACED EXCAVATIONS

Sheet piling or timbering is normally used to support the sides of deep, narrow excavations, stability being maintained by means of struts acting across the excavation, as shown in Figure 6.33(a). The piling is usually driven first, the struts being installed in stages as excavation proceeds. When the first row of struts is installed the depth of excavation is small and no significant yielding of the soil mass will have taken place. As the depth of excavation increases, significant yielding of the soil occurs before strut installation but the first row of struts prevents yielding near the surface. Deformation of the wall, therefore, will be of the form shown in Figure 6.33(a), being negligible at the top and increasing with depth. Thus the deformation condition of the Rankine theory is not satisfied and the theory cannot be used for this type of wall. Failure of the soil will take place along a surface of the form shown in Figure 6.33(a), only the lower part of the soil wedge within this surface reaching a state of plastic equilibrium, the upper part remaining in a state of elastic equilibrium.

Failure of a braced wall is normally due to the initial failure of one of the struts, resulting in the progressive failure of the whole system. The forces in the individual struts may differ widely because they depend on such random factors as the force with which

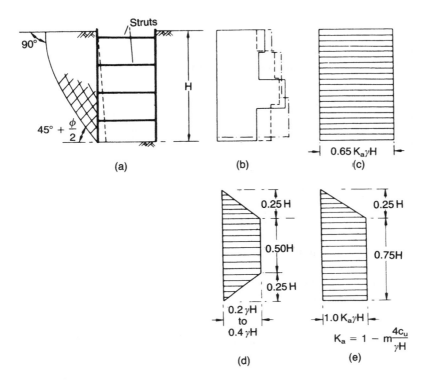

Figure 6.33 Braced excavations: Peck's envelopes.

the struts are wedged home and the time between excavation and installation of struts. The usual design procedure for braced walls is semi-empirical, being based on actual measurements of strut loads in excavations in sands and clays in a number of locations. For example, Figure 6.33(b) shows the apparent distributions of earth pressure derived from load measurements in the struts at three sections of a braced excavation in a dense sand. Since it is essential that no individual strut should fail, the pressure distribution assumed in design is taken as the envelope covering all the random distributions obtained from field measurements. Such an envelope should not be thought of as representing the actual distribution of earth pressure with depth but as a hypothetical pressure diagram from which the strut loads can be obtained with some degree of confidence. The pressure envelope proposed by Terzaghi and Peck for medium to dense sands is shown in Figure 6.33(c), being a uniform distribution of 0.65 times the Rankine active value.

According to Peck [17, 26], the behaviour of a braced excavation in clay depends on the value of the stability number $\gamma H / c_u$, where c_u is the average undrained shear strength of the clay adjacent to the excavation. It should be noted that the value of the stability number increases as the depth of excavation increases in a particular clay. If the stability number is less than 4, most of the clay adjacent to the excavation should be in a state of elastic equilibrium and for this condition Peck proposed that the envelope shown in Figure 6.33(d) should be used to estimate the strut loads. The lower limit of $0.2\gamma H$ should be used only if indicated by observations in similar conditions;

otherwise a lower limit of $0.3\gamma H$ is appropriate. If the stability number is greater than 4, plastic zones can be expected to develop near the bottom of the excavation and the envelope shown in Figure 6.33(e) should be used provided the abscissae are greater than those in Figure 6.33(d). If this is not the case, Figure 6.33(d) should be used regardless of the value of the stability number. In general, the value of m in Figure 6.33(e) should be taken as 1.0; however, in the case of soft, normally consolidated clays a value of m as low as 0.4 may be appropriate.

In the case of excavations in clay for which the stability number is greater than 7 there is a possibility that the base of the excavation will fail by heaving (see Section 8.2) and this should be analysed before the strut loads are considered. Due to base heave and the inward deformation of the clay there will be horizontal and vertical movement of the soil outside the excavation. Such movements may result in damage to adjacent structures and services and should be monitored during excavation; advance warning of excessive movement or possible instability can thus be obtained. In general, the greater the flexibility of the wall system and the longer the time before struts or anchors are installed, the greater will be the movements outside the excavation. It should be appreciated that all factors influencing wall behaviour cannot be represented by a stability number.

It should be noted that strut loads determined from the envelopes shown in Figure 6.33 (which are based on measured values) should be multiplied by an appropriate load factor to obtain the design load.

Much more data on strut loads in braced excavations is now available and, based on 81 case studies in a range of soils in the UK, Twine and Roscoe [27] have updated Peck's proposals, the revised envelopes being shown in Figure 6.35. In clays the envelopes take into account the increase in strut load which accompanies dissipation of the negative excess pore water pressure induced during excavation. The envelopes are based on characteristic strut loads, the appropriate partial factor then being applied to give the design load (i.e. the limit state approach, Case B). The envelopes also allow for a nominal surface surcharge of $10\,\mathrm{kN/m^2}$.

For soft and firm clays an envelope of the form shown in Figure 6.34(a) is proposed for flexible walls (i.e. sheet pile walls and timber sheeting) and tentatively for stiff walls (i.e. diaphragm and contiguous pile walls). The upper and lower pressure values are represented by $a\gamma H$ and $b\gamma H$, respectively, where γ is the total unit weight of the soil

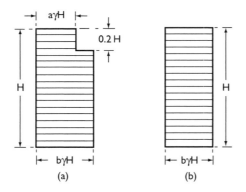

Figure 6.34 Braced excavations: Twine and Roscoe envelopes.

and H the depth of the excavation, including an allowance for over-excavation. The pressure values for soft clay also depend on whether the base of the excavation is stable, i.e. there is an adequate safety margin against failure by base heave. Base stability is enhanced if the wall extends below excavation level, reducing inward yielding of the wall but resulting in higher strut loads. For soft clay with a stable base, $a = 0.65$ and $b = 0.50$. For soft clay with enhanced base stability, $a = 1.15$ and $b = 0.65$. For firm clay the values of a and b are 0.3 and 0.2, respectively.

The envelopes for stiff and very stiff clay and for coarse soils are rectangular (Figure 6.34(b)). For stiff and very stiff clays the value of b for flexible walls is 0.3 and for stiff walls, $b = 0.5$. For coarse soils, $b = 0.2$ but below the water table the pressure is $b\gamma'H$ (γ' being the buoyant unit weight) with hydrostatic pressure acting in addition.

6.9 DIAPHRAGM WALLS

A diaphragm wall is a relatively thin reinforced concrete membrane cast in a trench, the sides of which are supported prior to casting by the hydrostatic pressure of a slurry of bentonite (a montmorillonite clay) in water. When mixed with water, bentonite readily disperses to form a colloidal suspension which exhibits thixotropic properties, i.e. it gels when left undisturbed but becomes fluid when agitated. The trench, the width of which is equal to that of the wall, is excavated progressively in suitable lengths from the ground surface, generally using a power-closing clamshell grab: shallow concrete guide walls are normally constructed as an aid to excavation. The trench is filled with the bentonite slurry as excavation proceeds: excavation thus takes place through the slurry already in place. The excavation process turns the gel into a fluid but the gel becomes re-established when disturbance ceases. The slurry tends to become contaminated with soil and cement in the course of construction but can be cleaned and re-used.

The bentonite particles form a skin of very low permeability, known as the filter cake, on the excavated soil faces. This occurs due to the fact that water filters from the slurry into the soil, leaving a layer of bentonite particles, a few millimetres thick, on the surface of the soil. Consequently, the full hydrostatic pressure of the slurry acts against the sides of the trench, enabling stability to be maintained. The filter cake will form only if the fluid pressure in the trench is greater than the pore water pressure in the soil; a high water table level can thus be a considerable impediment to diaphragm wall construction. In soils of low permeability, such as clays, there will be virtually no filtration of water into the soil and therefore no significant filter cake formation will take place; however, total stress conditions apply and slurry pressure will act against the clay. In soils of high permeability, such as sandy gravels, there may be excessive loss of bentonite into the soil, resulting in a layer of bentonite-impregnated soil and poor filter cake formation. However, if a small quantity of fine sand (around 1%) is added to the slurry the sealing mechanism in soils of high permeability can be improved, with a considerable reduction in bentonite loss. Trench stability depends on the presence of an efficient seal on the soil surface; the higher the permeability of the soil, the efficiency of the seal becomes more vital.

A slurry having a relatively high density is desirable from the points of view of trench stability, reduction of loss into soils of high permeability and the retention of contaminating particles in suspension. On the other hand, a slurry of relatively low density will be displaced more cleanly from the soil surfaces and the reinforcement, and will be more easily pumped and decontaminated. The specification for the slurry must reflect a compromise between these conflicting requirements. Slurry specifications are usually based on density, viscosity, gel strength and pH.

On completion of excavation the reinforcement is positioned and the length of trench is filled with wet concrete using a tremie pipe. The wet concrete (which has a density of approximately twice that of the slurry) displaces the slurry upwards from the bottom of the trench, the tremie being raised in stages as the level of concrete rises. Once the wall (constructed as a series of individual panels keyed together) has been completed and the concrete has achieved adequate strength, the soil on one side of the wall can be excavated. It is usual for ground anchors (Section 8.8) to be installed at appropriate levels, as excavation proceeds, to tie the wall back into the retained soil. The method is very convenient for the construction of deep basements and under-passes, an important advantage being that the wall can be constructed close to adjoining structures, provided that the soil is moderately compact and ground deform-ations are tolerable. Diaphragm walls are often preferred to sheet pile walls because of their relative rigidity and their ability to be incorporated as part of the structure. An alternative to a diaphragm wall is the contiguous pile wall comprising a line of bored piles, each being in contact with the adjacent piles.

The decision whether to use a triangular or a trapezoidal distribution of lateral pressure in the design of a diaphragm wall depends on the anticipated wall deform-ation. A triangular distribution would probably be indicated in the case of a single row of tie-backs near the top of the wall. In the case of multiple rows of tie-backs over the height of the wall, a trapezoidal distribution might be considered appropriate.

Trench stability

It is assumed that the full hydrostatic pressure of the slurry acts against the sides of the trench, the bentonite forming a thin, virtually impermeable layer on the surface of the soil. Consider a wedge of soil above a plane inclined at angle α to the horizontal, as shown in Figure 6.35. When the wedge is on the point of sliding into the trench, i.e. the soil within the wedge is in a condition of limiting equilibrium, the angle α can be assumed to be $45° + \phi/2$. The unit weight of the slurry is γ_s and that of the soil is γ. The depth of the slurry is nH and the height of the water table above the bottom of the

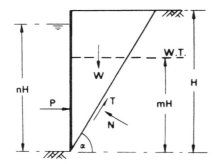

Figure 6.35 Stability of slurry trench.

trench is mH, where H is the depth of the trench. The normal and tangential components of the resultant force on the failure plane are N and T, respectively.

Considering force equilibrium,

$$P + T\cos\alpha - N\sin\alpha = 0 \tag{6.30}$$

$$W - T\sin\alpha - N\cos\alpha = 0 \tag{6.31}$$

Now

$$P = \frac{1}{2}\gamma_s(nH)^2$$

and

$$W = \frac{1}{2}\gamma H^2 \cot\alpha$$

In the case of a sand ($c' = 0$), an effective stress analysis is relevant. Hence

$$T = (N - U)\tan\phi'$$

where U, the boundary water force on the failure plane, is given by

$$U = \frac{1}{2}\gamma_w(mH)^2\operatorname{cosec}\alpha$$

and

$$\alpha = 45° + \frac{1}{2}\phi'$$

In the case of a saturated clay ($\phi_u = 0$), a total stress analysis is relevant. Hence

$$T = c_u H \operatorname{cosec}\alpha$$

where $\alpha = 45°$.

Using Equations 6.30 and 6.31, the minimum slurry density can be determined for a factor of safety of unity against shear failure (a factor of unity being assumed in the above expressions). Alternatively, for a slurry of given density, mobilized shear strength parameters $\tan^{-1}(\tan\phi'/F)$ or c_u/F must be used in the equations and the value of F obtained by iteration. The slurry density required for stability is very sensitive to the level of the water table.

6.10 REINFORCED SOIL

Reinforced soil consists of a compacted soil mass within which tensile reinforcing elements, typically in the form of horizontal steel strips, are embedded. (Patents for the technique were taken out by Henri Vidal and the Reinforced Earth Company, the term *reinforced earth* being their trademark.) Other forms of reinforcement include strips, rods, grids and meshes of metallic or polymeric materials and sheets of geotextiles. The mass is stabilized as a result of interaction between the soil and the elements, the lateral stresses within the soil being transferred to the elements which are thereby placed in tension. The soil used as fill material should be predominantly coarse-grained and must be adequately drained to prevent it from becoming saturated. In coarse fills the interaction is due to frictional forces which depend on the characteristics of the soil together with the type and surface texture of the elements. In the cases of grid and mesh reinforcement, interaction is enhanced by inter-locking between the soil and the apertures in the material.

In a reinforced-soil retaining structure (also referred to as a composite wall), a facing is attached to the outside ends of the elements to prevent the soil spilling out at the edge of the mass and to satisfy aesthetic requirements: the facing does *not* act as a retaining wall. The facing should be sufficiently flexible to withstand any deformation of the fill. The types of facing normally used are discrete precast concrete panels, full-height panels and pliant U-shaped sections aligned horizontally. The basic features of a reinforced-soil retaining wall are shown in Figure 6.36. Such structures possess considerable inherent flexibility and, consequently, can withstand relatively large differential settlement. The reinforced soil principle can also be employed in embankments, normally by means of geotextiles, and in slope stabilization by the insertion of steel rods, a technique known as soil nailing. There are many points of detail to be considered in reinforced soil construction and reference should be made to BS 8006: 1995 [6], the UK code of practice for strengthened/reinforced soils and other fills.

Both external and internal stability must be considered in design. The external stability of a reinforced soil structure should be analysed in the same way as a gravity wall (Section 6.6). The back of the wall should be taken as the vertical plane (FG) through the inner end of the lowest element, the total active thrust (P_a) due to the backfill behind this plane then being calculated as for a gravity wall. The ultimate limit states for external stability are: (1) bearing resistance failure of the underlying soil, resulting in tilting of the structure, (2) sliding between the reinforced fill and the underlying soil and (3) development of a deep slip surface which envelops the structure as a whole. Serviceability limit states are excessive values of settlement and wall deformation. For design according to the recommendations of BS 8006, partial factors are applied to representative values of dead loads (including the unit weight of the soil), imposed loads, the shear strength of the soil and the tensile strength of the reinforcing material. An additional factor is introduced to take account of the economic implications of failure. Appropriate values of the factors are given in the code.

In considering the internal stability of the structure the principal limit states are tensile failure of the elements and slipping between the elements and the soil. Tensile failure of one of the elements could lead to the progressive collapse of the whole structure (an ultimate limit state). Local slipping due to inadequate frictional resistance would result in a redistribution of tensile stress and the gradual deformation of the structure, not necessarily leading to collapse (a serviceability limit state).

Consider a reinforcing element at depth z below the surface of a soil mass. The tensile force in the element due to the transfer of lateral stress from the soil is given by

$$T_p = K\sigma_z S_x S_z \qquad (6.32)$$

where K is the appropriate earth pressure coefficient at depth z, σ_z the vertical stress, S_x the horizontal spacing of the elements and S_z the vertical spacing. If the reinforcement consists of a continuous layer, such as a grid, then the value of S_x is unity and T_p is the tensile force per unit length of wall. The vertical stress σ_z is due to the overburden pressure at depth z plus the stresses due to any surcharge loading and external bending moment (including that due to the total active thrust on the part of plane FG between the surface and depth z). The average vertical stress can be expressed as

$$\sigma_z = \frac{V}{L - 2e}$$

where V is the vertical component of the resultant force at depth z, e the eccentricity of the force and L the length of the reinforcing element at that depth. Given the design tensile strength of the material, the required cross-sectional area or thickness of the element can be obtained from the value of T_p. The addition of surface loading will cause an increase in vertical stress which can be calculated from elastic theory.

There are two procedures for the design of retaining structures. One approach is the *tie-back wedge method* which is applicable to structures with reinforcement of relatively high extensibility such as grids, meshes and geotextile sheets. The active state is assumed to be reached throughout the soil mass because of the relatively large strains possible at the interface between the soil and the reinforcement, therefore the earth pressure coefficient in Equation 6.32 is taken as K_a at all depths and the failure surface at collapse will be a plane AB inclined at $45° + \phi/2$ to the horizontal, as shown in Figure 6.36(a), dividing the reinforced mass into an active zone within which shear stresses on elements act outwards towards the face of the structure and a resistant zone within which shear stresses act inwards. The frictional resistance available on the top and bottom surfaces of an element is then given by

$$T_r = 2bL_e\sigma_z \tan\delta \tag{6.33}$$

where b is the width of the element, L_e the length of the element in the resistant zone and δ the angle of friction between soil and element. Slipping between elements and soil (referred to as bond failure) will not occur if T_r is greater than or equal to T_p. The value of δ can be determined by means of direct shear tests or full-scale pullout tests.

The stability of the wedge ABC is checked in addition to the external and internal stability of the structure as a whole. The forces acting on the wedge, as shown in Figure 6.36(a), are the weight of the wedge (W), the reaction on the failure plane (R) acting at angle ϕ to the normal (being the resultant of the normal and shear forces) and the total tensile force resisted by all the reinforcing elements (T_w). The value of T_w can thus be determined. In effect the force T_w replaces the reaction P of a retaining wall (as, for example, in Figure 6.13(a)). Any external forces must be included in the analysis, in which case the inclination of the failure plane will not be equal to $45° + \phi/2$ and a series of trial wedges must be analysed to obtain the maximum value of T_w. The design requirement is that the factored sum of the tensile forces in all the elements, calculated from Equation 6.33, must be greater than or equal to T_w.

The second design procedure is the *coherent gravity method*, due to Juran and Schlosser [12], and is applicable to structures with elements of relatively low

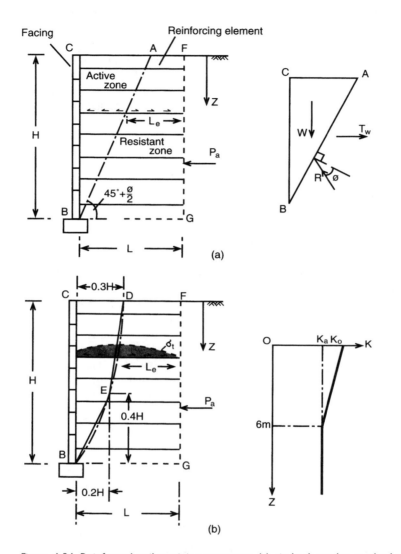

Figure 6.36 Reinforced soil-retaining structure: (a) tie-back wedge method and (b) coherent gravity method.

extensibility such as steel strips. Experimental work indicated that the distribution of tensile stress (σ_t) along such an element was of the general form shown in Figure 6.36(b), the maximum value occurring not at the face of the structure but at a point within the reinforced soil, the position of this point varying with depth as indicated by the curve DB in Figure 6.36(b). This curve again divides the reinforced mass into an active zone and a resistant zone, the method being based on the stability analysis of the active zone. The assumed mode of failure is that the reinforcing elements fracture progressively at the points of maximum tensile stress and, consequently, that conditions of plastic equilibrium develop in a thin layer of soil along the path of fracture. The curve of maximum tensile stress therefore defines the potential failure surface. If it is assumed that the soil becomes perfectly plastic, the failure surface will be a logarithmic spiral. The spiral is assumed to pass through the bottom of the facing and to intersect the surface of the fill at right angles, at a point approximately $0.3\,H$ behind the facing, as shown in Figure 6.36(b). A simplified analysis can be made by assuming that the curve of maximum tensile stress is represented by the bilinear approximation DEB shown in Figure 6.36(b), where $CD = 0.3\,H$. Equations 6.32 and 6.33 are then applied. The earth pressure coefficient in Equation 6.32 is assumed to be equal to K_0 (the at-rest coefficient) at the top of the wall reducing linearly to K_a at a depth of 6 m. The addition of surface loading would result in the modification of the line of maximum tensile stress and for this situation an amended bilinear approximaion is proposed in BS 8006.

PROBLEMS

6.1   The backfill behind a retaining wall, located above the water table, consists of a sand of unit weight $17\,\text{kN/m}^3$. The height of the wall is 6 m and the surface of the backfill is horizontal. Determine the total active thrust on the wall according to the Rankine theory if $c' = 0$ and $\phi' = 37°$. If the wall is prevented from yielding, what is the approximate value of the thrust on the wall?

6.2   Plot the distribution of active pressure on the wall surface shown in Figure 6.37. Calculate the total thrust on the wall (active + hydrostatic) and determine its point of application. Assume $\delta = 0$ and $c_w = 0$.

6.3   A line of sheet piling is driven 4 m into a firm clay and retains, on one side, a 3 m depth of fill on top of the clay. Water table level is at the surface of the clay. The unit weight of the fill is $18\,\text{kN/m}^3$ and the saturated unit weight of the clay is $20\,\text{kN/m}^3$. Calculate the active and passive pressures at the lower end of the sheet piling (a) if $c_u = 50\,\text{kN/m}^2$, $c_w = 25\,\text{kN/m}^2$ and $\phi_u = \delta = 0$ and (b) if $c' = 10\,\text{kN/m}^2$, $c_w = 5\,\text{kN/m}^2$, $\phi' = 26°$ and $\delta = 13°$, for the clay.

6.4   Details of a reinforced concrete cantilever retaining wall are shown in Figure 6.38, the unit weight of concrete being $23.5\,\text{kN/m}^3$. Due to inadequate drainage the water table has risen to the level indicated. Above the water table the unit weight of the retained soil is $17\,\text{kN/m}^3$ and below the water table the saturated unit weight is $20\,\text{kN/m}^3$. Characteristic values of the shear strength parameters are $c' = 0$ and $\phi' = 38°$. The angle of friction between the base of the wall and the foundation soil is $25°$. (a) Using the traditional approach, determine the maximum and minimum pressures under the base and the factor of safety against sliding. (b) Using the limit state approach, check whether or not the overturning and sliding limit states have been satisfied.

6.5   The sides of an excavation 3.0 m deep in sand are to be supported by a cantilever sheet pile wall. The water table is 1.5 m below the bottom of the excavation. The sand has a saturated unit weight of $20\,\text{kN/m}^3$, a unit weight above the water table of $17\,\text{kN/m}^3$ and the characteristic value of ϕ' is $36°$. Using the traditional method, determine the required depth of embedment of the piling below the bottom of the excavation to give a factor of safety of 2.0 with respect to passive resistance.

6.6   The section through a gravity retaining wall is shown in Figure 6.39, the unit weight of the wall material being $23.5\,\text{kN/m}^3$. The unit weight of the backfill is

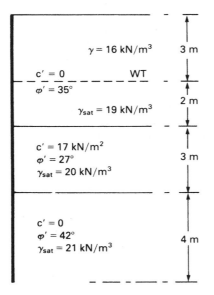

Figure 6.37

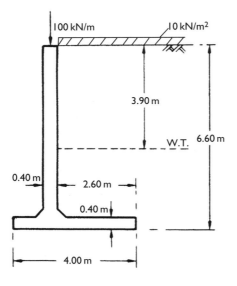

Figure 6.38

$19\,\text{kN/m}^3$ and design values of the shear strength parameters are $c' = 0$ and $\phi' = 36°$. The value of δ between wall and backfill and between base and foundation soil is $25°$. The ultimate bearing capacity of the foundation soil is $250\,\text{kN/m}^2$. Use the limit state method to determine if the design of the wall is satisfactory with respect to the overturning, bearing resistance and sliding limit states.

6.7   An anchored sheet pile wall is constructed by driving a line of piling into a soil for which the saturated unit weight is $21\,\text{kN/m}^3$ and the characteristic shear strength parameters are $c' = 10\,\text{kN/m}^2$ and $\phi' = 27°$. Backfill is placed to a depth of 8.00 m behind the piling, the backfill having a saturated unit weight of $20\,\text{kN/m}^3$, a unit weight above the water table of $17\,\text{kN/m}^3$ and characteristic shear strength parameters of $c' = 0$ and $\phi' = 35°$. Tie rods, spaced at 2.5 m centres, are located 1.5 m below the surface of the backfill. The water level in front of the wall and the water table behind the wall are both 5.00 m below the surface of the backfill. Determine the required depth of embedment for a factor of safety of 2.0 with respect to gross passive resistance and the force in each tie rod.

6.8   The soil on both sides of the anchored sheet pile wall detailed in Figure 6.40 has a saturated unit weight of $21\,\text{kN/m}^3$ and a unit weight above the water table of $18\,\text{kN/m}^3$. Characteristic parameters for the soil are $c' = 0$, $\phi' = 36°$ and $\delta = 0°$. There is a lag of 1.5 m between the water table behind the wall and tidal level in front. Determine (a) the factor of safety with respect to gross passive resistance and the force in each tie rod using the traditional method and (b) the required depth of embedment using the limit state method.

6.9   Details of a propped cantilever wall are shown in Figure 6.41. The saturated unit weight of the soil is $20\,\text{kN/m}^3$ and the unit weight above the water table is $17\,\text{kN/m}^3$. Characteristic parameters for the soil are $c' = 0$, $\phi' = 30°$ and $\delta = 15°$. Using the Burland, Potts and Walsh method, determine the factor of safety with respect to net passive resistance.

6.10   The struts in a braced excavation 9 m deep in a dense sand are placed at 1.5 m centres vertically and 3.0 m centres horizontally, the bottom of the excavation being above the water table. The unit weight of the sand is $19\,\text{kN/m}^3$. Based on shear strength parameters $c' = 0$ and $\phi' = 40°$, what load should each strut be designed to carry?

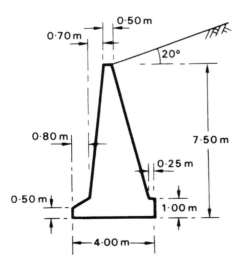

Figure 6.39

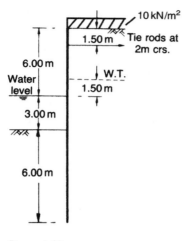

Figure 6.40

6.11 A diaphragm wall is to be constructed in a soil having a unit weight of $18\,kN/m^3$ and design shear strength parameters $c' = 0$ and $\phi' = 34°$. The depth of the trench is 3.50 m and the water table is 1.85 m above the bottom of the trench. Determine the factor of safety with respect to shear strength if the unit weight of the slurry is $10.6\,kN/m^3$ and the depth of slurry in the trench is 3.35 m.

6.12 A reinforced soil wall is 5.2 m high. The reinforcing elements, which are spaced at 0.65 m vertically and 1.20 m horizontally, measure 65×3 mm in section and are 5.0 m in length. The ultimate tensile strength of the reinforcing material is $340\,N/mm^2$. Design values to be used are as follows: unit weight of the selected fill $= 18\,kN/m^3$; angle of shearing resistance of selected fill $= 36°$; angle of friction between fill and elements $= 30°$. Using (a) the tie-back wedge method and (b) the coherent gravity method, check that an element 3.6 m below the top of the wall will not suffer tensile failure and that slipping between the element and the fill will not occur. The value of K_a for the material retained by the reinforced fill is 0.30, the unit weight of this material also being $18\,kN/m^3$.

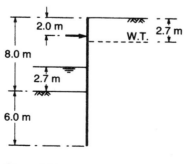

Figure 6.41

REFERENCES

1　Atkinson, J.H. (1981) *Foundations and Slopes: An Introduction to Applications of Critical State Soil Mechanics*, McGraw-Hill, Berkshire.

2　Barden, L. (1974) *Sheet Pile Wall Design Based on Rowe's Method*, CIRIA Technical Note 54, London.

3　Bjerrum, L. and Andersen, K. (1972) *In-situ* measurement of lateral pressures in clay, in *Proc. 5th European Conference SMFE, Madrid*, Vol. 1, Spanish Society SMFE, Madrid, pp. 11–20.

4　Bolton, M.D., Powrie, W. and Symons, I.F. (1989) The design of stiff *in-situ* walls retaining overconsolidated clay, *Ground Engineering*, **22** (8), 44–8; **22** (9), 34–40, **23** (2), 22–8.

5　British Standard 8002 (1994) *Code of Practice for Earth Retaining Structures*, BSI, London.

6　British Standard 8006 (1995) *Code of Practice for Strengthened Reinforced Soils and Other Fills*, BSI, London.

7　Burland, J.B., Potts, D.M. and Walsh, N.M. (1981) The overall stability of free and propped embedded cantilever retaining walls, *Ground Engineering*, **14** (5), 28–38.

8　Caquot, A. and Kerisel, J. (1966) *Traite de Mecanique des Sols, 4ᵉ edn*, Gauthier-Villars, Paris.

9　European Committee for Standardization (1994) *Eurocode 7: Geotechnical Design – Part 1: General Rules (ENV 1997–1)*, Brussels.

10　Ingold, T.S. (1979) The effects of compaction on retaining walls, *Geotechnique*, **29**, 265–83.

11　Ingold, T.S. (1982) *Reinforced Earth*, Thomas Telford, London.

12　Juran, I. and Schlosser, F. (1978) Theoretical analysis of failure in reinforced earth structures, in *Proc. Symposium on Earth Reinforcement*, ASCE Convention, Pittsburgh, ASCE, New York, pp. 528–55.

13　Kerisel, J. and Absi, E. (1990) *Active and Passive Earth Pressure Tables*, 3rd edn, A.A. Balkema, Rotterdam.

14　Mayne, P.W. and Kulhawy, F.H. (1982) K_0–OCR relationships in soil, *Journal ASCE*, **108** (GT6), 851–72.

15　Padfield, C.J. and Mair, R.J. (1984) Design of retaining walls embedded in stiff clay, *CIRIA Report* 104.

16　Parry, R.H.G. (1995) *Mohr Circles, Stress Paths and Geotechnics*, E. & F.N. Spon, London.

17　Peck, R.B. (1969) Deep excavations and tunnelling in soft ground, in *Proc. 7th International Conference SMFE, Mexico* (State of the Art Volume), Mexican Society SMFE, pp. 225–81.

18　Potts, D.M. and Fourie, A.B. (1984) The behaviour of a propped retaining wall: results of a numerical experiment, *Geotechnique*, **34**, 383–404.

19　Potts, D.M. and Fourie, A.B. (1985) The effect of wall stiffness on the behaviour of a propped retaining wall, *Geotechnique*, **35**, 347–52.

20　Puller, M. and Lee, C.K.T. (1996) A comparison between the design methods for earth retaining structures recommended by BS 8002: 1994 and previously used methods, *Proc. Institution of Civil Engineers, Geotechnical Engineering*, **119**, 29–34.

21　Rowe, P.W. (1952) Anchored Sheet Pile Walls, *Proc. Institution of Civil Engineers*, Part 1, **1**, 27–70.

22　Rowe, P.W. (1957) Sheet pile walls in clay, *Proc. Institution of Civil Engineers*, Part 1, **7**, 629–54.

23　Rowe, P.W. and Peaker, K. (1965) Passive earth pressure measurements, *Geotechnique*, **15**, 57–78.

24　Sokolovski, V.V. (1965) *Statics of Granular Media*, Pergamon Press, Oxford.

25　Terzaghi, K. (1943) *Theoretical Soil Mechanics*, John Wiley & Sons, New York.

26 Terzaghi, K., Peck, R.B. and Mesri, G. (1996) *Soil Mechanics in Engineering Practice*, 3rd edn, John Wiley & Sons, New York.

27 Twine, D. and Roscoe, H. (1999) Temporary propping of deep excavations: guidance on design, *CIRIA Report C517*, London.

28 Wroth, C.P. and Hughes, J.M.O. (1973) An instrument for the *in-situ* measurement of the properties of soft clays, in *Proc. 8th International Conference SMFE, Moscow*, Vol. 1 (2), 487–94.

Chapter 7

Consolidation theory

7.1 INTRODUCTION

As explained in Chapter 3, consolidation is the gradual reduction in volume of a fully saturated soil of low permeability due to drainage of some of the pore water, the process continuing until the excess pore water pressure set up by an increase in total stress has completely dissipated; the simplest case is that of one-dimensional consolidation, in which a condition of zero lateral strain is implicit. The process of swelling, the reverse of consolidation, is the gradual increase in volume of a soil under negative excess pore water pressure.

Consolidation settlement is the vertical displacement of the surface corresponding to the volume change at any stage of the consolidation process. Consolidation settlement will result, for example, if a structure is built over a layer of saturated clay or if the water table is lowered permanently in a stratum overlying a clay layer. On the other hand, if an excavation is made in a saturated clay, heaving (the reverse of settlement) will result in the bottom of the excavation due to swelling of the clay. In cases in which significant lateral strain takes place, there will be an immediate settlement due to deformation of the soil under undrained conditions, in addition to consolidation settlement. Immediate settlement can be estimated using the results from elastic theory given in Chapter 5. This chapter is concerned with the prediction of both the magnitude and the rate of consolidation settlement (required to ensure that serviceability limit states are satisfied).

The progress of consolidation *in situ* can be monitored by installing piezometers to record the change in pore water pressure with time. The magnitude of settlement can be measured by recording the levels of suitable reference points on a structure or in the ground: precise levelling is essential, working from a benchmark which is not subject to even the slightest settlement. Every opportunity should be taken of obtaining settlement data, as it is only through such measurements that the adequacy of theoretical methods can be assessed.

7.2 THE OEDOMETER TEST

The characteristics of a soil during one-dimensional consolidation or swelling can be determined by means of the oedometer test. Figure 7.1 shows diagrammatically a cross-section through an oedometer. The test specimen is in the form of a disc,

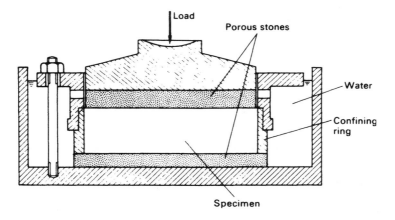

Figure 7.1 The oedometer.

held inside a metal ring and lying between two porous stones. The upper porous stone, which can move inside the ring with a small clearance, is fixed below a metal loading cap through which pressure can be applied to the specimen. The whole assembly sits in an open cell of water to which the pore water in the specimen has free access. The ring confining the specimen may be either fixed (clamped to the body of the cell) or floating (being free to move vertically): the inside of the ring should have a smooth polished surface to reduce side friction. The confining ring imposes a condition of zero lateral strain on the specimen, the ratio of lateral to vertical effective stress being K_0, the coefficient of earth pressure at-rest. The compression of the specimen under pressure is measured by means of a dial gauge operating on the loading cap.

The test procedure has been standardized in BS 1377 (Part 5) [4] which specifies that the oedometer shall be of the fixed ring type. The initial pressure will depend on the type of soil, then a sequence of pressures is applied to the specimen, each being double the previous value. Each pressure is normally maintained for a period of 24 h (in exceptional cases a period of 48 h may be required), compression readings being observed at suitable intervals during this period. At the end of the increment period, when the excess pore water pressure has completely dissipated, the applied pressure equals the effective vertical stress in the specimen. The results are presented by plotting the thickness (or percentage change in thickness) of the specimen or the void ratio at the end of each increment period against the corresponding effective stress. The effective stress may be plotted to either a natural or a logarithmic scale. If desired, the expansion of the specimen can be measured under successive decreases in applied pressure. However, even if the swelling characteristics of the soil are not required, the expansion of the specimen due to the removal of the final pressure should be measured.

The void ratio at the end of each increment period can be calculated from the dial gauge readings and either the water content or the dry weight of the specimen at the end of the test. Referring to the phase diagram in Figure 7.2, the two methods of calculation are as follows.

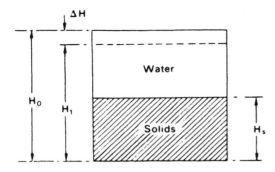

Figure 7.2 Phase diagram.

1   Water content measured at end of test $= w_1$
Void ratio at end of test $= e_1 = w_1 G_s$ (assuming $S_r = 100\%$)
Thickness of specimen at start of test $= H_0$
Change in thickness during test $= \Delta H$
Void ratio at start of test $= e_0 = e_1 + \Delta e$
where

$$\frac{\Delta e}{\Delta H} = \frac{1 + e_0}{H_0} \tag{7.1}$$

In the same way Δe can be calculated up to the end of any increment period.

2   Dry weight measured at end of test $= M_s$ (i.e. mass of solids)
Thickness at end of any increment period $= H_1$
Area of specimen $= A$
Equivalent thickness of solids $= H_s = M_s/AG_s\rho_w$
Void ratio,

$$e_1 = \frac{H_1 - H_s}{H_s} = \frac{H_1}{H_s} - 1 \tag{7.2}$$

Compressibility characteristics

Typical plots of void ratio (e) after consolidation against effective stress (σ') for a saturated clay are shown in Figure 7.3, the plots showing an initial compression followed by expansion and recompression (cf. Figure 4.9 for isotropic consolidation). The shapes of the curves are related to the stress history of the clay. The e–$\log \sigma'$ relationship for a normally consolidated clay is linear (or nearly so) and is called the virgin compression line. If a clay is overconsolidated, its state will be represented by a point on the expansion or recompression parts of the e–$\log \sigma'$ plot. The recompression curve ultimately joins the virgin compression line: further compression then occurs along the virgin line. During compression, changes in soil structure continuously

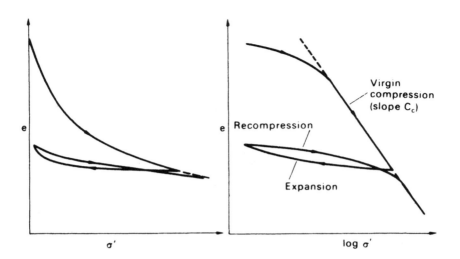

Figure 7.3 Void ratio–effective stress relationship.

take place and the clay does not revert to the original structure during expansion. The plots show that a clay in the overconsolidated state will be much less compressible than that in a normally consolidated state.

The compressibility of the clay can be represented by one of the following coefficients.

1   *The coefficient of volume compressibility* (m_v), defined as the volume change per unit volume per unit increase in effective stress. The units of m_v are the inverse of pressure (m²/MN). The volume change may be expressed in terms of either void ratio or specimen thickness. If, for an increase in effective stress from σ'_0 to σ'_1, the void ratio decreases from e_0 to e_1, then

$$m_v = \frac{1}{1+e_0}\left(\frac{e_0 - e_1}{\sigma'_1 - \sigma'_0}\right) \tag{7.3}$$

$$m_v = \frac{1}{H_0}\left(\frac{H_0 - H_1}{\sigma'_1 - \sigma'_0}\right) \tag{7.4}$$

The value of m_v for a particular soil is not constant but depends on the stress range over which it is calculated. BS 1377 specifies the use of the coefficient m_v calculated for a stress increment of 100 kN/m² in excess of the effective overburden pressure of the *in-situ* soil at the depth of interest, although the coefficient may also be calculated, if required, for any other stress range.

2   *The compression index* (C_c) is the slope of the linear portion of the e–log σ' plot and is dimensionless. For any two points on the linear portion of the plot

$$C_c = \frac{e_0 - e_1}{\log(\sigma'_1/\sigma'_0)} \tag{7.5}$$

The expansion part of the e–$\log \sigma'$ plot can be approximated to a straight line, the slope of which is referred to as the *expansion index* $C_{\rm e}$.

Preconsolidation pressure

Casagrande proposed an empirical construction to obtain, from the e–$\log \sigma'$ curve for an overconsolidated clay, the maximum effective vertical stress that has acted on the clay in the past, referred to as the *preconsolidation pressure* ($\sigma'_{\rm c}$). Figure 7.4 shows a typical e–$\log \sigma'$ curve for a specimen of clay, initially overconsolidated. The initial curve indicates that the clay is undergoing recompression in the oedometer, having at some stage in its history undergone expansion. Expansion of the clay *in situ* may, for example, have been due to melting of ice sheets, erosion of overburden or a rise in water table level. The construction for estimating the preconsolidation pressure consists of the following steps:

1   Produce back the straight-line part (BC) of the curve.
2   Determine the point (D) of maximum curvature on the recompression part (AB) of the curve.
3   Draw the tangent to the curve at D and bisect the angle between the tangent and the horizontal through D.
4   The vertical through the point of intersection of the bisector and CB produced gives the approximate value of the preconsolidation pressure.

Whenever possible the preconsolidation pressure for an overconsolidated clay should not be exceeded in construction. Compression will not usually be great if the effective vertical stress remains below $\sigma'_{\rm c}$: only if $\sigma'_{\rm c}$ is exceeded will compression be large.

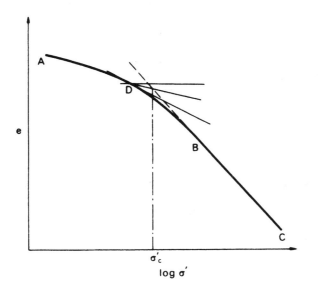

Figure 7.4 Determination of preconsolidation pressure.

In-situ e–log σ′ curve

Due to the effects of sampling and preparation, the specimen in an oedometer test will be slightly disturbed. It has been shown that an increase in the degree of specimen disturbance results in a slight decrease in the slope of the virgin compression line. It can therefore be expected that the slope of the line representing virgin compression of the *in-situ* soil will be slightly greater than the slope of the virgin line obtained in a laboratory test.

No appreciable error will be involved in taking the *in-situ* void ratio as being equal to the void ratio (e_0) at the start of the laboratory test. Schmertmann [18] pointed out that the laboratory virgin line may be expected to intersect the *in-situ* virgin line at a void ratio of approximately 0.42 times the initial void ratio. Thus the *in-situ* virgin line can be taken as the line EF in Figure 7.5 where the coordinates of E are log σ'_c and e_0, and F is the point on the laboratory virgin line at a void ratio of $0.42e_0$.

In the case of overconsolidated clays the *in-situ* condition is represented by the point (G) having coordinates σ'_0 and e_0, where σ'_0 is the present effective overburden pressure. The *in-situ* recompression curve can be approximated to the straight line GH parallel to the mean slope of the laboratory recompression curve.

Example 7.1

The following compression readings were obtained in an oedometer test on a specimen of saturated clay ($G_s = 2.73$):

Pressure (kN/m^2)	0	54	107	214	429	858	1716	3432	0
Dial gauge after 24 h (mm)	5.000	4.747	4.493	4.108	3.449	2.608	1.676	0.737	1.480

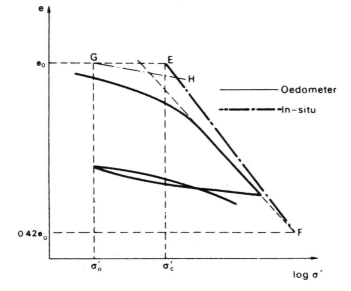

Figure 7.5 In-situ e–log σ′ curve.

The initial thickness of the specimen was 19.0 mm and at the end of the test the water content was 19.8%. Plot the e–$\log \sigma'$ curve and determine the preconsolidation pressure. Determine the values of m_v for the stress increments 100–200 and 1000–1500 kN/m². What is the value of C_c for the latter increment?

Void ratio at end of test $= e_1 = w_1 G_s = 0.198 \times 2.73 = 0.541$

Void ratio at start of test $= e_0 = e_1 + \Delta e$

Now

$$\frac{\Delta e}{\Delta H} = \frac{1 + e_0}{H_0} = \frac{1 + e_1 + \Delta e}{H_0}$$

i.e.

$$\frac{\Delta e}{3.520} = \frac{1.541 + \Delta e}{19.0}$$

$$\Delta e = 0.350$$

$$e_0 = 0.541 + 0.350 = 0.891$$

In general, the relationship between Δe and ΔH is given by

$$\frac{\Delta e}{\Delta H} = \frac{1.891}{19.0}$$

i.e. $\Delta e = 0.0996 \, \Delta H$, and can be used to obtain the void ratio at the end of each increment period (see Table 7.1). The e–$\log \sigma'$ curve using these values is shown in Figure 7.6. Using Casagrande's construction, the value of the preconsolidation pressure is 325 kN/m².

$$m_v = \frac{1}{1 + e_0} \cdot \frac{e_0 - e_1}{\sigma'_1 - \sigma'_0}$$

Table 7.1

Pressure (kN/m²)	ΔH (mm)	Δe	e
0	0	0	0.891
54	0.253	0.025	0.866
107	0.507	0.050	0.841
214	0.892	0.089	0.802
429	1.551	0.154	0.737
858	2.392	0.238	0.653
1716	3.324	0.331	0.560
3432	4.263	0.424	0.467
0	3.520	0.350	0.541

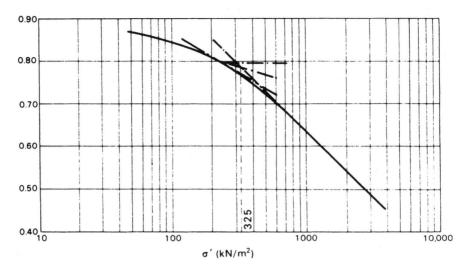

Figure 7.6 Example 7.1.

For $\sigma'_0 = 100\,\text{kN/m}^2$ and $\sigma'_1 = 200\,\text{kN/m}^2$,

$$e_0 = 0.845 \quad \text{and} \quad e_1 = 0.808$$

and therefore

$$m_\text{v} = \frac{1}{1.845} \times \frac{0.037}{100} = 2.0 \times 10^{-4}\,\text{m}^2/\text{kN} = 0.20\,\text{m}^2/\text{MN}$$

For $\sigma'_0 = 1000\,\text{kN/m}^2$ and $\sigma'_1 = 1500\,\text{kN/m}^2$,

$$e_0 = 0.632 \quad \text{and} \quad e_1 = 0.577$$

and therefore

$$m_\text{v} = \frac{1}{1.632} \times \frac{0.055}{500} = 6.7 \times 10^{-5}\,\text{m}^2/\text{kN} = 0.067\,\text{m}^2/\text{MN}$$

and

$$C_\text{c} = \frac{0.632 - 0.577}{\log(1500/1000)} = \frac{0.055}{0.176} = 0.31$$

Note that C_c will be the same for any stress range on the linear part of the e–$\log \sigma'$ curve; m_v will vary according to the stress range, even for ranges on the linear part of the curve.

7.3 CONSOLIDATION SETTLEMENT: ONE-DIMENSIONAL METHOD

In order to estimate consolidation settlement, the value of either the coefficient of volume compressibility or the compression index is required. Consider a layer of saturated clay of thickness H. Due to construction the total vertical stress in an elemental layer of thickness dz at depth z is increased by $\Delta\sigma$ (Figure 7.7). It is assumed that the condition of zero lateral strain applies within the clay layer. After the completion of consolidation, an equal increase $\Delta\sigma'$ in effective vertical stress will have taken place corresponding to a stress increase from σ'_0 to σ'_1 and a reduction in void ratio from e_0 to e_1 on the e–σ' curve. The reduction in volume per unit volume of clay can be written in terms of void ratio:

$$\frac{\Delta V}{V_0} = \frac{e_0 - e_1}{1 + e_0}$$

Since the lateral strain is zero, the reduction in volume per unit volume is equal to the reduction in thickness per unit thickness, i.e. the settlement per unit depth. Therefore, by proportion, the settlement of the layer of thickness dz will be given by

$$\begin{aligned}
ds_c &= \frac{e_0 - e_1}{1 + e_0} dz \\
&= \left(\frac{e_0 - e_1}{\sigma'_1 - \sigma'_0}\right)\left(\frac{\sigma'_1 - \sigma'_0}{1 + e_0}\right) dz \\
&= m_v \Delta\sigma' dz
\end{aligned}$$

where s_c = consolidation settlement.

The settlement of the layer of thickness H is given by

$$s_c = \int_0^H m_v \Delta\sigma' dz$$

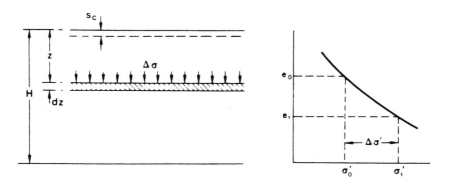

Figure 7.7 Consolidation settlement.

If m_v and $\Delta\sigma'$ are assumed constant with depth, then

$$s_c = m_v \Delta\sigma' H \tag{7.6}$$

or

$$s_c = \frac{e_0 - e_1}{1 + e_0} H \tag{7.7}$$

or, in the case of a normally consolidated clay,

$$s_c = \frac{C_c \log(\sigma_1'/\sigma_0')}{1 + e_0} H \tag{7.8}$$

In order to take into account the variation of m_v and/or $\Delta\sigma'$ with depth, the graphical procedure shown in Figure 7.8 can be used to determine s_c.

The variations of initial effective vertical stress (σ_0') and effective vertical stress increment ($\Delta\sigma'$) over the depth of the layer are represented in Figure 7.8(a); the variation of m_v is represented in Figure 7.8(b). The curve in Figure 7.8(c) represents the variation with depth of the dimensionless product $m_v \Delta\sigma'$ and the area under this curve is the settlement of the layer. Alternatively the layer can be divided into a suitable number of sublayers and the product $m_v \Delta\sigma'$ evaluated at the centre of each sublayer: each product $m_v \Delta\sigma'$ is then multiplied by the appropriate sublayer thickness to give the sublayer settlement. The settlement of the whole layer is equal to the sum of the sublayer settlements.

Example 7.2

A building is supported on a raft $45 \times 30\,\text{m}$, the net foundation pressure (assumed to be uniformly distributed) being $125\,\text{kN/m}^2$. The soil profile is as shown in Figure 7.9. The value of m_v for the clay is $0.35\,\text{m}^2/\text{MN}$. Determine the final settlement under the centre of the raft due to consolidation of the clay.

The clay layer is thin relative to the dimensions of the raft, therefore it can be assumed that consolidation is approximately one-dimensional. In this case it will be

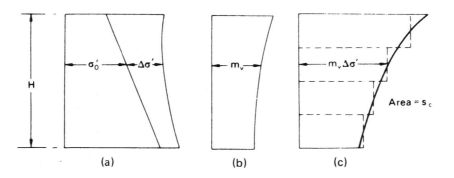

Figure 7.8 Consolidation settlement: graphical procedure.

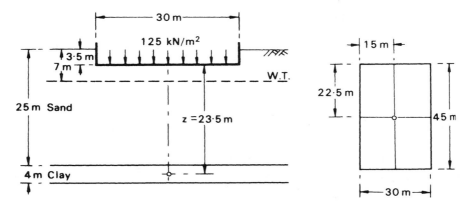

Figure 7.9 Example 7.2.

sufficiently accurate to consider the clay layer as a whole. Because the consolidation settlement is to be calculated in terms of m_v, only the effective stress *increment* at mid-depth of the layer is required (the increment being assumed constant over the depth of the layer). Also, $\Delta\sigma' = \Delta\sigma$ for one-dimensional consolidation and can be evaluated from Figure 5.10.

At mid-depth of the layer, $z = 23.5\,\mathrm{m}$. Below the centre of the raft:

$$m = \frac{22.5}{23.5} = 0.96$$

$$n = \frac{15}{23.5} = 0.64$$

$$I_r = 0.140$$

$$\Delta\sigma' = 4 \times 0.140 \times 125 = 70\,\mathrm{kN/m^2}$$

$$s_c = m_v\Delta\sigma'H = 0.35 \times 70 \times 4 = 98\,\mathrm{mm}$$

7.4   SETTLEMENT BY THE SKEMPTON–BJERRUM METHOD

Predictions of consolidation settlement using the one-dimensional method are based on the results of oedometer tests using representative samples of the clay. Due to the confining ring the net lateral strain in the test specimen is zero and for this condition the initial excess pore water pressure is equal theoretically to the increase in total vertical stress, i.e. the pore pressure coefficient A is equal to unity.

In practice the condition of zero lateral strain is satisfied approximately in the cases of thin clay layers and of layers under loaded areas which are large compared with the layer thickness. In many practical situations, however, significant lateral strain will occur and the initial excess pore water pressure will depend on the *in-situ* stress conditions and the value of the pore pressure coefficient A (which will not be equal to unity).

In cases in which the lateral strain is not zero, there will be an immediate settlement, under undrained conditions, in addition to the consolidation settlement. Immediate settlement is zero if the lateral strain is zero, as assumed in the one-dimensional method of calculating settlement. In the Skempton–Bjerrum method [22], the total settlement (s) of a foundation on clay is given by

$$s = s_i + s_c$$

where s_i = immediate settlement, occurring under undrained conditions, and s_c = consolidation settlement, due to the volume reduction accompanying the gradual dissipation of excess pore water pressure.

The immediate settlement (s_i) can be estimated from the results of elastic theory (Section 5.3). The value of Poisson's ratio (ν) relevant to undrained conditions in a fully saturated clay is taken to be 0.5. The undrained Young's modulus (E_u) must be estimated from the results of laboratory tests, *in-situ* load tests or pressuremeter tests. However, for most soils, the modulus increases with depth. Use of a constant value of E_u overestimates immediate settlement. An approximate analysis has been proposed by Butler [5] for the case of increasing modulus with depth.

If there is no change in static pore water pressure, the *initial* value of excess pore water pressure (denoted by u_i) at a point in the clay layer is given by Equation 4.25, with $B = 1$ for a fully saturated soil. Thus

$$
\begin{aligned}
u_i &= \Delta\sigma_3 + A(\Delta\sigma_1 - \Delta\sigma_3) \\
&= \Delta\sigma_1 \left[A + \frac{\Delta\sigma_3}{\Delta\sigma_1}(1 - A) \right]
\end{aligned}
\tag{7.9}
$$

where $\Delta\sigma_1$ and $\Delta\sigma_3$ are the total principal stress increments due to surface loading. From Equation 7.9 it is seen that

$$u_i > \Delta\sigma_3$$

if A is positive. Note also that $u_i = \Delta\sigma_1$ if $A = 1$. The value of A depends on the type of clay, the stress levels and the stress system.

The *in-situ* effective stresses before loading, immediately after loading and after consolidation are represented in Figure 7.10 and the corresponding Mohr circles (A, B and C, respectively) in Figure 7.11. In Figure 7.11, *abc* is the ESP for *in-situ* loading and consolidation, *ab* representing an immediate change of stress and *bc* a gradual change of

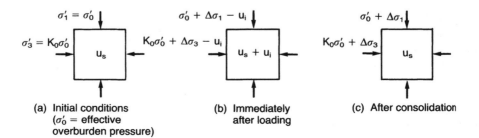

(a) Initial conditions
(σ'_0 = effective
overburden pressure)

(b) Immediately
after loading

(c) After consolidation

Figure 7.10 In-situ effective stresses.

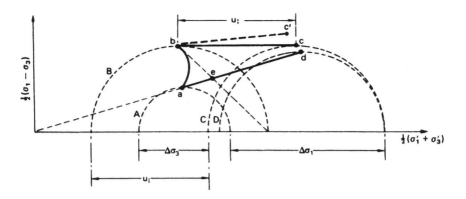

Figure 7.11 Stress paths.

stress as the excess pore water pressure dissipates. Immediately after loading there is a reduction in σ'_3 due to u_i being greater than $\Delta\sigma_3$ and lateral expansion will occur. Subsequent consolidation will therefore involve lateral recompression. Circle D in Figure 7.11 represents the corresponding stresses in the oedometer test after consolidation and *ad* is the corresponding ESP for the oedometer test. As the excess pore water pressure dissipates, Poisson's ratio decreases from the undrained value (0.5) to the drained value at the end of consolidation. The decrease in Poisson's ratio does not significantly affect the vertical stress but results in a small decrease in horizontal stress (point *c* would become *c'* in Figure 7.11): this decrease is neglected in the Skempton–Bjerrum method.

Skempton and Bjerrum proposed that the effect of lateral strain be neglected in the calculation of consolidation settlement (s_c), thus enabling the oedometer test to be maintained as the basis of the method. It was admitted, however, that this simplification could involve errors of up to 20% in vertical settlements. However, the value of excess pore water pressure given by Equation 7.9 is used in the method.

By the one-dimensional method, consolidation settlement (equal to the total settlement) is given as

$$s_{od} = \int_0^H m_v \Delta\sigma_1 dz \quad (\text{i.e. } \Delta\sigma' = \Delta\sigma_1)$$

where H is the thickness of the clay layer. By the Skempton–Bjerrum method, consolidation settlement is expressed in the form

$$s_c = \int_0^H m_v u_i dz$$

$$= \int_0^H m_v \Delta\sigma_1 \left[A + \frac{\Delta\sigma_3}{\Delta\sigma_1}(1 - A) \right] dz$$

A settlement coefficient μ is introduced such that

$$s_c = \mu s_{od} \tag{7.10}$$

where

$$\mu = \frac{\int_0^H m_v \Delta \sigma_1 [A + (\Delta \sigma_3 / \Delta \sigma_1)(1 - A)] dz}{\int_0^H m_v \Delta \sigma_1 dz}$$

If it can be assumed that m_v and A are constant with depth (sublayers can be used in analysis) then μ can be expressed as

$$\mu = A + (1 - A)\alpha \tag{7.11}$$

where

$$\alpha = \frac{\int_0^H \Delta \sigma_3 dz}{\int_0^H \Delta \sigma_1 dz}$$

Taking Poisson's ratio (ν) as 0.5 for a saturated clay during loading under undrained conditions, the value of α depends only on the shape of the loaded area and the thickness of the clay layer in relation to the dimensions of the loaded area; thus α can be estimated from elastic theory.

The value of initial excess pore water pressure (u_i) should, in general, correspond to the *in-situ* stress conditions. The use of a value of pore pressure coefficient A obtained from the results of a triaxial test on a cylindrical clay specimen is strictly applicable only for the condition of axial symmetry, i.e. for the case of settlement under the centre of a circular footing. However, a value of A so obtained will serve as a good approximation for the case of settlement under the centre of a square footing (using a circular footing of the same area). Under a strip footing, however, plane strain conditions apply and the intermediate principal stress increment $\Delta \sigma_2$, in the direction of the longitudinal axis, is equal to $0.5(\Delta \sigma_1 + \Delta \sigma_3)$. Scott [20] has shown that the value of u_i appropriate in the case of a strip footing can be obtained by using a pore pressure coefficient A_s, where

$$A_s = 0.866A + 0.211$$

The coefficient A_s replaces A (the coefficient for conditions of axial symmetry) in Equation 7.11 for the case of a strip footing, with the expression for α being unchanged.

Values of the settlement coefficient μ, for circular and strip footings, in terms of A and the ratio of layer thickness/breadth of footing (H/B) are given in Figure 7.12.

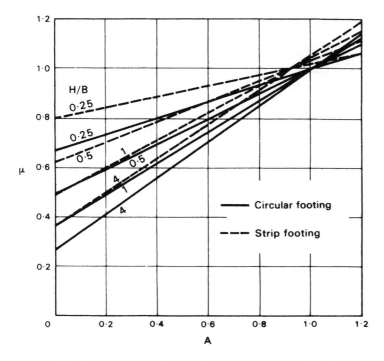

Values of μ are typically within the following ranges:

Soft, sensitive clays	1.0–1.2
Normally consolidated clays	0.7–1.0
Lightly overconsolidated clays	0.5–0.7
Heavily overconsolidated clays	0.2–0.5

Example 7.3

A footing 6 m square, carrying a net pressure of $160 \, \text{kN/m}^2$, is located at a depth of 2 m in a deposit of stiff clay 17 m thick: a firm stratum lies immediately below the clay. From oedometer tests on specimens of the clay the value of m_v was found to be $0.13 \, \text{m}^2/\text{MN}$ and from triaxial tests the value of A was found to be 0.35. The undrained Young's modulus for the clay is estimated to be $55 \, \text{MN/m}^2$. Determine the total settlement under the centre of the footing.

In this case there will be significant lateral strain in the clay beneath the footing (resulting in immediate settlement) and it is appropriate to use the Skempton–Bjerrum method. The section is shown in Figure 7.13.

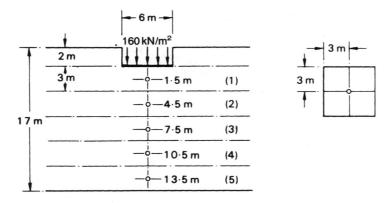

Figure 7.13 Example 7.3.

(a) *Immediate settlement.* The influence factors are obtained from Figure 5.15. Now

$$\frac{H}{B} = \frac{15}{6} = 2.5$$
$$\frac{D}{B} = \frac{2}{6} = 0.33$$
$$\frac{L}{B} = 1$$
$$\therefore \mu_0 = 0.95 \quad \text{and} \quad \mu_1 = 0.55$$

Hence

$$s_i = \mu_0 \mu_1 \frac{qB}{E_u}$$
$$= 0.95 \times 0.55 \times \frac{160 \times 6}{55} = 9\,\text{mm}$$

(b) *Consolidation settlement.* In Table 7.2,

$$\Delta\sigma' = 4 \times 160 \times I_r \ (\text{kN/m}^2)$$
$$s_{od} = 0.13 \times \Delta\sigma' \times 3 = 0.39\Delta\sigma' \ (\text{mm})$$

Now

$$\frac{H}{B} = \frac{15}{6.77} = 2.2$$

(equivalent diameter = 6.77 m)

$$A = 0.35$$

Table 7.2

Layer	z (m)	m, n	I_r	$\Delta\sigma'$ (kN/m^2)	s_{od} (mm)
1	1.5	2.00	0.233	149	58.1
2	4.5	0.67	0.121	78	30.4
3	7.5	0.40	0.060	38	14.8
4	10.5	0.285	0.033	21	8.2
5	13.5	0.222	0.021	13	5.1
					116.6

Hence, from Figure 7.12,

$$\mu = 0.55$$

Then

$$s_c = 0.55 \times 116.6 = 64 \, \text{mm}$$
$$\text{Total settlement} = s_i + s_c$$
$$= 9 + 64$$
$$= 73 \, \text{mm}$$

7.5   THE STRESS PATH METHOD

In this method it is recognized that soil deformation is dependent on the stress path followed prior to the final state of stress. The stress path for a soil element subjected to undrained loading followed by consolidation (neglecting the decrease in Poisson's ratio) is *abc* in Figure 7.11, while the stress paths for consolidation only according to the one-dimensional and Skempton–Bjerrum methods are *ad* and *ed*, respectively. In the stress path method, due to Lambe [12], the actual stress paths for a number of 'average' *in-situ* elements are estimated and laboratory tests are run, using the hydraulic triaxial apparatus, as closely as possible along the same stress paths, beginning at the initial stresses prior to construction. The measured vertical strains (ε_1) during the test are then used to obtain the settlement, i.e. for a layer of thickness H:

$$s = \int_0^H \varepsilon_1 \, dz \tag{7.12}$$

In-situ pore water pressure conditions and partial drainage during the construction period can be simulated if desired. As an example, Figure 7.14 shows a soil element under a circular storage tank and the ESP and corresponding vertical strains for a triaxial specimen in which undrained loading (*ab*), consolidation (*bc*), undrained unloading (*cd*) and swelling (*de*) are simulated.

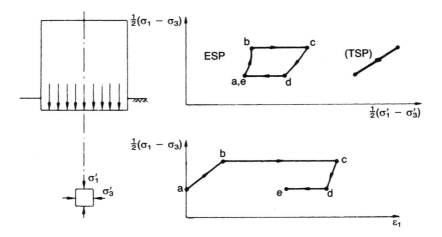

Figure 7.14 The stress path method. (Reproduced from T.W. Lambe (1967) *Journal ASCE,* **93,** No. SM6, by permission of the American Society of Civil Engineers.)

Although sound in principle, the method depends on the correct selection of typical soil elements and on the test specimens being truly representative of the *in-situ* material. In addition, the triaxial techniques involved in running the correct stress paths are complex and time-consuming unless computer-controlled equipment is available. Knowledge of the value of K_0 is also required.

Simons and Som [21] investigated the effect of stress path on axial and volumetric compressibility and proposed a method of calculating settlement based on the relationship between the ratio of vertical to volumetric strain ($\varepsilon_1/\varepsilon_v$) and the ratio $\Delta\sigma'_3/\Delta\sigma'_1$.

7.6 DEGREE OF CONSOLIDATION

For an element of soil at a particular depth z in a clay layer the progress of the consolidation process under a particular total stress increment can be expressed in terms of void ratio as follows:

$$U_z = \frac{e_0 - e}{e_0 - e_1}$$

where U_z is defined as the degree of consolidation, at a particular instant of time, at depth z ($0 \leq U_z \leq 1$), and $e_0 =$ void ratio before the start of consolidation, $e_1 =$ void ratio at the end of consolidation and $e =$ void ratio, at the time in question, during consolidation.

If the e–σ' curve is assumed to be linear over the stress range in question, as shown in Figure 7.15, the degree of consolidation can be expressed in terms of σ':

$$U_z = \frac{\sigma' - \sigma'_0}{\sigma'_1 - \sigma'_0}$$

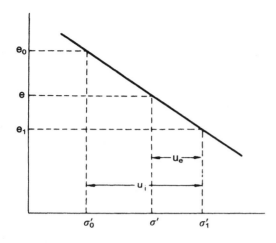

Figure 7.15 Assumed linear e–σ' relationship.

Suppose that the total vertical stress in the soil at the depth z is increased from σ_0 to σ_1 and there is *no lateral strain*. Immediately after the increase takes place, although the total stress has increased to σ_1, the effective vertical stress will still be σ'_0; only after the completion of consolidation will the effective stress become σ'_1. During consolidation the increase in effective vertical stress is numerically equal to the decrease in excess pore water pressure. If σ' and u_e are, respectively, the values of effective stress and excess pore water pressure at any time during the consolidation process and if u_i is the initial excess pore water pressure (i.e. the value immediately after the increase in total stress) then, referring to Figure 7.15:

$$\sigma'_1 = \sigma'_0 + u_i = \sigma' + u_e$$

The degree of consolidation can then be expressed as

$$U_z = \frac{u_i - u_e}{u_i} = 1 - \frac{u_e}{u_i} \tag{7.13}$$

7.7   TERZAGHI'S THEORY OF ONE-DIMENSIONAL CONSOLIDATION

The assumptions made in the theory are:

1   The soil is homogeneous.
2   The soil is fully saturated.
3   The solid particles and water are incompressible.
4   Compression and flow are one-dimensional (vertical).
5   Strains are small.

6 Darcy's law is valid at all hydraulic gradients.
7 The coefficient of permeability and the coefficient of volume compressibility remain constant throughout the process.
8 There is a unique relationship, independent of time, between void ratio and effective stress.

Regarding assumption 6, there is evidence of deviation from Darcy's law at low hydraulic gradients. Regarding assumption 7, the coefficient of permeability decreases as the void ratio decreases during consolidation. The coefficient of volume compressibility also decreases during consolidation since the e–σ' relationship is nonlinear. However, for small stress increments assumption 7 is reasonable. The main limitations of Terzaghi's theory (apart from its one-dimensional nature) arise from assumption 8. Experimental results show that the relationship between void ratio and effective stress is not independent of time.

The theory relates the following three quantities.

1 The *excess* pore water pressure (u_e).
2 The depth (z) below the top of the clay layer.
3 The time (t) from the instantaneous application of a total stress increment.

Consider an element having dimensions dx, dy and dz within a clay layer of thickness $2d$, as shown in Figure 7.16. An increment of total vertical stress $\Delta\sigma$ is applied to the element.

The flow velocity through the element is given by Darcy's law as

$$v_z = ki_z = -k\frac{\partial h}{\partial z}$$

Since any change in total head (h) is due only to a change in pore water pressure:

$$v_z = -\frac{k}{\gamma_w}\frac{\partial u_e}{\partial z}$$

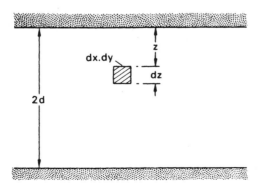

Figure 7.16 Element within a clay layer.

The condition of continuity (Equation 2.7) can therefore be expressed as

$$-\frac{k}{\gamma_w}\frac{\partial^2 u_e}{\partial z^2}\,dx\,dy\,dz = \frac{dV}{dt} \tag{7.14}$$

The rate of volume change can be expressed in terms of m_v:

$$\frac{dV}{dt} = m_v\frac{\partial\sigma'}{\partial t}\,dx\,dy\,dz$$

The total stress increment is gradually transferred to the soil skeleton, increasing effective stress, as the excess pore water pressure decreases. Hence the rate of volume change can be expressed as

$$\frac{dV}{dt} = -m_v\frac{\partial u_e}{\partial t}\,dx\,dy\,dz \tag{7.15}$$

Combining Equations 7.14 and 7.15,

$$m_v\frac{\partial u_e}{\partial t} = \frac{k}{\gamma_w}\frac{\partial^2 u_e}{\partial z^2}$$

or

$$\frac{\partial u_e}{\partial t} = c_v\frac{\partial^2 u_e}{\partial z^2} \tag{7.16}$$

This is the differential equation of consolidation, in which

$$c_v = \frac{k}{m_v\gamma_w} \tag{7.17}$$

c_v being defined as the *coefficient of consolidation*, suitable unit being m^2/year. Since k and m_v are assumed as constants, c_v is constant during consolidation.

Solution of the consolidation equation

The total stress increment is assumed to be applied instantaneously. At zero time, therefore, the increment will be carried entirely by the pore water, i.e. the initial value of excess pore water pressure (u_i) is equal to $\Delta\sigma$ and the initial condition is

$$u_e = u_i \quad \text{for} \quad 0 \le z \le 2d \quad \text{when} \quad t = 0$$

The upper and lower boundaries of the clay layer are assumed to be free-draining, the permeability of the soil adjacent to each boundary being very high compared to that of the clay. Thus the boundary conditions at any time after the application of $\Delta\sigma$ are

$$u_e = 0 \quad \text{for} \quad z = 0 \quad \text{and} \quad z = 2d \quad \text{when} \quad t > 0$$

The solution for the excess pore water pressure at depth z after time t is

$$u_e = \sum_{n=1}^{n=\infty} \left(\frac{1}{d} \int_0^{2d} u_i \sin \frac{n\pi z}{2d} \, dz \right) \left(\sin \frac{n\pi z}{2d} \right) \exp\left(-\frac{n^2 \pi^2 c_v t}{4d^2} \right) \tag{7.18}$$

where $d = $ length of the longest drainage path and $u_i = $ initial excess pore water pressure, in general a function of z.

For the particular case in which u_i is constant throughout the clay layer:

$$u_e = \sum_{n=1}^{n=\infty} \frac{2u_i}{n\pi} (1 - \cos n\pi) \left(\sin \frac{n\pi z}{2d} \right) \exp\left(-\frac{n^2 \pi^2 c_v t}{4d^2} \right) \tag{7.19}$$

When n is even, $(1 - \cos n\pi) = 0$ and when n is odd, $(1 - \cos n\pi) = 2$. Only odd values of n are therefore relevant and it is convenient to make the substitutions

$$n = 2m + 1$$

and

$$M = \frac{\pi}{2}(2m + 1)$$

It is also convenient to substitute

$$T_v = \frac{c_v t}{d^2} \tag{7.20}$$

a dimensionless number called the *time factor*. Equation 7.19 then becomes

$$u_e = \sum_{m=0}^{m=\infty} \frac{2u_i}{M} \left(\sin \frac{Mz}{d} \right) \exp(-M^2 T_v) \tag{7.21}$$

The progress of consolidation can be shown by plotting a series of curves of u_e against z for different values of t. Such curves are called *isochrones* and their form will depend on the initial distribution of excess pore water pressure and the drainage conditions at the boundaries of the clay layer. A layer for which both the upper and lower boundaries are free-draining is described as an *open* layer; a layer for which only one boundary is free-draining is a *half-closed* layer. Examples of isochrones are shown in Figure 7.17. In part (a) of the figure the initial distribution of u_i is constant and for an open layer of thickness $2d$ the isochrones are symmetrical about the centre line. The upper half of this diagram also represents the case of a half-closed layer of thickness d. The slope of an isochrone at any depth gives the hydraulic gradient and also indicates the direction of flow. In parts (b) and (c) of the figure, with a triangular distribution of u_i, the direction of flow changes over certain parts of the layer. In part (c) the lower boundary is impermeable and for a time swelling takes place in the lower part of the layer.

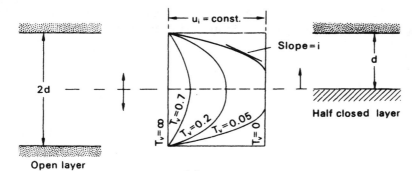

Open layer

(a)

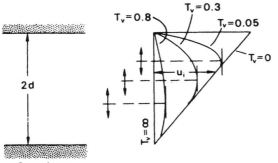

Open layer

(b)

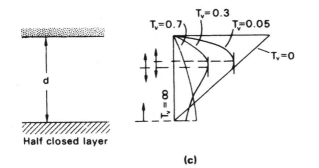

Half closed layer

(c)

Figure 7.17 Isochrones.

The degree of consolidation at depth z and time t can be obtained by substituting the value of u_e (Equation 7.21) in Equation 7.13 giving

$$U_z = 1 - \sum_{m=0}^{m=\infty} \frac{2}{M} \left(\sin \frac{Mz}{d} \right) \exp(-M^2 T_v) \tag{7.22}$$

In practical problems it is the *average* degree of consolidation (U) over the depth of the layer as a whole that is of interest, the consolidation settlement at time t being given by the product of U and the final settlement. The average degree of consolidation at time t for constant u_i is given by

$$U = 1 - \frac{(1/2d) \int_0^{2d} u_e \, dz}{u_i}$$

$$= 1 - \sum_{m=0}^{m=\infty} \frac{2}{M^2} \exp(-M^2 T_v) \tag{7.23}$$

The relationship between U and T_v given by Equation 7.23 is represented by curve 1 in Figure 7.18. Equation 7.23 can be represented almost exactly by the following empirical equations:

$$\text{for} \quad U < 0.60, \quad T_v = \frac{\pi}{4} U^2 \tag{7.24a}$$

$$\text{for} \quad U > 0.60, \quad T_v = -0.933 \log(1 - U) - 0.085 \tag{7.24b}$$

If u_i is not constant the average degree of consolidation is given by

$$U = 1 - \frac{\int_0^{2d} u_e \, dz}{\int_0^{2d} u_i \, dz} \tag{7.25}$$

where

$$\int_0^{2d} u_e \, dz = \text{area under isochrone at the time in question}$$

and

$$\int_0^{2d} u_i \, dz = \text{area under initial isochrone}$$

(For a half-closed layer the limits of integration are 0 and d in the above equations.)

The initial variation of excess pore water pressure in a clay layer can usually be approximated in practice to a linear distribution. Curves 1, 2 and 3 in Figure 7.18 represent the solution of the consolidation equation for the cases shown in Figure 7.19.

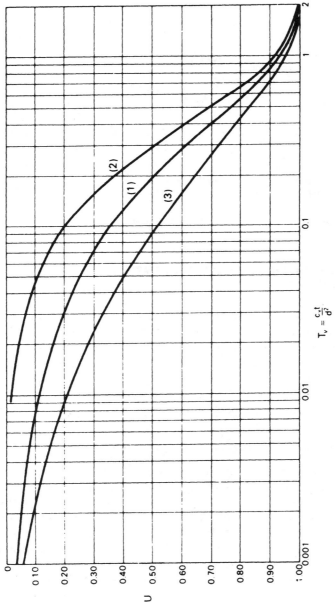

$$T_v = \frac{c_v t}{d^2}$$

Figure 7.18 Relationships between average degree of consolidation and time factor.

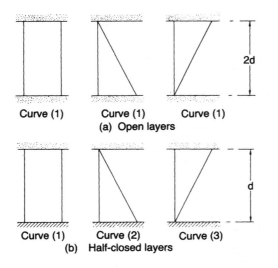

Figure 7.19 Initial variations of excess pore water pressure.

7.8   DETERMINATION OF COEFFICIENT OF CONSOLIDATION

The value of c_v for a particular pressure increment in the oedometer test can be determined by comparing the characteristics of the experimental and theoretical consolidation curves, the procedure being referred to as *curve-fitting*. The characteristics of the curves are brought out clearly if time is plotted to a square root or a logarithmic scale. It should be noted that once the value of c_v has been determined, the coefficient of permeability can be calculated from Equation 7.17, the oedometer test being a useful method for obtaining the permeability of a clay.

The log time method (due to Casagrande)

The forms of the experimental and theoretical curves are shown in Figure 7.20. The experimental curve is obtained by plotting the dial gauge readings in the oedometer test against the logarithm of time in minutes. The theoretical curve is given as the plot of the average degree of consolidation against the logarithm of the time factor. The theoretical curve consists of three parts: an initial curve which approximates closely to a parabolic relationship, a part which is linear and a final curve to which the horizontal axis is an asymptote at $U = 1.0$ (or 100%). In the experimental curve the point corresponding to $U = 0$ can be determined by using the fact that the initial part of the curve represents an approximately parabolic relationship between compression and time. Two points on the curve are selected (A and B in Figure 7.20) for which the values of t are in the ratio of 4:1, and the vertical distance between them is measured. An equal distance set off above the first point fixes the point (a_s) corresponding to $U = 0$. As a check the procedure should be repeated using different pairs of points. The point corresponding to $U = 0$

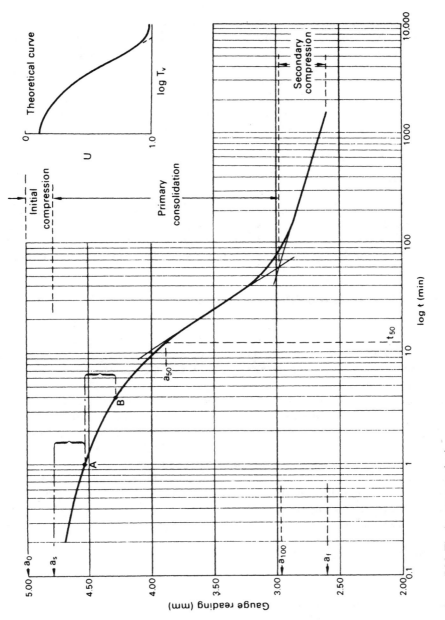

Figure 7.20 The log time method.

will not generally correspond to the point (a_0) representing the initial dial gauge reading, the difference being due mainly to the compression of small quantities of air in the soil, the degree of saturation being marginally below 100%: this compression is called *initial compression*. The final part of the experimental curve is linear but not horizontal and the point (a_{100}) corresponding to $U = 100\%$ is taken as the intersection of the two linear parts of the curve. The compression between the a_s and a_{100} points is called *primary consolidation* and represents that part of the process accounted for by Terzaghi's theory. Beyond the point of intersection, compression of the soil continues at a very slow rate for an indefinite period of time and is called *secondary compression*.

The point corresponding to $U = 50\%$ can be located midway between the a_s and a_{100} points and the corresponding time t_{50} obtained. The value of T_v corresponding to $U = 50\%$ is 0.196 and the coefficient of consolidation is given by

$$c_v = \frac{0.196d^2}{t_{50}} \tag{7.26}$$

the value of d being taken as half the average thickness of the specimen for the particular pressure increment. In BS 1377 it is stated that if the average temperature of the soil *in situ* is known and differs from the average test temperature, a correction should be applied to the value of c_v, correction factors being given in the standard.

The root time method (due to Taylor)

Figure 7.21 shows the forms of the experimental and theoretical curves, the dial gauge readings being plotted against the square root of time in minutes and the average degree of consolidation against the square root of time factor. The theoretical curve is linear up to about 60% consolidation and at 90% consolidation the abscissa (AC) is 1.15 times the abscissa (AB) of the production of the linear part of the curve. This characteristic is used to determine the point on the experimental curve corresponding to $U = 90\%$.

The experimental curve usually consists of a short curve representing initial compression, a linear part and a second curve. The point (D) corresponding to $U = 0$ is obtained by producing back the linear part of the curve to the ordinate at zero time. A straight line (DE) is then drawn having abscissae 1.15 times the corresponding abscissae on the linear part of the experimental curve. The intersection of the line DE with the experimental curve locates the point (a_{90}) corresponding to $U = 90\%$ and the corresponding value $\sqrt{t_{90}}$ can be obtained. The value of T_v corresponding to $U = 90\%$ is 0.848 and the coefficient of consolidation is given by

$$c_v = \frac{0.848d^2}{t_{90}} \tag{7.27}$$

If required, the point (a_{100}) on the experimental curve corresponding to $U = 100\%$, the limit of primary consolidation, can be obtained by proportion. As in the log time plot the curve extends beyond the 100% point into the secondary compression range. The root time method requires compression readings covering a much shorter period of time compared with the log time method, which requires the accurate definition of

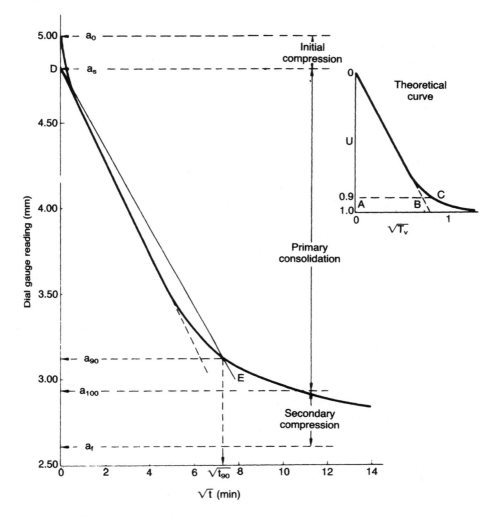

Figure 7.21 The root time method.

the second linear part of the curve well into the secondary compression range. On the other hand, a straight-line portion is not always obtained on the root time plot and in such cases the log time method should be used.

Other methods of determining c_v have been proposed by Naylor and Doran [14], Scott [19] and Cour [6].

The compression ratios

The relative magnitudes of the initial compression, the compression due to primary consolidation and the secondary compression can be expressed by the following ratios (refer Figures 7.20 and 7.21).

Initial compression ratio : $r_0 = \dfrac{a_0 - a_s}{a_0 - a_f}$ (7.28)

Primary compression ratio (log time) : $r_p = \dfrac{a_s - a_{100}}{a_0 - a_f}$ (7.29)

Primary compression ratio (root time) : $r_p = \dfrac{10(a_s - a_{90})}{9(a_0 - a_f)}$ (7.30)

Secondary compression ratio : $r_s = 1 - (r_0 + r_p)$ (7.31)

In-situ value of c_v

Settlement observations have indicated that the rates of settlement of full-scale structures are generally much greater than those predicted using values of c_v obtained from the results of oedometer tests on small specimens (e.g. 75 mm diameter × 20 mm thick). Rowe [16] has shown that such discrepancies are due to the influence of the clay macro-fabric on drainage behaviour. Features such as laminations, layers of silt and fine sand, silt-filled fissures, organic inclusions and root-holes, if they reach a major permeable stratum, have the effect of increasing the overall permeability of the clay mass. In general, the macro-fabric of a clay is not represented accurately in a small oedometer specimen and the permeability of such a specimen will be lower than the mass permeability.

In cases where fabric effects are significant, more realistic values of c_v can be obtained by means of the hydraulic oedometer developed by Rowe and Barden [17] and manufactured for a range of specimen sizes. Specimens 250 mm in diameter by 100 mm in thick are considered large enough to represent the natural macro-fabric of most clays: values of c_v obtained from tests on specimens of this size have been shown to be consistent with observed settlement rates.

Details of a hydraulic oedometer are shown in Figure 7.22. Vertical pressure is applied to the specimen by means of water pressure acting across a convoluted rubber

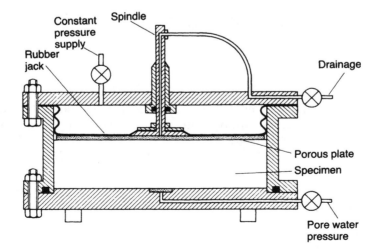

Figure 7.22 Hydraulic oedometer.

jack. The system used to apply the pressure must be capable of compensating for pressure changes due to leakage and specimen volume change. Compression of the specimen is measured by means of a central spindle passing through a sealed housing in the top plate of the oedometer. Drainage from the specimen can be either vertical or radial. Pore water pressure can be measured during the test and back-pressure may be applied to the specimen. The apparatus can also be used for flow tests, from which the coefficient of permeability can be determined directly. Methods of test using the hydraulic oedometer are specified in BS 1377 (Part 6).

Piezometers can be used for the *in-situ* determination of c_v but the method requires the use of three-dimensional consolidation theory. The most satisfactory procedure is to maintain a constant head (above or below the ambient pore water pressure in the clay) at the piezometer tip and measure the rate of flow into or out of the system. If the rate of flow is measured at various times the value of c_v (and of the coefficient of permeability k) can be deduced. Details have been given by Gibson [8, 9] and Wilkinson [25].

Another method of determining c_v is to combine laboratory values of m_v (which from experience are known to be more reliable than laboratory values of c_v) with *in-situ* measurements of k, using Equation 7.17.

Secondary compression

In the Terzaghi theory it is implied by assumption 8 that a change in void ratio is due entirely to a change in effective stress brought about by the dissipation of excess pore water pressure, with permeability alone governing the time dependency of the process. However, experimental results show that compression does not cease when the excess pore water pressure has dissipated to zero but continues at a gradually decreasing rate under constant effective stress. Secondary compression is thought to be due to the gradual readjustment of the clay particles into a more stable configuration following the structural disturbance caused by the decrease in void ratio, especially if the clay is laterally confined. An additional factor is the gradual lateral displacements which take place in thick clay layers subjected to shear stresses. The rate of secondary compression is thought to be controlled by the highly viscous film of adsorbed water surrounding the clay mineral particles in the soil. A very slow viscous flow of adsorbed water takes place from the zones of film contact, allowing the solid particles to move closer together. The viscosity of the film increases as the particles move closer, resulting in a decrease in the rate of compression of the soil. It is presumed that primary consolidation and secondary compression proceed simultaneously from the time of loading.

The rate of secondary compression in the oedometer test can be defined by the slope (C_α) of the final part of the compression–log time curve, measured as the unit compression over one decade on the log time scale. The magnitude of secondary compression in a given time is generally greater in normally consolidated clays than in overconsolidated clays. In overconsolidated clays, strains are mainly elastic but in normally consolidated clays significant plastic strains occur. For certain highly plastic clays and organic clays the secondary compression part of the compression–log time curve may completely mask the primary consolidation part. For a particular soil the magnitude of secondary compression over a given time, as a percentage of the total compression, increases as the ratio of pressure increment to initial pressure decreases; the magnitude of secondary compression also increases as the thickness of the oedometer specimen decreases and as temperature increases. Thus the secondary compression characteristics of an oedometer specimen cannot normally be extrapolated to the case of a full-scale foundation.

In a small number of normally consolidated clays it has been found that secondary compression forms the greater part of the total compression under applied pressure. Bjerrum [3] showed that such clays have gradually developed a reserve resistance against further compression as a result of the considerable decrease in void ratio which has occurred, under constant effective stress, over the hundreds or thousands of years since sedimentation. These clays, although normally consolidated, exhibit a quasi-preconsolidation pressure. It has been shown that provided any additional applied pressure is less than approximately 50% of the difference between the quasi-preconsolidation pressure and the effective overburden pressure the resultant settlement will be relatively small.

Example 7.4

The following compression readings were taken during an oedometer test on a saturated clay specimen ($G_s = 2.73$) when the applied pressure was increased from 214 to 429 kN/m^2:

Time (min)	0	$\frac{1}{4}$	$\frac{1}{2}$	1	$2\frac{1}{4}$	4	9	16	25
Gauge (mm)	5.00	4.67	4.62	4.53	4.41	4.28	4.01	3.75	3.49

Time (min)	36	49	64	81	100	200	400	1440
Gauge (mm)	3.28	3.15	3.06	3.00	2.96	2.84	2.76	2.61

After 1440 min the thickness of the specimen was 13.60 mm and the water content was 35.9%. Determine the coefficient of consolidation from both the log time and the root time plots and the values of the three compression ratios. Determine also the value of the coefficient of permeability.

$$\text{Total change in thickness during increment} = 5.00 - 2.61 = 2.39\,\text{mm}$$

$$\text{Average thickness during increment} = 13.60 + \frac{2.39}{2} = 14.80\,\text{mm}$$

$$\text{Length of drainage path, } d = \frac{14.80}{2} = 7.40\,\text{mm}$$

From the log time plot (Figure 7.20),

$$t_{50} = 12.5\,\text{min}$$

$$c_v = \frac{0.196d^2}{t_{50}} = \frac{0.196 \times 7.40^2}{12.5} \times \frac{1440 \times 365}{10^6} = 0.45\,\text{m}^2/\text{year}$$

$$r_0 = \frac{5.00 - 4.79}{5.00 - 2.61} = 0.088$$

$$r_p = \frac{4.79 - 2.98}{5.00 - 2.61} = 0.757$$

$$r_s = 1 - (0.088 + 0.757) = 0.155$$

From the root time plot (Figure 7.21) $\sqrt{t_{90}} = 7.30$, and therefore

$$t_{90} = 53.3\,\text{min}$$

$$c_v = \frac{0.848d^2}{t_{90}} = \frac{0.848 \times 7.40^2}{53.3} \times \frac{1440 \times 365}{10^6} = 0.46\,\text{m}^2/\text{year}$$

$$r_0 = \frac{5.00 - 4.81}{5.00 - 2.61} = 0.080$$

$$r_p = \frac{10(4.81 - 3.12)}{9(5.00 - 2.61)} = 0.785$$

$$r_s = 1 - (0.080 + 0.785) = 0.135$$

In order to determine the permeability, the value of m_v must be calculated.

Final void ratio, $e_1 = w_1 G_s = 0.359 \times 2.73 = 0.98$
Initial void ratio, $e_0 = e_1 + \Delta e$

Now

$$\frac{\Delta e}{\Delta H} = \frac{1 + e_0}{H_0}$$

i.e.

$$\frac{\Delta e}{2.39} = \frac{1.98 + \Delta e}{15.99}$$

Therefore

$$\Delta e = 0.35 \quad \text{and} \quad e_0 = 1.33$$

Now

$$m_v = \frac{1}{1 + e_0} \cdot \frac{e_0 - e_1}{\sigma_1' - \sigma_0'}$$

$$= \frac{1}{2.33} \times \frac{0.35}{215} = 7.0 \times 10^{-4}\,\text{m}^2/\text{kN}$$

$$= 0.70\,\text{m}^2/\text{MN}$$

Coefficient of permeability:

$$k = c_v m_v \gamma_w$$

$$= \frac{0.45 \times 0.70 \times 9.8}{60 \times 1440 \times 365 \times 10^3}$$

$$= 1.0 \times 10^{-10}\,\text{m/s}$$

7.9   CORRECTION FOR CONSTRUCTION PERIOD

In practice, structural loads are applied to the soil not instantaneously but over a period of time. Initially there is usually a reduction in net load due to excavation, resulting in swelling of the clay: settlement will not begin until the applied load exceeds the weight of the excavated soil. Terzaghi proposed an empirical method of correcting the instantaneous time–settlement curve to allow for the construction period.

The net load (P') is the gross load less the weight of soil excavated, and the effective construction period (t_c) is measured from the time when P' is zero. It is assumed that the net load is applied uniformly over the time t_c (Figure 7.23) and that the degree of consolidation at time t_c is the same as if the load P' had been acting as a constant load for the period $\frac{1}{2}t_c$. Thus the settlement at any time during the construction period is equal to that occurring for instantaneous loading at half that time; however, since the load then acting is not the total load, the value of settlement so obtained must be reduced in the proportion of that load to the total load.

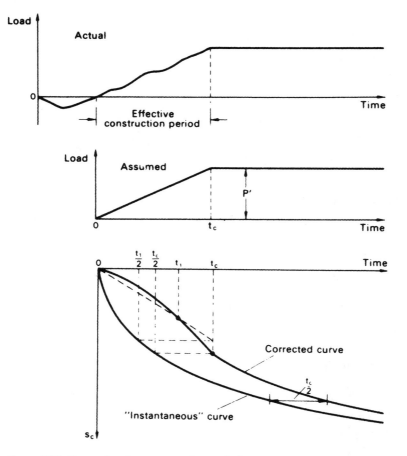

Figure 7.23 Correction for construction period.

For the period subsequent to the completion of construction, the settlement curve will be the instantaneous curve offset by half the effective construction period. Thus at any time after the end of construction the corrected time corresponding to any value of settlement is equal to the time from the start of loading less half the effective construction period. After a long period of time the magnitude of settlement is not appreciably affected by the construction time.

Alternatively, a numerical solution (Section 7.10) can be used, successive increments of excess pore water pressure being applied over the construction period.

Example 7.5

A layer of clay 8 m thick lies between two layers of sand. The upper sand layer extends from ground level to a depth of 4 m, the water table being at a depth of 2 m. The lower sand layer is under artesian pressure, the piezometric level being 6 m above ground level. For the clay $m_v = 0.94 \, \text{m}^2/\text{MN}$ and $c_v = 1.4 \, \text{m}^2/\text{year}$. As a result of pumping from the artesian layer the piezometric level falls by 3 m over a period of 2 years. Draw the time–settlement curve due to consolidation of the clay for a period of 5 years from the start of pumping.

In this case, consolidation is due only to the change in pore water pressure at the lower boundary of the clay: there is no change in total vertical stress. The effective vertical stress remains unchanged at the top of the clay layer but will be increased by $3\gamma_w$ at the bottom of the layer due to the decrease in pore water pressure in the adjacent artesian layer. The distribution of $\Delta\sigma'$ is shown in Figure 7.24. The problem is one-dimensional since the increase in effective vertical stress is the same over the

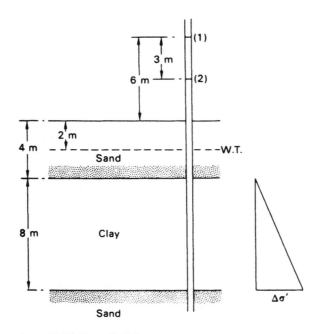

Figure 7.24 Example 7.5.

Table 7.3

U	T_v	t (years)	s_c (mm)
0.10	0.008	0.09	11
0.20	0.031	0.35	22
0.30	0.070	0.79	33
0.40	0.126	1.42	44
0.50	0.196	2.21	55
0.60	0.285	3.22	66
0.73	0.437	5.00	80

entire area in question. In calculating the consolidation settlement it is necessary to consider only the value of $\Delta\sigma'$ at the centre of the layer. Note that in order to obtain the value of m_v it would have been necessary to calculate the initial and final values of effective vertical stress in the clay.

At the centre of the clay layer, $\Delta\sigma' = 1.5\gamma_w = 14.7\,\mathrm{kN/m^2}$. The *final* consolidation settlement is given by

$$s_{cf} = m_v \Delta\sigma' H$$

$$= 0.94 \times 14.7 \times 8$$

$$= 110\,\mathrm{mm}$$

The clay layer is open, and therefore $d = 4\,\mathrm{m}$. For $t = 5$ years,

$$T_v = \frac{c_v t}{d^2}$$

$$= \frac{1.4 \times 5}{4^2}$$

$$= 0.437$$

From curve 1 of Figure 7.18, the corresponding value of U is 0.73. To obtain the time–settlement relationship a series of values of U is selected up to 0.73 and the corresponding times calculated from the time factor equation: the corresponding values of settlement (s_c) are given by the product of U and s_{cf} (see Table 7.3). The plot of s_c against t gives the 'instantaneous' curve. Terzaghi's method of correction for the 2-year period over which pumping takes place is then carried out as shown in Figure 7.25.

Example 7.6

An 8 m depth of sand overlies a 6 m layer of clay, below which is an impermeable stratum (Figure 7.26); the water table is 2 m below the surface of the sand. Over a period of 1 year, a 3 m depth of fill (unit weight $20\,\mathrm{kN/m^3}$) is to be dumped on the surface over an extensive area. The saturated unit weight of the sand is $19\,\mathrm{kN/m^3}$ and that of the clay is $20\,\mathrm{kN/m^3}$; above the water table the unit weight of the sand is

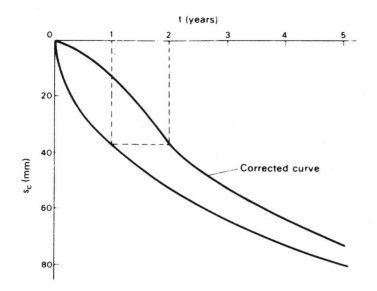

Figure 7.25 Example 7.5.

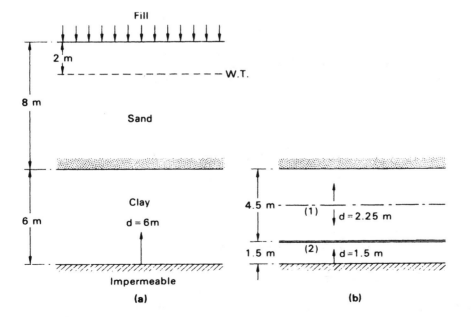

Figure 7.26 Example 7.6.

$17\,\text{kN/m}^3$. For the clay, the relationship between void ratio and effective stress (units kN/m^2) can be represented by the equation

$$e = 0.88 - 0.32\log\left(\frac{\sigma'}{100}\right)$$

and the coefficient of consolidation is $1.26\,\text{m}^2/\text{year}$.

(a)   Calculate the final settlement of the area due to consolidation of the clay and the settlement after a period of 3 years from the start of dumping.
(b)   If a very thin layer of sand, freely draining, existed $1.5\,\text{m}$ above the bottom of the clay layer, what would be the values of the final and 3-year settlements?

(a) Since the fill covers a wide area, the problem can be considered to be one-dimensional. The consolidation settlement will be calculated in terms of C_c, considering the clay layer as a whole, and therefore the initial and final values of effective vertical stress at the centre of the clay layer are required.

$$\sigma'_0 = (17 \times 2) + (9.2 \times 6) + (10.2 \times 3) = 119.8\,\text{kN/m}^2$$
$$e_0 = 0.88 - 0.32\log 1.198 = 0.88 - 0.025 = 0.855$$
$$\sigma'_1 = 119.8 + (3 \times 20) = 179.8\,\text{kN/m}^2$$
$$\log\left(\frac{179.8}{119.8}\right) = 0.176$$

The final settlement is calculated from Equation 7.8:

$$s_{\text{cf}} = \frac{0.32 \times 0.176 \times 6000}{1.855} = 182\,\text{mm}$$

In the calculation of the degree of consolidation 3 years after the start of dumping, the corrected value of time to allow for the 1-year dumping period is

$$t = 3 - \frac{1}{2} = 2.5\ \text{years}$$

The layer is half-closed, and therefore $d = 6\,\text{m}$. Then

$$T_{\text{v}} = \frac{c_{\text{v}}t}{d^2} = \frac{1.26 \times 2.5}{6^2}$$
$$= 0.0875$$

From curve 1 of Figure 7.18, $U = 0.335$. Settlement after 3 years:

$$s_{\text{c}} = 0.335 \times 182 = 61\,\text{mm}$$

(b) The final settlement will still be $182\,\text{mm}$ (ignoring the thickness of the drainage layer): only the rate of settlement will be affected. From the point of view of drainage there is now an open layer of thickness 4.5 m ($d = 2.25$ m) above a half-closed layer of thickness 1.5 m ($d = 1.5$ m): these layers are numbered 1 and 2, respectively.

By proportion

$$T_{v_1} = 0.0875 \times \frac{6^2}{2.25^2} = 0.622$$

$$\therefore U_1 = 0.825$$

and

$$T_{v_2} = 0.0875 \times \frac{6^2}{1.5^2} = 1.40$$

$$\therefore U_2 = 0.97$$

Now for each layer, $s_c = Us_{cf}$ which is proportional to UH. Hence if $\overline{U}$ is the overall degree of consolidation for the two layers combined:

$$4.5U_1 + 1.5U_2 = 6.0\overline{U}$$

i.e. $(4.5 \times 0.825) + (1.5 \times 0.97) = 6.0\overline{U}$.

Hence $\overline{U} = 0.86$ and the 3-year settlement is

$$s_c = 0.86 \times 182 = 157 \, \text{mm}$$

7.10 NUMERICAL SOLUTION

The one-dimensional consolidation equation can be solved numerically by the method of finite differences. The method has the advantage that any pattern of initial excess pore water pressure can be adopted and it is possible to consider problems in which the load is applied gradually over a period of time. The errors associated with the method are negligible and the solution is easily programmed for the computer.

The method is based on a depth–time grid as shown in Figure 7.27. The depth of the clay layer is divided into m equal parts of thickness Δz and any specified period of time is divided into n equal intervals Δt. Any point on the grid can be identified by the subscripts i and j, the depth position of the point being denoted by $i (0 \leq i \leq m)$ and the elapsed time by $j (0 \leq j \leq n)$. The value of excess pore water pressure at any depth after any time is therefore denoted by $u_{i,j}$. (In this section the subscript e is dropped from the symbol for excess pore water pressure, i.e. u represents u_e as defined in Section 3.3.)

The following finite difference approximations can be derived from Taylor's theorem:

$$\frac{\partial u}{\partial t} = \frac{1}{\Delta t}(u_{i,j+1} - u_{i,j})$$

$$\frac{\partial^2 u}{\partial z^2} = \frac{1}{(\Delta z)^2}(u_{i-1,j} + u_{i+1,j} - 2u_{i,j})$$

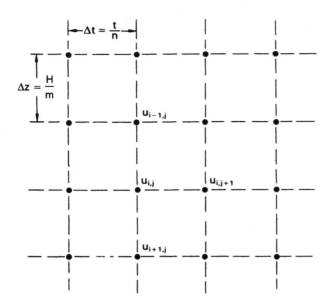

Figure 7.27 Depth–time grid.

Substituting these values in Equation 7.16 yields the finite difference approximation of the one-dimensional consolidation equation:

$$u_{i,j+1} = u_{i,j} + \frac{c_{\mathrm{v}}\Delta t}{(\Delta z)^2}\left(u_{i-1,j} + u_{i+1,j} - 2u_{i,j}\right) \tag{7.32}$$

It is convenient to write

$$\beta = \frac{c_{\mathrm{v}}\Delta t}{(\Delta z)^2} \tag{7.33}$$

this term being called the *operator* of Equation 7.32. It has been shown that for convergence the value of the operator must not exceed $\frac{1}{2}$. The errors due to neglecting higher-order derivatives in Taylor's theorem are reduced to a minimum when the value of the operator is $\frac{1}{6}$.

It is usual to specify the number of equal parts m into which the depth of the layer is to be divided and as the value of β is limited a restriction is thus placed on the value of Δt. For any specified period of time t in the case of an *open* layer:

$$
\begin{aligned}
T_{\mathrm{v}} &= \frac{c_{\mathrm{v}}(n\,\Delta t)}{(\frac{1}{2}m\Delta z)^2} \\
&= 4\frac{n}{m^2}\beta
\end{aligned}
\tag{7.34}
$$

In the case of a *half-closed* layer the denominator becomes $(m\Delta z)^2$ and

$$T_v = \frac{n}{m^2}\beta \tag{7.35}$$

A value of n must therefore be chosen such that the value of β in Equation 7.34 or 7.35 does not exceed $\frac{1}{2}$.

Equation 7.32 does not apply to points on an impermeable boundary. There can be no flow across an impermeable boundary, a condition represented by the equation:

$$\frac{\partial u}{\partial z} = 0$$

which can be represented by the finite difference approximation:

$$\frac{1}{2\Delta z}(u_{i-1,j} - u_{i+1,j}) = 0$$

the impermeable boundary being at a depth position denoted by subscript i, i.e.

$$u_{i-1,j} = u_{i+1,j}$$

For all points *on* an impermeable boundary, Equation 7.32 becomes

$$u_{i,j+1} = u_{i,j} + \frac{c_v\Delta t}{(\Delta z)^2}(2u_{i-1,j} - 2u_{i,j}) \tag{7.36}$$

The degree of consolidation at any time t can be obtained by determining the areas under the initial isochrone and the isochrone at time t as in Equation 7.25.

Example 7.7

A half-closed clay layer (free-draining at the upper boundary) is 10 m thick and the value of c_v is 7.9 m^2/year. The initial distribution of excess pore water pressure is as follows:

Depth (m)	0	2	4	6	8	10
Pressure (kN/m^2)	60	54	41	29	19	15

Obtain the values of excess pore water pressure after consolidation has been in progress for 1 year.

The layer is half-closed, and therefore $d = 10$ m. For $t = 1$ year,

$$T_v = \frac{c_v t}{d^2} = \frac{7.9 \times 1}{10^2} = 0.079$$

Table 7.4

i	j										
	0	1	2	3	4	5	6	7	8	9	10
0	0	0	0	0	0	0	0	0	0	0	0
1	54.0	40.6	32.6	27.3	23.5	20.7	18.5	16.7	15.3	14.1	13.1
2	41.0	41.2	38.7	35.7	32.9	30.4	28.2	26.3	24.6	23.2	21.9
3	29.0	29.4	29.9	30.0	29.6	29.0	28.3	27.5	26.7	26.0	25.3
4	19.0	20.2	21.3	22.4	23.3	24.0	24.5	24.9	25.1	25.2	25.2
5	15.0	16.6	18.0	19.4	20.6	21.7	22.6	23.4	24.0	24.4	24.7

The layer is divided into five equal parts, i.e. $m = 5$. Now

$$T_v = \frac{n}{m^2} \beta$$

Therefore

$$n\beta = 0.079 \times 5^5 = 1.98 \quad \text{(say 2.0)}$$

(This makes the actual value of $T_v = 0.080$ and $t = 1.01$ years.) The value of n will be taken as 10 (i.e. $\Delta t = \frac{1}{10}$ year), making $\beta = 0.2$. The finite difference equation then becomes

$$u_{i,j+1} = u_{i,j} + 0.2 \left(u_{i-1,j} + u_{i+1,j} - 2u_{i,j} \right)$$

but on the impermeable boundary:

$$u_{i,j+1} = u_{i,j} + 0.2 \left(2u_{i-1,j} - 2u_{i,j} \right)$$

On the permeable boundary, $u = 0$ for all values of t, assuming the initial pressure of $60 \, \text{kN/m}^2$ instantaneously becomes zero.

The computation is set out in Table 7.4. Conveniently, the computation can be performed using a spreadsheet.

7.11 VERTICAL DRAINS

The slow rate of consolidation in saturated clays of low permeability may be accelerated by means of vertical drains which shorten the drainage path within the clay. Consolidation is then due mainly to horizontal radial drainage, resulting in the faster dissipation of excess pore water pressure; vertical drainage becomes of minor importance. In theory the final magnitude of consolidation settlement is the same, only the *rate* of settlement being affected.

In the case of an embankment constructed over a highly compressible clay layer (Figure 7.28), vertical drains installed in the clay would enable the embankment to be brought into service much sooner and there would be a quicker increase in the shear strength of the clay. A degree of consolidation of the order of 80% would be desirable

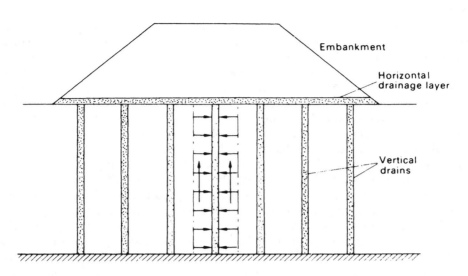

Figure 7.28 Vertical drains.

at the end of construction. Any advantages, of course, must be set against the additional cost of the installation.

The traditional method of installing vertical drains is by driving boreholes through the clay layer and backfilling with a suitably graded sand. Typical diameters are 200–400 mm and drains have been installed to depths of over 30 m. The sand must be capable of allowing the efficient flow of water while preventing fine soil particles from being washed in. Careful backfilling is essential to avoid discontinuities which could give rise to 'necking' and render a drain ineffective. Necking could also be caused by lateral soil displacement during consolidation.

Prefabricated drains are now generally used and tend to be more economic than backfilled drains for a given area of treatment. One type of drain (often referred to as a 'sandwick') consists of a filter stocking, usually of woven polypropylene, filled with sand. Compressed air is used to ensure that the stocking is completely filled with sand. This type of drain, a typical diameter being 65 mm, is very flexible and is generally unaffected by lateral soil displacement, the possibility of necking being virtually eliminated. The drains are installed either by insertion into pre-bored holes or, more commonly, by placing them inside a mandrel or casing which is then driven or vibrated into the ground.

Another type of prefabricated drain is the *band* drain, consisting of a flat plastic core indented with drainage channels, surrounded by a layer of filter fabric. The fabric must have sufficient strength to prevent it from being squeezed into the channels and the mesh size must be small enough to prevent the passage of soil particles which could clog the channels. Typical dimensions of a band drain are 100×4 mm and in design the equivalent diameter is assumed to be the perimeter divided by π. Band drains are installed by placing them inside a steel mandrel which is either pushed, driven or vibrated into the ground. An anchor is attached to the lower end of the drain to keep it in position as the mandrel is withdrawn. The anchor also prevents soil from entering the mandrel during installation.

Drains are normally installed in either a square or a triangular pattern. As the object is to reduce the length of the drainage path, the spacing of the drains is the most important design consideration. The spacing must obviously be less than the thickness of the clay layer and there is no point in using vertical drains in relatively thin layers. It is essential for a successful design that the coefficients of consolidation in both the horizontal and the vertical directions (c_h and c_v, respectively) are known as accurately as possible. In particular, the accuracy of c_h is the most crucial factor in design, more important, for example, than the effect of simplifying assumptions in the theory used. The ratio c_h/c_v is normally between 1 and 2; the higher the ratio, the more advantageous a drain installation will be. The values of the coefficients for the clay immediately surrounding the drains may be significantly reduced due to remoulding during installation, especially if boring is used, an effect known as *smear*. The smear effect can be taken into account either by assuming a reduced value of c_h or by using a reduced drain diameter. Another design complication in the case of large diameter sand drains is that the column of sand tends to act as a weak pile, reducing the vertical stress increment imposed on the clay layer by an unknown degree, resulting in lower excess pore water pressure and therefore reduced consolidation settlement. This effect is minimal in the case of prefabricated drains because of their flexibility.

Vertical drains may not be effective in overconsolidated clays if the vertical stress after consolidation remains less than the preconsolidation pressure. Indeed, disturbance of overconsolidated clay during drain installation might even result in increased final consolidation settlement. It should be realized that the rate of secondary compression cannot be controlled by vertical drains.

In polar coordinates the three-dimensional form of the consolidation equation, with different soil properties in the horizontal and vertical directions, is

$$\frac{\partial u_e}{\partial t} = c_h \left(\frac{\partial^2 u_e}{\partial r^2} + \frac{1}{r} \frac{\partial u_e}{\partial r} \right) + c_v \frac{\partial^2 u_e}{\partial z^2} \tag{7.37}$$

The vertical prismatic blocks of soil surrounding the drains are replaced by cylindrical blocks, of radius R, having the same cross-sectional area (Figure 7.29). The solution to Equation 7.37 can be written in two parts:

$$U_v = f(T_v)$$

and

$$U_r = f(T_r)$$

where U_v = average degree of consolidation due to vertical drainage only; U_r = average degree of consolidation due to horizontal (radial) drainage only;

$$T_v = \frac{c_v t}{d^2}$$

= time factor for consolidation due to vertical drainage only $\tag{7.38}$

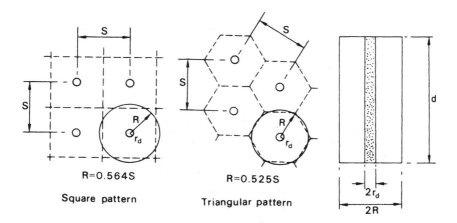

Figure 7.29 Cylindrical blocks.

$$T_r = \frac{c_h t}{4R^2}$$

= time factor for consolidation due to radial drainage only (7.39)

The expression for T_r confirms the fact that the closer the spacing of the drains, the quicker the consolidation process due to radial drainage proceeds. The solution for radial drainage, due to Barron, is given in Figure 7.30, the U_r/T_r relationship depending on the ratio $n = R/r_d$, where R is the radius of the equivalent cylindrical block and r_d the radius of the drain. It can also be shown that

$$(1 - U) = (1 - U_v)(1 - U_r)$$ (7.40)

where U is the average degree of consolidation under combined vertical and radial drainage.

Example 7.8

An embankment is to be constructed over a layer of clay 10 m thick, with an impermeable lower boundary. Construction of the embankment will increase the total vertical stress in the clay layer by 65 kN/m². For the clay, $c_v = 4.7\,\mathrm{m^2/year}$, $c_h = 7.9\,\mathrm{m^2/year}$ and $m_v = 0.25\,\mathrm{m^2/MN}$. The design requirement is that all but 25 mm of the settlement due to consolidation of the clay layer will have taken place after 6 months. Determine the spacing, in a square pattern, of 400 mm diameter sand drains to achieve the above requirement.

Final settlement $= m_v\,\Delta\sigma'\,H = 0.25 \times 65 \times 10$

$= 162\,\mathrm{mm}$

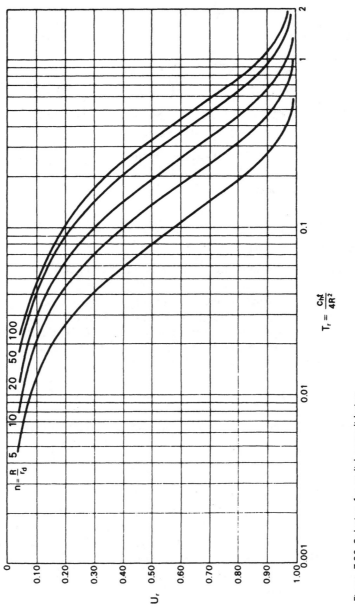

Figure 7.30 Solution for radial consolidation.

For $t = 6$ months,

$$U = \frac{162 - 25}{162} = 0.85$$

Diameter of sand drains is 0.4 m, i.e. $r_d = 0.2$ m.
  Radius of cylindrical block:

$$R = nr_d = 0.2n$$

The layer is half-closed, and therefore $d = 10$ m.

$$T_v = \frac{c_v t}{d^2} = \frac{4.7 \times 0.5}{10^2} = 0.0235$$

From curve 1 of Figure 7.18, $U_v = 0.17$:

$$T_r = \frac{c_h t}{4R^2} = \frac{7.9 \times 0.5}{4 \times 0.2^2 \times n^2} = \frac{24.7}{n^2}$$

i.e.

$$n = \sqrt{\frac{24.7}{T_r}}$$

Now $(1 - U) = (1 - U_v)(1 - U_r)$, and therefore

$$0.15 = 0.83(1 - U_r)$$
$$U_r = 0.82$$

A trial-and-error solution is necessary to obtain the value of n. Starting with a value of n corresponding to one of the curves in Figure 7.30 the value of T_r for $U_r = 0.82$ is obtained from that curve. Using this value of T_r the value of $\sqrt{24.7/T_r}$ is calculated and plotted against the selected value of n.

n	T_r	$\sqrt{24.7/T_r}$
5	0.20	11.1
10	0.33	8.6
15	0.42	7.7

From Figure 7.31 it is seen that $n = 9$. Therefore

$$R = 0.2 \times 9 = 1.8 \text{ m}$$

Spacing of drains in a square pattern is given by

$$S = \frac{R}{0.564} = \frac{1.8}{0.564} = 3.2 \text{ m}$$

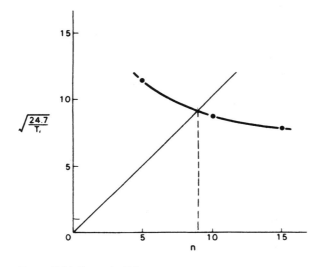

Figure 7.31 Example 7.8.

PROBLEMS

7.1   In an oedometer test on a specimen of saturated clay ($G_s = 2.72$) the applied pressure was increased from 107 to $214 \, \text{kN/m}^2$ and the following compression readings recorded:

Time (min)	0	$\frac{1}{4}$	$\frac{1}{2}$	1	$2\frac{1}{4}$	4	$6\frac{1}{4}$	9	16
Gauge (mm)	7.82	7.42	7.32	7.21	6.99	6.78	6.61	6.49	6.37

Time (min)	25	36	49	64	81	100	300	1440
Gauge (mm)	6.29	6.24	6.21	6.18	6.16	6.15	6.10	6.02

After 1440 min the thickness of the specimen was 15.30 mm and the water content was 23.2%. Determine the values of the coefficient of consolidation and the compression ratios from (a) the root time plot and (b) the log time plot. Determine also the values of the coefficient of volume compressibility and the coefficient of permeability.

7.2   The following results were obtained from an oedometer test on a specimen of saturated clay:

Pressure (kN/m²)	27	54	107	214	429	214	107	54
Void ratio	1.243	1.217	1.144	1.068	0.994	1.001	1.012	1.024

A layer of this clay 8 m thick lies below a 4 m depth of sand, the water table being at the surface. The saturated unit weight for both soils is $19 \, \text{kN/m}^3$. A 4 m depth of fill of unit weight $21 \, \text{kN/m}^3$ is placed on the sand over an extensive area. Determine the final settlement due to consolidation of the clay. If the fill were to be removed some time after the completion of consolidation, what heave would eventually take place due to swelling of the clay?

7.3 In an oedometer test a specimen of saturated clay 19 mm thick reaches 50% consolidation in 20 min. How long would it take a layer of this clay 5 m thick to reach the same degree of consolidation under the same stress and drainage conditions? How long would it take the layer to reach 30% consolidation?

7.4 Assuming the fill in Problem 7.2 is dumped very rapidly, what would be the value of excess pore water pressure at the centre of the clay layer after a period of 3 years? The layer is open and the value of c_v is 2.4 m^2/year.

7.5 An open clay layer is 6 m thick, the value of c_v being 1.0 m^2/year. The initial distribution of excess pore water pressure varies linearly from 60 kN/m^2 at the top of the layer to zero at the bottom. Using the finite difference approximation of the one-dimensional consolidation equation, plot the isochrone after consolidation has been in progress for a period of 3 years and from the isochrone determine the average degree of consolidation in the layer.

7.6 A 10 m depth of sand overlies an 8 m layer of clay, below which is a further depth of sand. For the clay, $m_v = 0.83$ m^2/MN and $c_v = 4.4$ m^2/year. The water table is at surface level but is to be lowered permanently by 4 m, the initial lowering taking place over a period of 40 weeks. Calculate the final settlement due to consolidation of the clay, assuming no change in the weight of the sand, and the settlement 2 years after the start of lowering.

7.7 A raft foundation 60 × 40 m carrying a net pressure of 145 kN/m^2 is located at a depth of 4.5 m below the surface in a deposit of dense sandy gravel 22 m deep: the water table is at a depth of 7 m. Below the sandy gravel is a layer of clay 5 m thick which, in turn, is underlain by dense sand. The value of m_v for the clay is 0.22 m^2/MN. Determine the settlement below the centre of the raft, the corner of the raft and the centre of each edge of the raft, due to consolidation of the clay.

7.8 An oil storage tank 35 m in diameter is located 2 m below the surface of a deposit of clay 32 m thick, the water table being at the surface: the net foundation pressure is 105 kN/m^2. A firm stratum underlies the clay. The average value of m_v for the clay is 0.14 m^2/MN and that of pore pressure coefficient A is 0.65. The undrained value of Young's modulus is estimated to be 40 MN/m^2. Determine the total settlement under the centre of the tank.

7.9 A half-closed clay layer is 8 m thick and it can be assumed that $c_v = c_h$. Vertical sand drains 300 mm in diameter, spaced at 3 m centres in a square pattern, are to be used to increase the rate of consolidation of the clay under the increased vertical stress due to the construction of an embankment. Without sand drains the degree of consolidation at the time the embankment is due to come into use has been calculated as 25%. What degree of consolidation would be reached with the sand drains at the same time?

7.10 A layer of saturated clay is 10 m thick, the lower boundary being impermeable; an embankment is to be constructed above the clay. Determine the time required for 90% consolidation of the clay layer. If 300 mm diameter sand drains at 4 m centres in a square pattern were installed in the clay, in what time would the same overall degree of consolidation be reached? The coefficients of consolidation in the vertical and horizontal directions, respectively, are 9.6 and 14.0 m^2/year.

REFERENCES

1   Atkinson, M.S. and Eldred, P.J.L. (1981) Consolidation of soil using vertical drains, *Geotechnique*, **31**, 33–43.

2   Barron, R.A. (1948) Consolidation of fine grained soils by drain wells, *Transactions ASCE*, **113**, 718–42.

3   Bjerrum, L. (1967) Engineering geology of Norwegian normally-consolidated marine clays as related to settlement of buildings, *Geotechnique*, **17**, 83–118.

4   British Standard 1377 (1990) *Methods of Test for Soils for Civil Engineering Purposes*, British Standards Institution, London.

5   Butler, F.G. (1974) Heavily overconsolidated clays, *Proc. Conf. on Settlement of Structures*, Pentech Press, Cambridge, pp. 531–78.

6   Cour, F.R. (1971) Inflection point method for computing c_v, Technical Note, *Journal ASCE*, **97** (SM5), 827–31.

7   Gibson, R.E. (1963) An analysis of system flexibility and its effects on time lag in pore water pressure measurements, *Geotechnique*, **13**, 1–11.

8   Gibson, R.E. (1966) A note on the constant head test to measure soil permeability *in-situ*, *Geotechnique*, **16**, 256–9.

9   Gibson, R.E. (1970) An extension to the theory of the constant head *in-situ* permeability test, *Geotechnique*, **20**, 193–7.

10   Holtz, R.D., Jamiolkowski, M.B., Lancellotta, R. and Pedroni, R. (1991) *Prefabricated Vertical Drains: Design and Performance*, CIRIA/Butterworth-Heinemann, London/Oxford.

11   Lambe, T.W. (1964) Methods of estimating settlement, *Journal ASCE*, **90** (SM5), 43–67.

12   Lambe, T.W. (1967) Stress path method, *Journal ASCE*, **93** (SM6), 309–31.

13   McGown, A. and Hughes, F.H. (1981) Practical aspects of the design and installation of deep vertical drains, *Geotechnique*, **31**, 3–17.

14   Naylor, A.H. and Doran, I.G. (1948) Precise determination of primary consolidation, in *Proceedings 2nd International Conference SMFE, Rotterdam*, Vol. 1, pp. 34–40.

15   Padfield, C.J. and Sharrock, M.J. (1983) *Settlement of Structures on Clay Soils*, PSA/CIRIA, London.

16   Rowe, P.W. (1968) The influence of geological features of clay deposits on the design and performance of sand drains, *Proceedings ICE*, Supplementary volume, paper 70585.

17   Rowe, P.W. and Barden, L. (1966) A new consolidation cell, *Geotechnique*, **16**, 162–70.

18   Schmertmann, J.H. (1953) Estimating the true consolidation behaviour of clay from laboratory test results, *Proceedings ASCE*, **79**, 1–26.

19   Scott, R.F. (1961) New method of consolidation coefficient evaluation, *Journal ASCE*, **87**, No. SM1.

20   Scott, R.E. (1963) *Principles of Soil Mechanics*, Addison-Wesley, Reading, MA.

21   Simons, N.E. and Som, N.N. (1969) The influence of lateral stresses on the stress deformation characteristics of London clay, in *Proceedings 7th International Conference SMFE, Mexico City*, Vol. 1.

22   Skempton, A.W. and Bjerrum, L. (1957) A contribution to the settlement analysis of foundations on clay, *Geotechnique*, **7**, 168–78.

23   Taylor, D.W. (1948) *Fundamentals of Soil Mechanics*, John Wiley & Sons, New York.

24   Terzaghi, K. (1943) *Theoretical Soil Mechanics*, John Wiley & Sons, New York.

25   Wilkinson, W.B. (1968) Constant head *in-situ* permeability tests in clay strata, *Geotechnique*, **18**, 172–94.

Chapter 8

Bearing capacity

8.1 FOUNDATION DESIGN

A foundation is that part of a structure which transmits loads directly to the underlying soil. This chapter is concerned with the bearing capacity of soils on which foundations are supported, allied to the general philosophy of foundation design. If a soil stratum near the surface is capable of adequately supporting the structural loads it is possible to use either *footings* or a *raft*, these being referred to in general as shallow foundations. A footing is a relatively small slab giving separate support to part of the structure. A footing supporting a single column is referred to as an individual footing or pad, one supporting a closely spaced group of columns as a combined footing and one supporting a load-bearing wall as a strip footing. A raft is a relatively large single slab, usually stiffened with cross members, supporting the structure as a whole. If the soil near the surface is incapable of adequately supporting the structural loads, *piles*, or other forms of deep foundations such as piers or caissons, are used to transmit the loads to suitable soil (or rock) at greater depth. In addition to being located within an adequate bearing stratum, a foundation should be below the depth which is subjected to frost action (around 0.5 m in the United Kingdom) and, where appropriate, the depth to which seasonal swelling and shrinkage of the soil takes place. Consideration must also be given to the problems arising from excavating below the water table if it is necessary to locate foundations below this level. The choice of foundation level may also be influenced by the possibility of future excavations for services close to the structure and by the effect of construction, particularly excavation, on existing structures and services.

Results from elastic theory (Figure 5.8) indicate that the increase in vertical stress in the soil below the centre of a strip footing of width B is approximately 20% of the foundation pressure at a depth of $3B$. In the case of a square footing the corresponding depth is $1.5B$. For practical purposes these depths can normally be accepted as the limits of the zones of influence of the respective foundations and are called the *significant depths*. An alternative approach is to take the significant depth as that at which the vertical stress is 20% of the effective overburden pressure. It is essential that the soil conditions are known within the significant depth of any foundation.

There are two possible approaches to foundation design, as follows.

(1) The *permissible stress* method, as used in BS 8004: 1986 [7], uses a lumped factor of safety to ensure that the pressure applied to a foundation element is significantly less than the value which would cause shear failure in the supporting soil. The applied pressure is

due to dead load and maximum imposed load. A relatively high factor of 2–3 (more often the latter) is specified to allow for uncertainties in soil conditions and analytical method, and to ensure that settlement is not excessive. The *allowable bearing capacity* (q_a) is defined as the maximum pressure which may be applied to the soil such that an adequate factor of safety against shear failure is ensured and that settlement (especially differential settlement) should not cause unacceptable damage nor interfere with the function of the structure. For preliminary design purposes, presumed bearing values (Table 8.1) are given in BS 8004, being pressure ranges which would normally result in an adequate factor of safety against shear failure for particular soil types, but without consideration of settlement.

(2) The *limit state* method, on which Eurocode 7 (EC7) is based, aims at ensuring that all relevant performance requirements are satisfied under all conceivable circumstances. Ultimate limit states are concerned with the avoidance of collapse or major damage. Serviceability limit states are aimed at the avoidance of unacceptable settlement which could give rise to minor damage or impairment of function. To these ends, design is based on partial safety factors which are applied to characteristic permanent (dead) and variable (imposed) loads, referred to as actions, and soil parameters, referred to as ground properties. Each action is multiplied by an appropriate partial factor and each ground property is divided by an appropriate factor. The margin of safety is thus derived from two sources.

The following limit states should be considered as appropriate. Both ultimate limit states and serviceability limit states must be satisfied.

Ultimate limit states

1   Bearing resistance failure caused by shear failure of the supporting soil.
2   Loss of overall stability due to the development of a deep slip surface within the supporting soil (analysed using the methods described in Chapter 9).
3   Failure by sliding under inclined loading.
4   Combined soil/structure failure or structural failure of the foundation element due to excessive foundation movement.

Table 8.1 Presumed bearing values (BS 8004: 1986)

Soil type	Bearing value (kN/m^2)	Remarks
Dense gravel or dense sand and gravel	>600	Width of
Medium-dense gravel or medium-dense sand and gravel	200–600	foundation (B) not less than
Loose gravel or loose sand and gravel	<200	1 m. Water table
Dense sand	>300	at least B below
Medium-dense sand	100–300	base of
Loose sand	<100	foundation
Very stiff boulder clays and hard clays	300–600	Susceptible to
Stiff clays	150–300	long-term
Firm clays	75–150	consolidation
Soft clays and silts	<75	settlement
Very soft clays and silts	–	

Serviceability limit states

5   Excessive settlement (or heaving): excessive angular distortion.
6   Vibration resulting in unacceptable effects such as settlement and soil liquefaction.

Three design cases are described in EC7. Case A concerns situations in which stability is governed by the weight of the structure, with ground properties such as shear strength being relatively insignificant. An example is the uplift on a foundation due to hydrostatic pressure. In this case, weight is a favourable permanent action to which a partial factor of 0.95 is applied while hydrostatic force is an unfavourable variable action to which a partial factor of 1.50 is applied. Another example is the overturning of a retaining wall. Case B concerns situations of uncertainty in unfavourable permanent and variable actions to which partial factors of 1.35 and 1.50, respectively, are applied while ground properties are unfactored. This case is normally relevant to the structural design of foundations and retaining walls. Case C deals with situations in which the values of ground properties are uncertain, therefore partial factors greater than unity are applied to the relevant soil parameters. For parameters c' and $\tan \phi'$ factors of 1.60 and 1.25, respectively, have been proposed. For parameter c_u the proposed factor is 1.40. In this case a factor of unity is applied to both favourable and unfavourable permanent actions and a factor of 1.30 to variable actions. In all cases, variable favourable actions are not considered, i.e. the partial factor is zero. Case C is normally critical in determining the dimensions of foundations, the depth of embedded retaining walls and in analysing the stability of slopes (Chapter 9). For settlement calculations all partial factors are 1.00.

In the limit state method *bearing resistance* (a load) is used in contrast to bearing capacity (a pressure) in the permissible stress method. The ultimate bearing resistance is developed when shear failure of the supporting soil is on the point of occurring. Bearing resistance is used in limit state design to be consistent with the resistance of other materials used in construction and because it takes account of the shape of the foundation and the nature of loading. Bearing resistance failure (and other ultimate states) should be checked for both Case B and Case C, as defined above. Both undrained and drained conditions should be checked in the case of clays, although the undrained condition is usually critical.

Settlement damage

Damage due to settlement may be classified as architectural, functional or structural. In the case of framed structures, settlement damage is usually confined to the cladding and finishes (i.e. architectural damage): such damage is due only to the settlement occurring subsequent to the application of the cladding and finishes. In some cases, structures can be designed and constructed in such a way that a certain degree of movement can be accommodated without damage. In other cases a certain degree of cracking may be inevitable if the structure is to be economic. It may be that damage to services, and not to the structure, will be the limiting criterion. Based on observations of damage in buildings, Skempton and MacDonald [43] proposed limits for maximum settlement at which damage could be expected and related maximum settlement to

Table 8.2 Angular distortion limits

1/150	Structural damage of general buildings expected
1/250	Tilting of high rigid buildings may be visible
1/300	Cracking in panel walls expected
	Difficulties with overhead cranes
1/500	Limit for buildings in which cracking is not permissible
1/600	Overstressing of structural frames with diagonals
1/750	Difficulties with machinery sensitive to settlement

angular distortion. The angular distortion (also known as relative rotation) between two points under a structure is equal to the differential settlement between the points divided by the distance between them. No damage was observed where the angular distortion was less than 1/300: for individual footings this figure corresponds roughly to a maximum settlement of 50 mm on sands and of 75 mm on clays. Angular distortion limits were subsequently proposed by Bjerrum [3] as a general guide for a number of structural situations (Table 8.2). It is recommended that the safe limit to avoid cracking in the panel walls of framed structures should be 1/500. In the case of load-bearing brickwork the criteria recommended by Polshin and Tokar [35] are generally used. These criteria are given in terms of the ratio of deflection to the length of the deflected part and depend on the length-to-height ratio of the building: recommended deflection ratios are within the range 0.3×10^{-3} to 0.7×10^{-3}. In the case of buildings subjected to hogging the criteria of Polshin and Tokar should be halved.

The above approach to settlement limits is empirical and is intended to be only a general guide for simple structures. A more fundamental damage criterion is the limiting tensile strain at which visible cracking occurs in a given material. Ideally the concept of limiting tensile strain should be used in conjunction with an elastic strain analysis using a simple idealization of the structure, including foundations, partitions and finishes. A comprehensive discussion of settlement damage in buildings has been presented by Burland and Wroth [13].

Geotechnical categories

Foundation problems, and geotechnical problems in general, can be divided into three categories, as follows, the scope of the ground investigation and design procedure depending on the category in question.

Category 1 comprises typically the foundations for relatively small and simple structures (e.g. lightly loaded 1–2 storey buildings) for which the ground conditions (confirmed by simple procedures such as trial pits) are known from experience to be uncomplicated with no loose or compressible strata and no significant groundwater problems. Design is normally based on experience and routine procedures such as the application of presumed bearing values. Stability and settlement calculations are not usually necessary.

Category 2 includes shallow or deep foundations for conventional structures, bridges, retaining walls and embankments for which abnormal risk, unusual loading conditions and difficult ground conditions are all absent. Details of ground conditions are determined from routine investigation procedures such as boreholes plus field and laboratory tests. No significant problems with groundwater conditions should occur either during or after construction. Design normally involves routine stability and settlement calculations.

Category 3 includes shallow and deep foundations for relatively large or unusual structures, deep excavations and embankments involving abnormal risk and/or exceptionally difficult ground conditions (including groundwater conditions). Extensive ground investigation supplemented by field and laboratory testing is normally required, often requiring the use of advanced techniques. Design usually necessitates extensive stability and settlement calculations, often involving the use of computer analyses such as the finite element method.

8.2 ULTIMATE BEARING CAPACITY

The ultimate bearing capacity (q_f) is defined as the pressure which would cause shear failure of the supporting soil immediately below and adjacent to a foundation.

Three distinct modes of failure have been identified and these are illustrated in Figure 8.1: they will be described with reference to a strip footing. In the case of *general shear failure*, continuous failure surfaces develop between the edges of the footing and the ground surface as shown in Figure 8.1. As the pressure is increased towards the value q_f the state of plastic equilibrium is reached initially in the soil around the edges of the footing, then gradually spreads downwards and outwards. Ultimately the state of plastic equilibrium is fully developed throughout the soil above the failure surfaces. Heaving of the ground surface occurs on both sides of the footing although the final slip movement would occur only on one side, accompanied by tilting of the footing. This mode of failure is typical of soils of low compressibility (i.e. dense or stiff soils) and the pressure–settlement curve is of the general form shown in Figure 8.1, the ultimate bearing capacity being well defined. In the mode of *local shear failure* there is significant compression of the soil under the footing and only partial development of the state of plastic equilibrium. The failure surfaces, therefore, do not reach

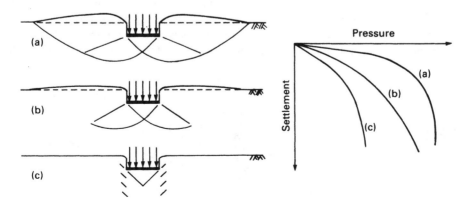

Figure 8.1 Modes of failure: (a) general shear, (b) local shear and (c) punching shear.

the ground surface and only slight heaving occurs. Tilting of the foundation would not be expected. Local shear failure is associated with soils of high compressibility and, as indicated in Figure 8.1, is characterized by the occurrence of relatively large settlements (which would be unacceptable in practice) and the fact that the ultimate bearing capacity is not clearly defined. *Punching shear failure* occurs when there is relatively high compression of the soil under the footing, accompanied by shearing in the vertical direction around the edges of the footing. There is no heaving of the ground surface away from the edges and no tilting of the footing. Relatively large settlements are also a characteristic of this mode and again the ultimate bearing capacity is not well defined. Punching shear failure will also occur in a soil of low compressibility if the foundation is located at considerable depth. In general the mode of failure depends on the compressibility of the soil and the depth of the foundation relative to its breadth.

The bearing capacity problem can be considered in terms of plasticity theory. The lower and upper bound theorems (Section 6.1) can be applied to give solutions for the ultimate bearing capacity of a soil. In certain cases, exact solutions can be obtained corresponding to the equality of the lower and upper bound solutions. However, such solutions are based on the assumption that the soil can be represented by a perfectly plastic stress–strain relationship, as shown in Figure 6.1. This approximation is only realistic for soils of low compressibility, i.e. soils corresponding to the general shear mode of failure. However, for the other modes, settlement and not shear failure is normally the limiting criterion.

A suitable failure mechanism for a strip footing is shown in Figure 8.2. The footing, of width B and infinite length, carries a uniform pressure q on the surface of a mass of homogeneous, isotropic soil. The shear strength parameters for the soil are denoted by the general symbols c and ϕ. As a simplifying assumption the unit weight of the soil is neglected (i.e. $\gamma = 0$). When the pressure becomes equal to the ultimate bearing capacity q_f the footing will have been pushed downwards into the soil mass, producing a state of plastic equilibrium, in the form of an active Rankine zone, below the footing, the angles ABC and BAC being $45° + \phi/2$. The downward movement of the wedge ABC forces the adjoining soil sideways, producing outward lateral forces on both sides of the wedge. Passive Rankine zones ADE and BGF therefore develop on both sides of the wedge ABC, the angles DEA and GFB being $45° - \phi/2$. The transition between the downward movement of the wedge ABC and the lateral movement of the wedges ADE and BGF takes place through zones of radial shear (also known as slip fans) ACD and BCG, the surfaces CD and CG being logarithmic spirals (or circular arcs if $\phi = 0$) to which BC and ED, or AC and FG, are tangential. A state of plastic equilibrium thus

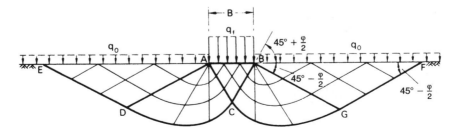

Figure 8.2 Failure under a strip footing.

exists above the surface EDCGF, the remainder of the soil mass being in a state of elastic equilibrium.

The following exact solution can be obtained, using plasticity theory, for the ultimate bearing capacity of a strip footing on the surface of a weightless soil, based on the mechanism described above. For the undrained condition ($\phi_u = 0$) in which the shear strength is given by c_u:

$$q_f = (2 + \pi)c_u = 5.14c_u \tag{8.1}$$

The derivation of this value has been given by Atkinson [1] and Parry [32].

For the general case in which the shear strength parameters are c and ϕ, it is necessary to consider a surcharge pressure q_o acting on the soil surface as shown in Figure 8.2; otherwise if $c = 0$ the bearing capacity of a soil for which the unit weight is neglected would be zero. The solution for this case, attributed to Prandtl and Reissner, is

$$q_f = c \cot \phi [\exp(\pi \tan \phi) \tan^2(45° + \phi/2) - 1]$$
$$+ q_o [\exp(\pi \tan \phi) \tan^2(45° + \phi/2)] \tag{8.2}$$

However, an additional term must be added to Equation 8.2 to take into account the component of bearing capacity due to the self-weight of the soil. This component can only be determined approximately, by numerical or graphical means, and is sensitive to the value assumed for the angles ABC and BAC in Figure 8.2.

Foundations are not normally located on the surface of a soil mass, as assumed in the above solutions, but at a depth D below the surface as shown in Figure 8.3. In applying these solutions in practice it is assumed that the shear strength of the soil between the surface and depth D is neglected, this soil being considered only as a surcharge imposing a uniform pressure $q_o = \gamma D$ on the horizontal plane at foundation level. This is a reasonable assumption for a *shallow* foundation (interpreted as a foundation for which the depth D is not greater than the breadth B). The soil above foundation level is normally weaker, especially if backfilled, than the soil at greater depth.

The ultimate bearing capacity of the soil under a shallow strip footing can be expressed by the following equation:

$$q_f = cN_c + \gamma D N_q + \frac{1}{2}\gamma B N_\gamma \tag{8.3}$$

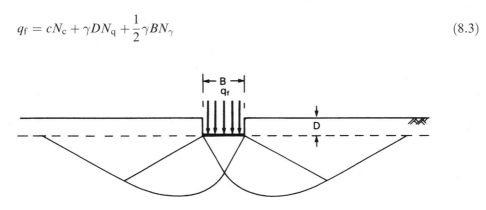

Figure 8.3 Footing at depth D below the surface.

where N_c, N_q and N_γ are bearing capacity factors depending only on the value of the shear strength parameter ϕ. The first term in Equation 8.3 is the contribution to bearing capacity due to the shear strength component represented by parameter c, the second term is the contribution due to the surcharge pressure and the third term is due to the weight of the soil below foundation level. The superposition of components of bearing capacity is theoretically incorrect for a plastic material but the resulting error is considered to be on the safe side.

The values of N_q and N_c implicit in Equation 8.2 should be used in bearing capacity calculations, i.e.

$$N_q = \exp(\pi \tan \phi) \tan^2 \left(45° + \frac{\phi}{2} \right)$$

$$N_c = (N_q - 1) \cot \phi$$

Values for factor N_γ have been obtained by Hansen [23] and Meyerhof [27], represented by the following approximations:

$$N_\gamma = 1.8(N_q - 1) \tan \phi \quad \text{(Hansen)}$$
$$N_\gamma = (N_q - 1) \tan (1.4\phi) \quad \text{(Meyerhof)}$$

In EC7 the following value is proposed:

$$N_\gamma = 2.0(N_q - 1) \tan \phi$$

Values of N_c, N_q and N_γ are plotted in terms of ϕ in Figure 8.4. Hansen's values of N_γ are used in the examples in this chapter.

The problems involved in extending the two-dimensional solution for a strip footing to three dimensions would be considerable. Accordingly, the bearing capacities of square, rectangular and circular footings are determined by means of semi-empirical shape factors applied to the solution for a strip footing. The bearing capacity factors N_c, N_q and N_γ should be multiplied by the respective shape factors s_c, s_q and s_γ. Various proposals for shape factors have been published. The following simplified values are sufficiently accurate for most cases in practice: $s_c = s_q = 1.2$ for both square and circular footings; $s_\gamma = 0.8$ for a square footing or 0.6 for a circular footing (i.e. the third term in Equation 8.3 becomes $0.4\gamma B N_\gamma$ or $0.3\gamma B N_\gamma$, respectively). For a rectangular footing of breadth B and length L, the shape factors are obtained by linear interpolation between the values for a strip footing ($B/L = 0$) and a square footing ($B/L = 1$), i.e. $s_c = s_q = 1 + 0.2B/L$ and $s_\gamma = 1 - 0.2B/L$. More detailed proposals for shape factors, as functions of ϕ', have been given by Hansen [23, 50] and DeBeer [17]. Detailed expressions are also given in EC7 [19]. Depth factors d_c, d_q and d_γ have also been given in terms of the ratio D/B but these should only be used if it is certain that the shear strength of the soil above foundation level is, and will remain, equal (or almost equal) to that below foundation level.

The effect of inclined loading on bearing capacity can be taken into account by means of inclination factors. If the angle of inclination of the resultant load to the

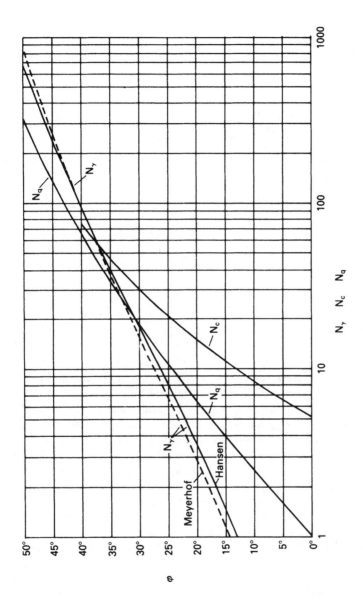

Figure 8.4 Bearing capacity factors for shallow foundations.

vertical is α then N_c, N_q and N_γ should be multiplied, respectively, by the following factors:

$$i_c = 1 - \frac{H}{2cB'L'} \tag{8.4}$$

$$i_q = 1 - \frac{1.5H}{V} \tag{8.5}$$

$$i_\gamma = i_q^2 \tag{8.6}$$

where V and H are the vertical and horizontal components of the resultant load, respectively. More detailed values of inclination factors, as functions of the shear strength parameters, have been given by Hansen [23, 50].

The bearing capacity equation can be written in general form by including the shape and inclination factors (plus depth factors if appropriate). Thus

$$q_f = cN_cs_ci_c + \gamma DN_qs_qi_q + \frac{1}{2}\gamma BN_\gamma s_\gamma i_\gamma \tag{8.7}$$

Footings may be subjected to eccentric and inclined loading resulting in a reduction in bearing capacity. If e is the eccentricity of the resultant load on the base of a footing of width B, an effective foundation width B' should be used in Equations 8.3 and 8.7, where

$$B' = B - 2e \tag{8.8}$$

The resultant load is assumed to be uniformly distributed over the effective width B'. If the resultant load is also eccentric in the length direction of a rectangular footing, a similar expression is used for the effective length L'.

It is vital that the appropriate values of unit weight are used in the bearing capacity equation. In an effective stress analysis three different situations should be considered. (1) If the water table is well below foundation level, the bulk unit weight (γ) is used in the second and third terms of Equation 8.3 or 8.7. (2) If the water table is at foundation level, the effective (buoyant) unit weight (γ') must be used in the third term (which represents the resistance due to the weight of the soil below foundation level), the bulk unit weight being used in the second term (representing the resistance due to the surcharge above foundation level). (3) If the water table is at the surface, the effective unit weight must be used in both the second and third terms. In the case of a sand ($c' = 0$) the first term is, of course, zero. In a total stress analysis of a foundation on fully saturated clay the saturated (i.e. total) unit weight (γ_{sat}) is used in the second term, the third term being zero ($N_\gamma = 0$ for $\phi_u = 0$).

For foundations under working load the maximum shear strain within the supporting soil will normally be less than that required to develop peak shear strength in dense sands or stiff clays. Strain must be low enough to ensure that the settlement of the foundation is acceptable. The allowable bearing capacity or the design bearing resistance should be calculated, therefore, using the peak strength parameters corresponding to the appropriate stress levels. It should be recognized, however, that the results of bearing capacity calculations are very sensitive to the values assumed for the shear strength parameters, especially the higher values of ϕ'. Due consideration must therefore be given to the probable degree of accuracy of the parameters.

Lumped and partial factors

The pressure applied to the soil at the base of a foundation due to all vertical loading above that level is referred to as the total or gross foundation pressure (q). The net foundation pressure (q_n) is the increase in pressure at foundation level being the total foundation pressure less the weight of soil per unit area permanently removed, i.e. the difference in pressure on the soil before and after construction. Thus

$$q_n = q - \gamma D$$

The unit weight γ may be either the total stress or the effective stress value, depending on the type of analysis.

According to the permissible stress method, the factor of safety (F) with respect to shear failure is defined in terms of net ultimate bearing capacity (q_{nf}), i.e.

$$F = \frac{q_{nf}}{q_n} = \frac{q_f - \gamma D}{q - \gamma D} \tag{8.9}$$

However in the case of shallow footings, if the value of ϕ is relatively high, there is no significant difference between the values of F in terms of net and total pressures.

In the limit state method the design bearing resistance (R_d) is calculated using the factored shear strength parameters. The vertical design action (V_d) is calculated from factored loads without subtracting the weight of overburden soil (because safety is not defined in terms of the ratio R_d/V_d). The bearing resistance limit state is satisfied if the design action is less than or equal to the design bearing resistance.

Skempton's values of N_c

In a review of bearing capacity theory, Skempton [40] concluded that in the case of saturated clays under undrained conditions ($\phi_u = 0$) the ultimate bearing capacity of a footing could be expressed by the equation:

$$q_f = c_u N_c + \gamma D \tag{8.10}$$

the factor N_c being a function of the shape of the footing and the depth/breadth ratio. Skempton's values of N_c are given in Figure 8.5. The factor for a rectangular footing of dimensions $B \times L$ (where $B < L$) is the value for a square footing multiplied by $0.84 + 0.16B/L$. The values of N_c may be used for stratified deposits provided the value of c_u for a particular stratum is not greater than nor less than the average value for all strata within the significant depth by more than 50% of that average value.

Base failure in excavations

Bearing capacity theory can be applied to the problem of base failure in braced excavations in clay under undrained conditions. The application is limited to the analysis of cases in which the bracing is adequate to prevent significant lateral deformation of the soil adjacent to the excavation. A simple failure mechanism, originally proposed by Terzaghi [45], is illustrated in Figure 8.6, the angle at *a* being

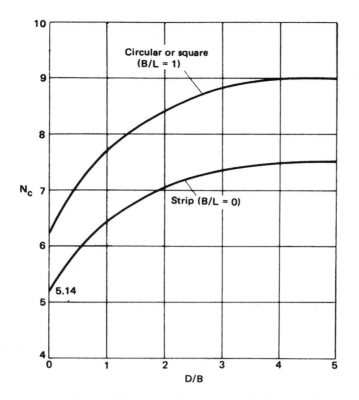

Figure 8.5 Skempton's values of N_c for $\phi_u = 0$. (Reproduced from A.W. Skempton (1951) *Proceedings of the Building Research Congress*, Division 1, p. 181, by permission of the Building Research Establishment, © *Crown copyright.*)

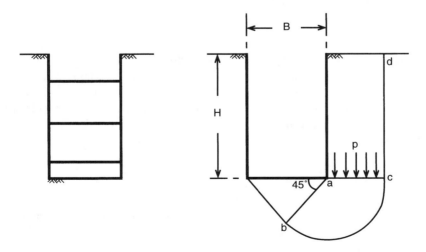

Figure 8.6 Base failure in excavation.

45° and bc being a circular arc if $\phi_u = 0$; therefore the length of ab is $(B/2)/\cos 45°$ (approximately $0.7B$). Failure occurs when the shear strength of the clay is insufficient to resist the average shear stress resulting from the vertical pressure (p) on ac due to the weight of the soil ($0.7\gamma BH$) reduced by the shear strength on cd (c_uH). Thus

$$p = \gamma H - \frac{c_u H}{0.7B}$$

The problem is essentially that of a bearing capacity analysis in reverse, the soil below the base being unloaded as excavation proceeds, there being zero pressure at the bottom of the excavation and p representing the overburden pressure. The shear strength available along the failure surface, acting in the opposite direction to that in the bearing capacity problem, can be expressed as $c_u N_c$, where N_c is the appropriate bearing capacity factor (taken to be 5.7 by Terzaghi). Thus for limiting equilibrium

$$c_u N_c = \gamma H - \frac{c_u H}{0.7B}$$

The factor of safety against base failure is given by

$$F_b = \frac{c_u N_c}{(\gamma - c_u/0.7B)H} \tag{8.11}$$

If a firm stratum were to exist at depth D_f below the base of the excavation, where $D_f < 0.7B$, then D_f replaces $0.7B$ in Equation 8.11.

Based on observations of actual base failures in Oslo, Bjerrum and Eide [5] concluded that Equation 8.11 gave reliable results only in the case of excavations with relatively low depth/breadth ratios. In the case of excavations with relatively large depth/breadth ratios, local failure occurred before shear failure on cd was fully mobilized up to surface level. Bjerrum and Eide proposed that Skempton's bearing capacity factors (Figure 8.5, with D being replaced by H) should be used in the analysis of base heave, the condition of limiting equilibrium being given by writing $q_f = 0$ in Equation 8.10. Thus, the factor of safety against base failure is given by

$$F_b = \frac{c_u N_c}{\gamma H} \tag{8.12}$$

If a surcharge pressure (q) acts on the surface adjacent to the excavation, the denominator in Equation 8.12 becomes $\gamma H + q$.

Example 8.1

A footing 2.25×2.25 m is located at a depth of 1.5 m in a sand, the shear strength parameters to be used in design being $c' = 0$ and $\phi' = 38°$. Determine the ultimate bearing capacity (a) if the water table is well below foundation level and (b) if the water table is at the surface. The unit weight of the sand above the water table is $18\,\mathrm{kN/m^3}$; the saturated unit weight is $20\,\mathrm{kN/m^3}$.

For a square footing the ultimate bearing capacity (with $c = 0$) is given by

$$q_f = \gamma D N_q + 0.4 \gamma B N_\gamma$$

For $\phi' = 38°$ the bearing capacity factors (Figure 8.4) are $N_\gamma = 67$ and $N_q = 49$. Therefore

$$q_f = (18 \times 1.5 \times 49) + (0.4 \times 18 \times 2.25 \times 67)$$

$$= 1323 + 1085$$

$$= 2408\,\mathrm{kN/m^2}$$

When the water table is at the surface, the ultimate bearing capacity is given by

$$q_f = \gamma' D N_q + 0.4 \gamma' B N_\gamma$$

$$= (10.2 \times 1.5 \times 49) + (0.4 \times 10.2 \times 2.25 \times 67)$$

$$= 750 + 615$$

$$= 1365\,\mathrm{kN/m^2}$$

Example 8.2

A strip footing is to be designed to support a dead load of 500 kN/m and an imposed load of 300 kN/m at a depth of 0.7 m in a gravelly sand. Characteristic values of the shear strength parameters are $c' = 0$ and $\phi' = 40°$. Determine the required width of the footing if a (lumped) factor of safety of 3.0 against shear failure is specified and assuming that the water table may rise to foundation level. Would a foundation of that width satisfy the bearing resistance limit state? The unit weight of the sand above the water table is $17\,\mathrm{kN/m^3}$ and below the water table the saturated unit weight is $20\,\mathrm{kN/m^3}$.

For $\phi' = 40°$ the bearing capacity factors (Figure 8.4) are $N_\gamma = 95$ and $N_q = 64$. For an analysis in terms of effective stress (the buoyant unit weight of the sand being relevant below the water table), the gross ultimate bearing capacity (units $\mathrm{kN/m^2}$) is

$$q_f = \gamma D N_q + \frac{1}{2}\gamma' B N_\gamma$$

$$= (17 \times 0.7 \times 64) + \left(\frac{1}{2} \times 10.2 \times B \times 95\right)$$

$$= 762 + 485B$$

Then the net ultimate bearing capacity is

$$q_{nf} = q_f - \gamma D = 750 + 485B$$

It should be noted that net ultimate bearing capacity would be obtained directly if $N_q - 1$ were used in the bearing capacity equation.

The net foundation pressure is

$$q_n = \frac{800}{B} - (17 \times 0.7)$$

Then, for a factor of safety of 3,

$$\frac{1}{3}(750 + 485B) = \frac{800}{B} - 12$$

Hence $B = 1.55\,\mathrm{m}$

In the limit state approach, Case C is relevant to the determination of footing width. Using the partial factors given in Section 8.1 the design value of ϕ' is $\tan^{-1}(\tan 40°/1.25) = 34°$ (i.e. $\phi'_d = \tan^{-1}(\tan \phi'_k/1.25)$) and the corresponding bearing capacity factors are $N_\gamma = 35$ and $N_q = 30$. Hence the design bearing resistance is

$$R_d = 1.55\left[(17 \times 0.7 \times 30) + \left(\frac{1}{2} \times 10.2 \times 1.55 \times 35\right)\right] = 982\,\mathrm{kN/m}$$

and the design action is

$$V_d = (500 \times 1.00) + (300 \times 1.30) = 890\,\mathrm{kN/m}$$

The design action is less than the design bearing resistance; therefore, the bearing resistance limit state is satisfied.

For Case B, which is relevant to the structural design of the footing, the design value of ϕ' is $40°$ (partial factor 1.00), then $N_q = 64$, $N_\gamma = 95$ and R_d is calculated to be $2344\,\mathrm{kN/m}$. The partial factors used in the calculation of the design action are 1.35 and 1.50, hence V_d is $(500 \times 1.35) + (300 \times 1.50) = 1125\,\mathrm{kN/m}$, i.e. considerably less than R_d. Case C clearly controls the design.

Example 8.3

A foundation $2.0 \times 2.0\,\mathrm{m}$ is located at a depth of $4.0\,\mathrm{m}$ in a stiff clay of saturated unit weight $21\,\mathrm{kN/m^3}$. The undrained shear strength at a depth of $4.0\,\mathrm{m}$ is given by the characteristic parameter $c_u = 120\,\mathrm{kN/m^2}$ ($\phi_u = 0$). The foundation supports a permanent load of $1200\,\mathrm{kN}$ and a variable load of $700\,\mathrm{kN}$. Is the foundation satisfactory with respect to the bearing resistance limit state? What is the factor of safety with respect to shear strength?

In the limit state approach the design action is

$$V_d = (1200 \times 1.00) + (700 \times 1.30) = 2110\,\mathrm{kN}$$

The design value of c_u is $120/1.40$, i.e. $85\,kN/m^2$. In this case $D/B = 2$, then from Figure 8.5 the value of coefficient N_c is 8.4. For $\phi_u = 0$, $N_q = 1$. Hence the design bearing resistance is

$$R_d = 2^2[(85 \times 8.4) + (21 \times 4)] = 3192\,kN$$

The design action is less than the design bearing resistance; therefore, the foundation is satisfactory with respect to the bearing resistance limit state.

The design should be checked for the drained condition using parameters c' and ϕ'.

In the permissible stress method unfactored loads are used, hence the net foundation pressure is

$$q_n = \left(\frac{1900}{2^2}\right) - (21 \times 4) = 391\,kN/m^2$$

The net ultimate bearing capacity is

$$q_{nf} = 120 \times 8.4 = 1008\,kN/m^2$$

The factor of safety is $1008/391 = 2.58$.

Example 8.4

The base of a long retaining wall is 3 m wide and is 1 m below the ground surface in front of the wall: the water table is well below base level. The vertical and horizontal components of the base reaction are 282 and $102\,kN/m$, respectively. The eccentricity of the base reaction is 0.36 m. Appropriate shear strength parameters for the foundation soil are $c' = 0$ and $\phi' = 35°$, and the unit weight of the soil is $18\,kN/m^3$. Determine the factor of safety against shear failure.

The effective width of the base is given by

$$B' = B - 2e = 2.28\,m$$

For $\phi' = 35°$ the bearing capacity factors (Figure 8.4) are $N_q = 33$ and $N_\gamma = 41$.

The inclination factors are:

$$i_q = 1 - \left(\frac{1.5 \times 102}{282}\right) = 0.46$$

$$i_\gamma = 0.46^2 = 0.21$$

The ultimate bearing capacity is given by

$$q_f = \gamma D N_q i_q + \frac{1}{2}\gamma B' N_\gamma i_\gamma$$

$$= (18 \times 1 \times 33 \times 0.46) + \left(\frac{1}{2} \times 18 \times 2.28 \times 41 \times 0.21\right)$$

$$= 273 + 177 = 450 \, \text{kN/m}^2$$

$$\therefore q_{nf} = q_f - \gamma D = 432 \, \text{kN/m}^2$$

The net base pressure is

$$q_n = \frac{282}{2.28} - 18 = 106 \, \text{kN/m}^2$$

Then the factor of safety is

$$F = \frac{q_{nf}}{q_n} = \frac{432}{106} = 4.0$$

8.3   ALLOWABLE BEARING CAPACITY OF CLAYS

The allowable bearing capacity of clays, silty clays and plastic silts may be limited either by the requirement of an adequate factor of safety against shear failure or by settlement considerations. Similarly, in the limit state approach, either the ultimate or the service-ability state may be the ruling criterion. Shear strength, and hence the factor of safety, will increase whenever consolidation takes place. For homogeneous clays with low mass permeability, the factor of safety, therefore, should be checked for the condition imme-diately after construction, using the undrained shear strength parameters. However, in the case of clays exhibiting significant macro-fabric features the mass permeability may be relatively high and the undrained condition may be overconservative at the end of construction. The methods of estimating the immediate settlement under undrained conditions and the long-term consolidation settlement are detailed in Chapters 5 and 7, respectively. For most cases in practice, simple settlement calculations are adequate provided that reliable values of soil parameters for the *in-situ* soil have been determined. It should be appreciated that the precision of settlement predictions is much more influenced by inaccuracies in the values of soil parameters than by shortcomings in the methods of analysis. Sampling disturbance can have a serious effect on the values of parameters determined in the laboratory. In settlement analysis the same degree of precision should not be expected as, for example, in structural calculations.

The factor of safety and immediate settlement should be estimated on the basis of dead load plus initial (short-term) imposed load. Estimates of consolidation settlement should be based on dead load plus the average imposed load expected over a long period of time.

Settlement on overconsolidated clays depends on whether the preconsolidation pres-sure is exceeded, and if so to what extent, for a given foundation. The bearing pressure should normally be limited so that the preconsolidation pressure is not exceeded. In the case of a series of footings, differential settlement may be reduced by increasing the size of the largest footings above that required by the allowable bearing capacity. Foundations are not usually supported on normally consolidated clays because the resulting con-solidation settlement would almost certainly be excessive.

If a stratum of soft clay lies below a firm stratum in which footings are located, there is a possibility that the footings may break through into the soft stratum. Such a possibility can be avoided if the vertical stress increments at the top level of the clay are less than the allowable bearing capacity of the clay by an adequate factor.

8.4   ALLOWABLE BEARING CAPACITY OF SANDS

In this section the term 'sand' includes gravelly sands, silty sands and non-plastic silts. Most sand deposits are non-homogeneous and the allowable bearing capacity of shallow foundations is limited by settlement considerations except possibly in the case of narrow footings. Thus in limit state design the serviceability limit state is normally the governing criterion and in the traditional approach the allowable settlement is reached at a pressure for which the factor of safety against shear failure is greater than 3. In the case of narrow footings, however, shear failure may be the limiting consideration. Other factors being equal, the pressure that will produce the allowable settlement in a dense sand will be greater than that needed to produce the allowable settlement in a loose sand. Settlement in sand is rapid and occurs almost entirely during construction and initial loading. Settlement, therefore, should be estimated using the dead load plus the maximum imposed load.

Differential settlement between a number of footings is governed mainly by variations in the homogeneity of the sand within the significant depth and to a lesser extent by variations in foundation pressure. Settlement records published over many years indicate that differential settlement between footings of approximately equal size carrying the same pressure is unlikely to exceed 50% of the maximum settlement. If the footings are of different size the differential settlement will be greater. For footings carrying the same pressure the maximum settlement increases with increasing footing size. There is no appreciable difference between the settlement of square and strip footings of the same width. For a given pressure and footing size the settlement decreases slightly with increasing footing depth below ground level due to the fact that the lateral confining pressure will be greater. In most cases, even under extreme variations of footing size and depth, it is unlikely that differential settlement will be greater than 75% of the maximum settlement. A few cases have been reported, however, in which the differential settlement was almost equal to the maximum settlement.

A reasonable design criterion for footings on sands is an allowable maximum settlement of 25 mm. The differential settlement between any two footings is then likely to be less than 20 mm. Differential settlement may be decreased by reducing the size of the smallest footings, provided the factor of safety with respect to shear failure remains above the specified value.

The settlement distribution under a raft is different from that for a series of footings. The settlement of footings is governed by the soil characteristics relatively near the surface and any one footing may be influenced by a loose pocket of soil. The settlement of a raft, on the other hand, is governed by the soil characteristics over a much greater depth. Loose pockets of soil may occur at random within this depth but they tend to be bridged over. The differential settlement of a raft as a percentage of the maximum settlement is roughly half the corresponding percentage for a series of footings. Thus for a differential settlement of 20 mm or less, the same as for a series of footings, the criterion for the maximum settlement of a raft is 50 mm.

The allowable bearing capacity of a sand depends primarily on the density index, the stress history, the position of the water table relative to foundation level and the size of the foundation. Of secondary importance are particle shape and grading. Both the magnitude of settlement and the value of the shear strength parameter ϕ' are strongly dependent on density index; the denser the sand, the less scope there is for particle rearrangement. However, the magnitude of settlement is also influenced by the stress history of the deposit, i.e. whether the sand is normally consolidated or overconsoli- dated and the previous stress path. If two sands having the same grading were to exist at the same density index but one were normally consolidated and the other over- consolidated, the settlement would be greater in the normally consolidated sand for identical loading conditions. The water table position affects both the settlement and the ultimate bearing capacity. If the sand within the significant depth is fully saturated the effective unit weight is roughly halved, resulting in a reduction in lateral confining pressure and a corresponding increase in settlement; the reduced effective unit weight will also result in a lower value of ultimate bearing capacity. The size of the foundation governs the depth to which the soil characteristics are relevant. It should be realized that unpredictable settlement can be caused by a reduction in density index due to disturbance of the sand during construction. Settlement can also be caused as a result of a reduction in lateral confining pressure, e.g. due to adjacent excavation. If a sand deposit is loose, vibration may result in volume decrease, causing appreciable settle- ment. Loose sands should be compacted prior to construction, e.g. by using the technique of *vibro-compaction* (Section 8.6), or else piles should be used.

Due to the extreme difficulty of obtaining undisturbed sand samples for laboratory testing and to the inherent heterogeneity of sand deposits, the allowable bearing capacity is normally estimated by means of correlations based on the results of *in-situ* tests. The tests in question are plate bearing tests and dynamic or static penetration tests.

The plate bearing test

In this test the sand is loaded through a steel plate at least 300 mm square, readings of load and settlement being observed up to failure or to at least 1.5 times the estimated allowable bearing capacity. The load increments should be approximately one-fifth of the estimated allowable bearing capacity. The test plate is generally located at founda- tion level in a pit at least 1.5 m square. The test is reliable only if the sand is reasonably uniform over the significant depth of the full-scale foundation. Minor local weaknesses near the surface will influence the results of the test while having no appreciable effect on the full-scale foundation. On the other hand, a weak stratum below the significant depth of the test plate but within the significant depth of the foundation, as shown in Figure 8.7, would have no influence on the test results; the weak stratum, however, would have an appreciable effect on the performance of the foundation.

Settlement in a sand increases as the size of the loaded area increases and the main problem with the use of plate bearing tests is the extrapolation of the settlement of a test plate to that of a full-scale foundation. The required correlation appears to depend on the density index, particle size distribution and stress history of the sand, and at

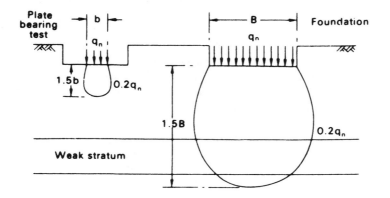

Figure 8.7 Influence of weak stratum.

present there is no reliable method of extrapolation. Bjerrum and Eggestad [4], for example, from a study of case records, showed that there is a considerable scatter in the relationship between settlement and the size of the loaded area for a given pressure. Ideally, plate bearing tests should be carried out at different depths and using plates of different sizes in order that extrapolations may be made, but this is generally ruled out on economic grounds; further problems would be introduced if the tests had to be carried out below water table level.

The screw-plate test is a form of bearing test in which no excavation is required. The plate penetrates the sand by rotation and can therefore be positioned, in turn, at a series of depths above or below the water table. Loading is carried out through the shaft of the screw plate.

The standard penetration test

This dynamic penetration test, specified in BS 1377 (Part 9), is used to assess the density index of a sand deposit. The test is performed using a split-barrel sampler (Figure 10.5(c)), 50 mm in external diameter, 35 mm in internal diameter and about 650 mm in length, connected to the end of boring rods. The sampler is driven into the sand at the bottom of a cased borehole by means of a 65 kg hammer falling freely through a height of 760 mm onto the top of the boring rods. In the UK a trip-release mechanism and guiding assembly are normally used to control the fall of the hammer, and an anvil at the lower end of the assembly is used to transmit the blow to the boring rods. However, different methods of releasing the hammer are used in different countries. The borehole must be cleaned out to the required depth, care being taken to ensure that the material to be tested is not disturbed: jetting as part of the boring operation is undesirable. The casing must not be driven below the level at which the test is to begin.

Initially the sampler is driven 150 mm into the sand to seat the device and to bypass any disturbed sand at the bottom of the borehole. The number of blows required to drive the sampler a further 300 mm is then recorded: this number is called the *standard penetration resistance* (N). The number of blows required for each 75 mm of penetration (including the initial drive) should be recorded separately. If 50 blows are reached before a penetration of 300 mm, no further blows should be applied but the actual penetration should be recorded. At the conclusion of a test the sampler is withdrawn and the sand extracted. Tests are normally carried out at intervals of between 0.75 and 1.50 m to a depth below foundation level at least equal to the width (B) of the foundation. If the test is to be carried out in gravelly soils the driving shoe is replaced by a solid 60° cone. There is evidence that slightly higher results are obtained in the same material when the normal driving shoe is replaced by the 60° cone.

When testing below the water table, care must be taken to avoid entry of water through the bottom of the borehole as this would tend to loosen the sand due to upward seepage pressure. Water should be added as necessary to maintain the water table level in the borehole (or at the level required to balance any excess pore water pressure). When the test is carried out in fine sand or silty sand below the water table the measured N value, if greater than 15, should be corrected for the increased resistance due to negative excess pore water pressure set up during driving and unable to dissipate immediately: the corrected value is given by

$$N' = 15 + \frac{1}{2}(N - 15) \tag{8.13}$$

Skempton [42] summarized the evidence regarding the influence of test procedure on the value of standard penetration resistance. Measured N values should be corrected to allow for the different methods of releasing the hammer, the type of anvil and the total length of boring rods. Only energy delivered to the sampler is applied in penetrating the sand, the ratio of the delivered energy to the free-fall energy of the hammer being referred to as the rod energy ratio. Rod energy ratios for the operating procedures used in several countries vary between 45 and 78%. For the trip-release mechanism, guiding assembly and anvil generally used in the UK the energy ratio for rod lengths exceeding 10 m is 60%. It has been recommended that a standard rod energy ratio of 60% should be adopted and that all measured N values should be normalized, by simple proportion of energy ratios, to this standard: the normalized values are denoted N_{60}. If a short length of boring rods (<10 m) is used in a test, a reflection of energy occurs and a further loss in delivered energy results. A further correction should therefore be applied to the measured N values if the total length of rods is less than 10 m, e.g. if a 3–4 m length is used a correction factor of 0.75 has been proposed. An additional effect relates to the borehole diameter, there being evidence that lower N values are obtained in 150 and 200 mm diameter boreholes than in those less than 115 mm in diameter. Tentative correction factors for 150 and 200 mm boreholes are 1.05 and 1.15, respectively.

A density classification for sands was proposed originally, in general terms, by Terzaghi and Peck, on the basis of standard penetration resistance, as shown in columns (1) and (2) of Table 8.3. Numerical values of density index, as shown in column (3), were subsequently added by Gibbs and Holtz [22]. However, standard penetration resistance depends not only on density index but also on the effective stresses at the depth of measurement; effective stresses can be represented to a first approximation by effective overburden pressure. This dependence was first demonstrated in the laboratory by Gibbs and Holtz and was later confirmed in the field. Sand at the same density index would thus give different values of standard penetration

Table 8.3 Density index of sands

(1) N value	(2) Classification	(3) I_D (%)	(4) $(N_1)_{60}$
0–4	Very loose	0–15	0–3
4–10	Loose	15–35	3–8
10–30	Medium dense	35–65	8–25
30–50	Dense	65–85	25–42
>50	Very dense	85–100	42–58

resistance at different depths. Several proposals have been made for the correction of measured N values following the work of Gibbs and Holtz. The corrected value (N_1) is related to the measured value (N) by the factor C_N, where

$$N_1 = C_N N \tag{8.14}$$

The relationship between C_N and effective overburden pressure shown in Figure 8.8 represents a consensus of published proposals.

The following relationship between standard penetration resistance (N), density index (I_D) and effective overburden pressure (σ'_0 kN/m^2) was proposed by Meyerhof:

$$\frac{N}{I_D^2} = a + b\frac{\sigma'_0}{100} \tag{8.15}$$

Values of the parameters a and b for a number of sands have been given by Skempton [42]. The characteristics of a sand can be represented by $(N_1)_{60}$ and $(N_1)_{60}/I_D^2$ where $(N_1)_{60}$ is the standard penetration resistance normalized to a rod energy ratio of 60%

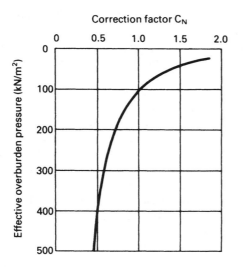

Figure 8.8 Correction of measured values of standard penetration resistance.

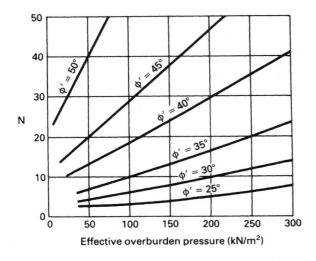

Figure 8.9 Correlation between shear strength parameter ϕ', standard penetration resistance and effective overburden pressure. (Reproduced from J.H. Schmertmann (1975) *Proceedings of Conference on In-Situ Measurement of Soil Properties*, by permission of the American Society of Civil Engineers.)

and an effective overburden pressure of $100 \, \text{kN/m}^2$. Appropriate values of $(N_1)_{60}$ were added to the Terzaghi and Peck classification of density index by Skempton, as shown in column (4) of Table 8.3. Table 8.3 should be considered to apply to normally consolidated sands.

There is evidence that standard penetration resistance is also influenced by the grading and shape of the particles, the degree of overconsolidation and the time during which the sand has been undergoing consolidation (referred to as the ageing effect). All other factors being equal, the evidence indicates that standard penetration resistance increases with increasing particle size, increasing OCR and ageing.

A correlation between the shear strength parameter ϕ', standard penetration resistance and effective overburden pressure, published by Schmertmann [38], but based on previous work by DeMello, is shown in Figure 8.9. It must be appreciated that this chart provides only a rough estimate of the value of ϕ' and should not be used for very shallow depths.

Associated design methods

The development of procedures to obtain the allowable bearing capacity of sands is described historically in this section. Initially, in 1948, Terzaghi and Peck [46], in their first edition, presented empirical correlations between standard penetration resistance, width of footing and the bearing pressure limiting maximum settlement to 25 mm (and differential settlement to 75% of maximum settlement). According to Terzaghi and Peck, the correlations, represented in Figure 8.10, are applicable to situations in which the water table is not less than $2B$ below the footing, where B is the width of the footing. If the sand at foundation level is saturated, the pressures obtained from Figure 8.10 should

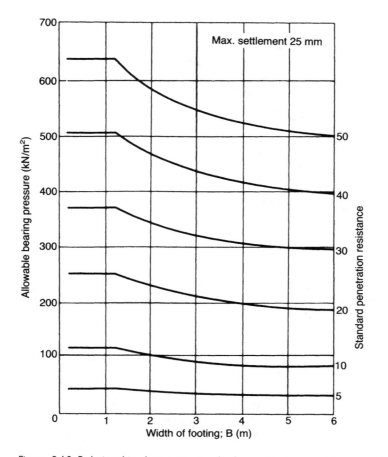

Figure 8.10 Relationship between standard penetration resistance and allowable bearing pressure. (Reproduced from K. Terzaghi and R.B. Peck (1967) *Soil Mechanics in Engineering Practice*, by permission of John Wiley & Sons, Inc.)

be reduced by one-half if the depth/breadth ratio of the footing is zero, and reduced by one-third if the depth/breadth ratio is unity. For intermediate positions of the water table and intermediate values of depth/breadth ratio the appropriate value of bearing pressure may be obtained by linear interpolation. However, these recommendations subsequently were considered to produce too severe a reduction in allowable pressure, and a correction should be made only if the water table is within a depth B below the foundation. Peck *et al.* [34] proposed that linear interpolation should be used between a reduction of 50% if the water table is at ground level and zero reduction if the water table is at depth B below the foundation. Thus the provisional value of allowable bearing pressure obtained from Figure 8.10 should be multiplied by a factor C_w, given by

$$C_w = 0.5 + 0.5 \frac{D_w}{D + B} \tag{8.16}$$

where D_w is the depth of the water table below the surface and D the depth of the foundation.

Terzaghi and Peck stated that their correlations were a conservative basis for the design of shallow footings. It was intended that the largest footing should not settle by more than 25 mm even if it were situated on the most compressible pocket of sand. It should be realized that the Terzaghi and Peck correlations were not intended to yield actual settlement values for particular footings but only to ensure that the maximum settlement *would not exceed* 25 mm.

In using the correlations the average N value (corrected as appropriate), to depth B below the foundation, is determined for each borehole and the lowest average is then used in design. For a series of footings the bearing pressure is obtained for the largest footing; this value of pressure is then used to calculate the dimensions of all other footings, subject to a check on the factor of safety against shear failure. In the case of rafts the allowable bearing pressure obtained from the design chart should be doubled because a maximum settlement of 50 mm is considered acceptable.

Settlement measurements on actual structures have shown that the Terzaghi and Peck method (incorporating the water table correction as originally proposed) is excessively conservative. Meyerhof [28] recommended that the allowable bearing pressure given by the Terzaghi and Peck method should be increased by 50% and that no correction should be applied for the position of the water table, arguing that its effect is reflected in the measured N values.

The influence of effective overburden pressure was not considered in the original Terzaghi and Peck correlations and it is now recognized that corrected values of standard penetration resistance (N_1), determined from Figure 8.8, should be used in determining allowable bearing pressures. It should be noted that the stress history of the sand is not taken into account in the Terzaghi and Peck design procedure.

Burland *et al.* [10] collated settlement data from a number of sources and plotted settlement per unit pressure (s/q) against foundation breadth (B): lines representing upper limits were then drawn for dense and medium-dense sands (as defined in Table 8.3).

This graph, shown in Figure 8.11, may be adequate for routine work. The 'probable' settlement could perhaps be taken as 50% of the upper limit value. In most cases the maximum settlement would be unlikely to exceed 75% of the upper limit value. Factors such as foundation depth and water table position are not considered. Use of the graph implies that the settlement–pressure relationship remains approximately linear.

Burland and Burbidge [11] subsequently carried out a statistical analysis of over 200 settlement records of foundations on sands and gravels. A relationship was established between the compressibility of the soil (a_f), the width of the foundation (B) and the average value of standard penetration resistance ($\overline{N}$) over the depth of influence of the foundation. The compressibility is given by the slope of the pressure–settlement plot, in mm/(kN/m^2), over the working range of pressure. Evidence was presented which indicated that if N tends to increase with depth or is approximately constant with depth then the ratio of the depth of influence to foundation width (z_I/B) decreases with increasing foundation width; values of z_I obtained from

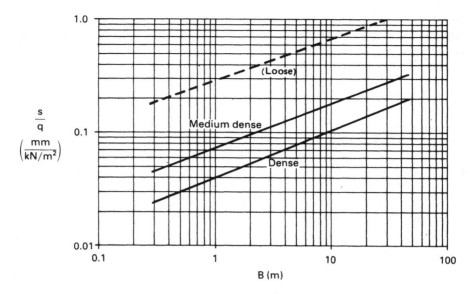

Figure 8.11 Envelopes of settlement per unit pressure. (Reproduced from J.B. Burland, B.B. Broms and V.F.B. De Mello (1977) *Proceedings 9th International Conference SMFE*, Tokyo, Vol. 2, by permission of the Japanese Society of Soil Mechanics and Foundation Engineering.)

Figure 8.12 can be used as a guide in design. However, if N tends to decrease with depth, the value of z_I should be taken as $2B$, provided the stratum thickness exceeds this value. The compressibility is related to foundation width by a compressibility index (I_c), where

$$I_c = \frac{a_f}{B^{0.7}} \tag{8.17}$$

The compressibility index, in turn, is related to the average value of standard penetration resistance ($\overline{N}$) by the expression

$$I_c = \frac{1.71}{\overline{N}^{1.4}} \tag{8.18}$$

The N values should not be corrected for effective overburden pressure as this has a major influence on both standard penetration resistance and compressibility; this influence, therefore, should not be eliminated from the correlation. The results of the analysis tend to confirm Meyerhof's conclusion that the influence of water table level is reflected in the measured N values. However, the position of the water table does influence *settlement* and if the level were to fall subsequent to the determination of the N values then a greater settlement would be expected. Equation 8.13 should be applied in the case of fine sands and silty sands below the water table. It was further

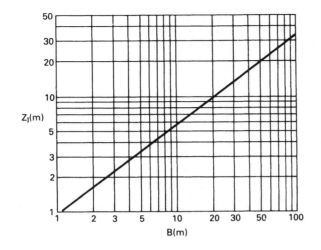

Figure 8.12 Relationship between depth of influence and foundation width. (Reproduced from J.B. Burland and M.C. Burbidge (1985) *Proceedings Institution of Civil Engineers*, Part I, Vol. 78, by permission of Thomas Telford Ltd.)

proposed that in the case of gravels or sandy gravels the measured N values should be increased by 25%.

In a normally consolidated sand the average settlement s_i (mm) at the end of construction for a foundation of width B (m) carrying a foundation pressure q (kN/m^2) is given by

$$s_i = qB^{0.7}I_c \qquad (8.19a)$$

If it can be established that the sand is overconsolidated and an estimate of preconsolidation pressure (σ'_c) can be made, the settlement is given by one or other of the following expressions:

$$s_i = \left(q - \frac{2}{3}\sigma'_c\right)B^{0.7}I_c \quad (\text{if } q > \sigma'_c) \qquad (8.19b)$$

$$s_i = qB^{0.7}\frac{I_c}{3} \quad (\text{if } q < \sigma'_c) \qquad (8.19c)$$

The analysis indicated that foundation depth had no significant influence on settlement for depth/breadth ratios less than 3. However, a significant correlation was found between settlement and the length/breadth ratio (L/B) of the foundation; accordingly the settlement given by Equations 8.19a–8.19c should be multiplied by a shape factor f_s, where

$$f_s = \left(\frac{1.25L/B}{L/B + 0.25}\right)^2 \qquad (8.20)$$

Table 8.4 Compressibility classification for normally consolidated sands and gravels (Burland and Burbidge [11])

N value	Compressibility grade
0–4	VII
4–8	VI
9–15	V
16–25	IV
26–40	III
41–60	II
>60	I

It was tentatively proposed that if the thickness (H) of the sand stratum below foundation level is less than the depth of influence (z_I), the settlement should be multiplied by a factor f_1, where

$$f_1 = \frac{H}{z_I}\left(2 - \frac{H}{z_I}\right) \tag{8.21}$$

Although it is normally assumed that settlement in sands is virtually complete by the end of construction and initial loading, the records indicated that continuing settlement can occur and it was proposed that the settlement should be multiplied by a factor f_t for time in excess of 3 years after the end of construction, where

$$f_t = \left(1 + R_3 + R_t \, \log\left(\frac{t}{3}\right)\right) \tag{8.22}$$

where R_3 is the time-dependent settlement, as a proportion of s_i, occurring during the first 3 years after construction and R_t the settlement occurring during each log cycle of time in excess of 3 years. A conservative interpretation of the data indicates that after 30 years f_t can reach 1.5 for static loads and 2.5 for fluctuating loads.

It should be noted that, unlike the Terzaghi and Peck procedure, the Burland and Burbidge method enables a specific value of settlement to be predicted for a given foundation pressure. The Burland and Burbidge procedure is now the recommended method in design, provided sufficient data are available. In the absence of data, the use of Figure 8.11 may be adequate for routine design.

Burland and Burbidge also introduced the concept of compressibility grades, based on uncorrected N values, as detailed in Table 8.4, these grades being a function of both density index and overburden pressure. Charts for the assessment of compressibility grades from the results of plate bearing tests were also presented.

Example 8.5

A footing 3×3 m is to be located at a depth of 1.5 m in a sand deposit, the water table being 3.5 m below the surface. Values of standard penetration resistance were determined as detailed in Table 8.5. Determine the allowable bearing capacity using the various design methods.

Table 8.5

Depth (m)	N	σ_v' (kN/m^2)	C_N	N_1
0.75	8	–	–	–
1.55	7	26	2.0	14
2.30	9	39	1.6	14
3.00	13	51	1.4	18
3.70	12	65	1.25	15
4.45	16	70	1.2	19
5.20	20	–	–	–
				(av. 16)

Terzaghi and Peck recommended that N values should be determined between foundation level and a depth of approximately B below the foundation; in this case between depths of 1.5 and 4.5 m: the values at depths of 0.75 and 5.20 m are therefore superfluous. The measured N values are corrected using Equation 8.14. Values of effective overburden pressure are calculated (using $\gamma = 17\,\mathrm{kN/m^3}$ above the water table and $\gamma' = 10\,\mathrm{kN/m^3}$ below the water table) and the corresponding values of C_N determined from Figure 8.8. The average of the corrected values (N_1) is 16. Then referring to Figure 8.10, for $B = 3\,\mathrm{m}$ and $N = 16$, the provisional value of allowable bearing capacity is $165\,\mathrm{kN/m^2}$. For the given water table level the provisional value should be multiplied by the factor C_w (Equation 8.16), where

$$C_w = 0.5 + \frac{0.5 \times 3.5}{4.5} = 0.89$$

Hence the allowable bearing capacity is given by

$$q_a = 0.89 \times 165 = 150\,\mathrm{kN/m^2}$$

Using Meyerhof's method the average of the measured N values between depths of 1.5 and 4.5 m is 11. For $B = 3\,\mathrm{m}$ and $N = 11$ the provisional value of allowable bearing capacity is $100\,\mathrm{kN/m^2}$. This value is increased by 50% with no correction being made for the position of the water table. Thus

$$q_a = 1.5 \times 100 = 150\,\mathrm{kN/m^2}$$

Using the Burland and Burbidge method, and assuming that the sand is normally consolidated, the depth of influence (Figure 8.12) for $B = 3\,\mathrm{m}$ is 2.2 m, i.e. 3.7 m below the surface. The average of the measured N values between depths of 1.5 and 3.7 m is 10, hence the compressibility index (Equation 8.17) is given by

$$I_c = \frac{1.71}{10^{1.4}} = 0.068$$

Then the allowable bearing capacity for a settlement of 25 mm at the end of construction is given by

$$q_a = \frac{s_i}{B^{0.7}I_c} = \frac{25}{3^{0.7} \times 0.068} = 170 \, \text{kN/m}^2$$

In design the usual requirement is to determine the foundation dimensions for the support of a given load and an iterative technique is necessary.

The static cone penetration test

The device used in this test consists of a cylindrical penetrometer (or probe), the lower end of which is fitted with a cone having an apex angle of 60° (Figure 8.13(a)). The diameter of the device is 35.7 mm, the projected area of the cone thus being 1000 mm². The penetrometer is attached to the lower end of a string of hollow boring rods, the diameter of the rods also being 35.7 mm. The rig is usually mounted on a trailer or vehicle. The test (specified in BS 1377, Part 9) consists of pushing the penetrometer directly into the ground (i.e. no boring is involved) at a rate of 20 mm/s by means of static thrust, usually applied by hydraulic jacking, and measuring the *cone penetration resistance* (q_c), defined as the force required to advance the cone divided by the projected area. The penetrometer normally incorporates a slightly roughened sleeve 1337 mm in length, giving a surface area of 150 cm². The sleeve enables the local frictional resistance (f_s) between sleeve and soil, defined as the frictional force on the sleeve divided by the surface area, to be measured. The resistances are measured by

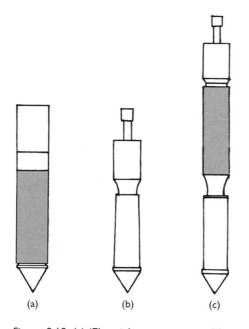

(a)   (b)   (c)

Figure 8.13 (a) 'Electric' penetrometer, (b) mantle cone and (c) friction jacket cone.

load sensors in the cone and sleeve, the cables from the sensors passing through the hollow rods to monitoring and recording equipment at the surface, enabling plots of the variation of q_c and f_s with depth to be produced. Some models, referred to as piezocones, also incorporate a piezometer (Section 11.2) consisting of a porous tip and a pressure transducer for the measurement of pore water pressure. During a test it is possible that the penetrometer may be deflected from the vertical by large soil particles such as cobbles. It is advisable, therefore, that the penetrometer should incorporate an inclinometer (Section 11.2) to detect loss of alignment and to give accurate values of vertical depth. The device also serves as a ground investigation tool in that it can detect thin layers of soil that can easily be missed by conventional procedures (Chapter 10) and indicate strata variations between boreholes. Soil types can be broadly identified from values of cone resistance (q_c) and the 'friction ratio' (f_s/q_c); see, e.g. Ref. [25].

The 'electrical' penetrometer described above has largely superseded mechanical devices. However, mechanical penetrometers are still used in some areas and are useful for preliminary tests to assess whether or not the more sensitive electrical models could be damaged by the particular ground conditions. There are two forms of mechanical penetrometer, the mantle cone and the friction jacket cone.

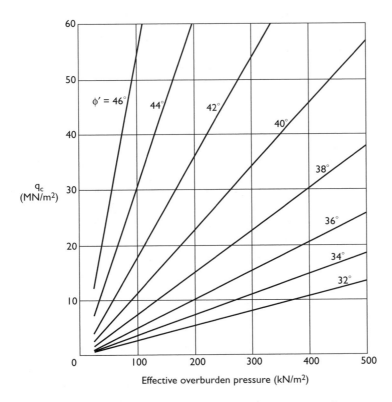

Figure 8.14 Correlation between shear strength parameter ϕ', cone penetration resistance and effective overburden pressure. (Reproduced from H.T. Durgunoglu and J.K. Mitchell (1975) *Proceedings of Conference on In-Situ Measurement of Soil Properties*, by permission of the American Society of Civil Engineers.)

The mantle cone (Figure 8.13(b)), which measures only cone resistance, is attached to a string of solid rods running inside the hollow outer rods. The outer rods are attached to a union sleeve, the lower end of which has a reduced diameter and runs inside the body of the cone. The cone is pushed 80 mm into the soil at a rate of 20 mm/s by means of the inner rods, the union sleeve remaining stationary. The thrust is usually measured by a hydraulic load cell at the surface. After the penetration resistance has been determined the outer rods are pushed downwards. The cone and sleeve are thus advanced together after the travel inside the body of the device has been taken up. The test is then repeated, cone penetration resistance normally being determined at depth intervals of 200 mm. The friction jacket cone (Figure 8.13(c)) incorporates a sleeve of equal diameter to that of the cone. Measurements are made of cone resistance alone and combined cone and sleeve resistances (sleeve resistance then being obtained by subtraction).

A correlation between the shear strength parameter ϕ', cone penetration resistance and effective overburden pressure, shown in Figure 8.14, was obtained by Durgunoglu and Mitchell [18].

Settlement by Schmertmann's method

This method of settlement estimation is based on a simplified distribution of vertical strain under the centre, or centre-line, of a shallow foundation, expressed in the form of a strain influence factor I_z. The vertical strain ε_z is written as

$$\varepsilon_z = \frac{q_n}{E_s} I_z$$

where q_n is the net pressure on the foundation and E_s the appropriate value of deformation modulus. The assumed distributions of strain influence factor with depth for square ($L/B = 1$) and long ($L/B \geq 10$) or strip foundations are shown in Figure 8.15, depth being expressed in terms of the width of the foundations. The two cases correspond to conditions of axial symmetry and plane strain, respectively. These are simplified distributions, based on both theoretical and experimental results, in which it is assumed that strains become insignificant at depths of $2B$ and $4B$, respectively, below the foundations. The peak value of strain influence factor I_{zp} in each case is given by the expression

$$I_{zp} = 0.5 + 0.1 \left(\frac{q_n}{\sigma'_p}\right)^{0.5}$$

where σ_p is the effective overburden pressure at the depth of I_{zp}. For rectangular foundations with L/B ratios between 1 and 10, distributions of strain influence factor are obtained by interpolation. It should be noted that the maximum vertical strains do not occur immediately below the foundations, as is the case with vertical stress. Corrections can be applied to the strain distributions for the depth of the foundation below the surface and for creep. Although it is usually assumed that settlement in sands is virtually complete by the end of construction, some case records indicate

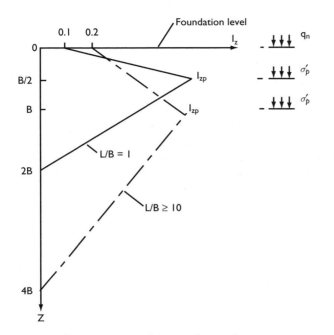

Figure 8.15 Distribution of strain influence factor.

continued settlement with time, thus suggesting a creep effect; however, the creep correction is often omitted. The correction factor for footing depth is given by

$$C_1 = 1 - 0.5\frac{\sigma'_0}{q_n} \qquad (8.23)$$

where σ'_0 = effective overburden pressure at foundation level and q_n = net foundation pressure. The correction factor for creep is given by

$$C_2 = 1 + 0.2\log\left(\frac{t}{0.1}\right) \qquad (8.24)$$

where t is the time in years at which the settlement is required.

The settlement of a footing carrying a net pressure q_n is written as

$$s = \int_0^{2B} \varepsilon_z \, \mathrm{d}z$$

or, approximately,

$$s = C_1 C_2 q_n \sum_0^{2B} \frac{I_z}{E} \Delta z \qquad (8.25)$$

Schmertmann obtained correlations, based on *in-situ* load tests, between deformation modulus and cone penetration resistance for normally consolidated sands as follows:

$$E_s = 2.5q_c \quad \text{for square foundations } (L/B = 1)$$
$$E_s = 3.5q_c \quad \text{for long foundations } (L/B \geq 10)$$

For overconsolidated sands the above values should be doubled.

The q_c/depth profile, to a depth of either $2B$ or $4B$ (or an interpolated depth) below the foundation, is divided into suitable layers of thicknesses Δz within each of which the value of q_c is assumed to be constant. The value of I_z at the centre of each layer is obtained from Figure 8.15. Equation 8.25 is then evaluated to give the settlement of the foundation.

Example 8.6

A footing 2.5×2.5 m supports a net foundation pressure of $150\,\text{kN/m}^2$ at a depth of 1.0 m in a deep deposit of normally consolidated fine sand of unit weight $17\,\text{kN/m}^3$. The variation of cone penetration resistance with depth is given in Figure 8.16. Estimate the settlement of the footing using Schmertmann's method.

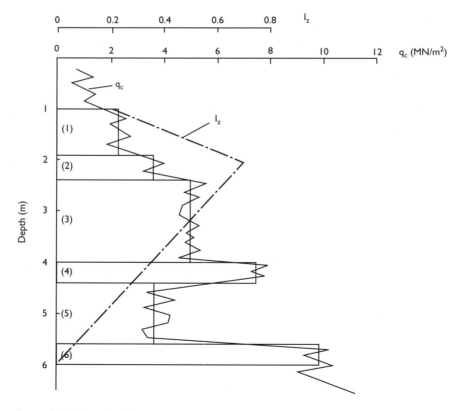

Figure 8.16 Example 8.6.

Table 8.6 Schmertmann's method

	Δz (m)	q_c (MN/m^2)	E_s (MN/m^2)	I_z	$I_z \Delta z / E_s$ (m^3/MN)
1	0.90	2.3	5.75	0.41	0.064
2	0.50	3.6	9.00	0.68	0.038
3	1.60	5.0	12.50	0.50	0.064
4	0.40	7.5	18.75	0.33	0.007
5	1.20	3.3	8.25	0.18	0.026
6	0.40	9.9	24.75	0.04	0.001
					0.200

The q_c/z plot below foundation level is divided into a number of layers, of thicknesses Δz, for each of which the value of q_c can be assumed constant.

The peak value of strain influence factor occurs at a depth 2.25 m (i.e. $B/2$ below foundation level) and is given by

$$I_{zp} = 0.5 + 0.1 \left(\frac{150}{17 \times 2.25} \right)^{0.5}$$

$$= 0.70$$

The distribution of strain influence factor with depth is superimposed on the q_c/z plot as shown in Figure 8.16 and the value of I_z is determined at the centre of each layer. The value of E_s for each layer is equal to $2.5q_c$.

The correction factor for foundation depth (Equation 8.23) is

$$C_1 = 1 - \frac{0.5 \times 17}{150} = 0.94$$

The correction factor for creep (C_2) will be taken as unity.

The calculations are set out in Table 8.6. The settlement is then given by Equation 8.25:

$$s = 0.94 \times 1.0 \times 150 \times 0.200$$

$$= 28 \, \text{mm}$$

8.5   BEARING CAPACITY OF PILES

Piles may be divided into two main categories according to their method of installation. The first category consists of driven piles of steel or precast concrete and piles formed by driving tubes or shells which are fitted with a driving shoe: the tubes or shells are filled with concrete after driving. Also included in this category are piles formed by placing concrete as the driven tubes are withdrawn. The installation of any type of driven pile causes displacement and disturbance of the soil around the pile.

However, in the case of steel H piles and tubes without a driving shoe, soil displacement is small. The second category consists of piles which are installed without soil displacement. Soil is removed by boring or drilling to form a shaft, concrete then being cast in the shaft to form the pile: the shaft may be cased or uncased depending on the type of soil. In clays the shaft may be enlarged at its base by a process known as under-reaming: the resultant pile then has a larger base area in contact with the soil. The principal types of pile are illustrated in Figure 8.17.

The bearing resistance of a pile can be determined by either analytical or semi-empirical methods, it being desirable to calibrate the results obtained against those from *in-situ* load tests. The ultimate bearing capacity is equal to the sum of the (ultimate) base and shaft resistances. The base resistance is the product of the base area (A_b) and the pressure (q_b) which would cause shear failure of the supporting soil immediately below and adjacent to the base of the pile. The shaft resistance is the product of the perimeter area of the shaft (A_s) and the average value of ultimate shearing resistance per unit

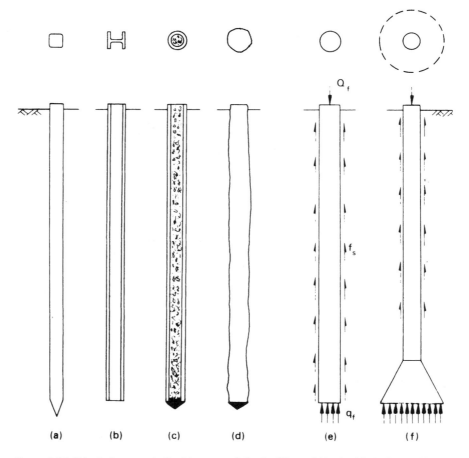

Figure 8.17 Principal types of pile: (a) precast RC pile, (b) steel H pile, (c) shell pile, (d) concrete pile cast as driven tube withdrawn, (e) bored pile (cast *in situ*) and (f) under-reamed bored pile (cast *in situ*).

area (q_s), generally referred to as the 'skin friction', between the pile and the soil. Thus the ultimate bearing resistance (Q_f) of a pile in compression is given by the equation

$$Q_f = A_b q_b + A_s q_s \qquad (8.26)$$

The weight of the soil displaced or removed in installation is generally assumed to be equal to the weight of the pile. For a pile subjected to uplift or tension, only the shaft resistance ($A_s q_s$) is relevant. Methods of determining the values of q_b and q_s are described in the following sections.

Evidence from load tests on instrumented piles indicates that in the initial stages of loading, most of the load is supported by skin friction on the upper part of the pile. Subsequently, as the load is increased, further mobilization of skin friction takes place but gradually a greater proportion of the load is supported by base resistance. The vertical displacement of the pile required for full mobilization of base resistance is significantly greater than that for full shaft resistance. It should be realized that other limit states (e.g. excessive settlement) may be reached before that of bearing resistance.

In the traditional method of design, either an overall load factor is applied to Q_f to obtain the allowable resistance or different factors are applied to the base and shaft components. In the case of large-diameter bored piles, including under-reamed piles, the shaft resistance may be fully mobilized at the design load and it is advisable to ensure a load factor of 3.0 for base resistance, with a factor of 1.0 for shaft resistance, in addition to an appropriate overall load factor, typically 2.0.

In the limit state method described in EC7, the 'design bearing resistance' (R_{cd}) of a pile in compression is expressed as

$$R_{cd} = \frac{R_{bk}}{\gamma_b} + \frac{R_{sk}}{\gamma_s} \qquad (8.27)$$

where γ_b and γ_s are partial factors for base and shaft resistance, respectively. A partial factor of 1.50 is applied to q_b and q_s, the base and shaft resistances per unit area, rather than to ground properties. Values of q_b and q_s derived from the results of laboratory or *in-situ* tests are thus divided by 1.50 to give characteristic values q_{bk} and q_{sk}, which are then multiplied by the base and shaft areas, respectively, to give the characteristic base and shaft resistances R_{bk} and R_{sk}, denoted in Equation 8.27. For bored piles the characteristic resistances are finally divided by partial factors of 1.60 and 1.30, respectively, to give the design values of base and shaft resistances (R_{bd} and R_{sd}, respectively); alternatively the sum of the characteristic resistances can be divided by a factor (γ_t) of 1.50 to give the total design resistance of the pile in compression (R_{cd}). If the pile shaft is excavated by a continuous-flight auger, the partial factors for base and shaft resistances are 1.45 and 1.30, respectively; for total resistance the factor is 1.40. For driven piles, a partial factor of 1.30 is applied to the characteristic values of both base and shaft resistances.

Piles in sands

The ultimate bearing capacity and settlement of a pile depend mainly on the density index of the sand. However, if a pile is driven into sand the density index adjoining the pile is increased by compaction due to soil displacement (except in dense sands, which may be loosened). The soil characteristics governing ultimate bearing capacity and settlement, therefore, are different from the original characteristics prior to driving. This fact, in addition to the heterogeneous nature of sand deposits, makes the prediction of pile behaviour by analytical methods extremely difficult.

The ultimate bearing capacity at base level can be expressed as

$$q_b = \sigma'_0 N_q \tag{8.28}$$

where σ'_0 is the effective overburden pressure at base level. (It should be noted that the N_γ term for a pile is negligible because the width B is small compared with the length L.)

Berezantzev *et al.* [2] developed a theory for the ultimate bearing capacity of piles in which failure is assumed to have taken place when the failure surfaces reach the level of the base, as shown in Figure 8.18. The surcharge at base level consists of the pressure due to the weight of an annulus of soil surrounding the pile, reduced by the frictional force on the outer surface of the annulus. The resulting factor N_q depends on the shear strength parameter ϕ' and the ratio L/B. For a given value of ϕ' the value of N_q decreases with increasing L/B ratio. Values of N_q for an L/B ratio of 25 are given in Table 8.7: extrapolated values for an L/B ratio of 50 are shown in brackets.

The average value of skin friction over the length of pile embedded in sand can be expressed as

$$q_s = K_s \overline{\sigma}'_0 \tan \delta \tag{8.29}$$

where K_s = the average coefficient of earth pressure along the embedded length, $\overline{\sigma}'_0$ = the average effective overburden pressure along the embedded length and δ = the

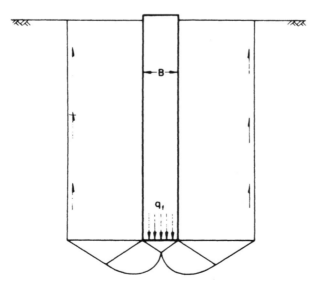

Figure 8.18 Failure mechanism in theory of Berezantzev *et al.* [2].

Table 8.7 Berezantzev *et al.* theory: relationship between ϕ' and N_q

ϕ' (°)	28	30	32	34	36	38	40
N_q	12 (9)	17 (14)	25 (22)	40 (37)	58 (56)	89 (88)	137 (136)

angle of friction between the pile and the sand. For concrete piles driven in sand, values of K_s of 1.0 and 2.0 for loose and dense sand, respectively, have been suggested for use in design. These values should be halved for steel H piles. Suggested values of δ are $0.75\phi'$ for concrete piles and 20° for steel piles.

Equations 8.28 and 8.29 represent a linear increase with depth of both q_b and q_s. However, tests on full-scale and model piles have indicated that these equations are valid only above a critical depth of roughly $15B$. Below the critical depth both q_b and q_s remain approximately constant at limiting values in uniform soil conditions. This is thought to be due to arching of the soil around the lower part of the pile when the soil yields below the base. Another possible explanation of the development of limiting values is that the peak value of ϕ' decreases with increasing confining pressure until the limiting critical-state value is reached.

Due to the critical depth limitation and the problem of obtaining values of the required parameters, the above equations are difficult to apply in practice. It is preferable to use empirical correlations, based on the results of pile loading tests and dynamic or static penetration tests, to estimate the values of q_b and q_s. The following correlations have been proposed by Meyerhof [29] for piles driven into a sand stratum:

$$q_b = 40N\frac{D_b}{B} \leq 400N \ (\text{kN/m}^2) \tag{8.30}$$

where N is the value of standard penetration resistance in the vicinity of the pile base and D_b the length of the pile embedded in the sand. For piles driven into non-plastic silts an upper limit of $300N$ is appropriate. Also

$$q_s = 2\overline{N}(\text{kN/m}^2) \tag{8.31}$$

where $\overline{N}$ is the average value of standard penetration resistance over the embedded length of the pile within the sand stratum. The value of q_s given by Equation 8.31 should be halved in the case of small-displacement piles such as steel H piles. For bored piles the values of q_b and q_s are approximately $\frac{1}{3}$ and $\frac{1}{2}$, respectively, of the corresponding values for driven piles.

The results of static cone penetration tests can also be used in pile design. The end bearing capacity (q_b) can be taken to be equal to the average value of cone penetration resistance ($\overline{q}_c$) in the vicinity of the pile base. Different procedures have been suggested for determining this average. One approach, described by Tomlinson [50], is to plot all relevant q_c/depth profiles together and draw an average line for the section around the pile base. A load factor of 2.0–2.5 is then applied to the base resistance ($A_b q_b$) depending on the scatter of the profile. A lower bound line should also be drawn and a check made to ensure that a load factor in excess of 1.0 for base resistance is achieved using the lower bound average. Another method, based on practice in the Netherlands, uses the mean of two averages q_{c1} and q_{c2}, for single profiles, determined (1) between $0.7B$ and $4B$ below the pile base and (2) $8B$ above the base, respectively. If q_c increases steadily below the pile, the average is determined only to depth $0.7B$. If a pronounced decrease in q_c occurs between $0.7B$ and $4B$ the lowest value within that range is taken as q_{c1}. The average value q_{c2} above the base should be determined, working upwards from the base, using only values which decrease from or are equal to that at the base. The value of end bearing capacity (q_b) should be restricted to $15\,MN/m^2$.

Shaft resistance per unit area (q_s) can be determined from values of local sleeve resistance (f_s). However, f_s must be multiplied by a factor to allow for the effect of pile installation on the density of the sand. The factor depends on the material and end shape of the pile, suggested values being 1.1 for a concrete pile with a pointed end and 0.7 for a steel H pile. Shaft resistance can also be determined from direct correlations with cone resistance, e.g. $q_s = 0.012q_c$ for timber, precast concrete and steel displacement piles. The value of q_s should be restricted to $0.12\,MN/m^2$.

In the limit state method both the values of q_b and q_s determined by the above procedures are divided by 1.50 to give the characteristic values q_{bk} and q_{sk}, respectively.

Piles in clays

In the case of driven piles, the clay adjacent to the pile is displaced both laterally and vertically. Upward displacement of the clay results in heaving of the ground surface around the pile and can cause a reduction in the bearing capacity of adjacent piles already installed. The clay in the disturbed zone around the pile is completely remoulded during driving. The excess pore water pressure set up by the driving stresses dissipates within a few months as the disturbed zone is relatively narrow (of the order of B): in general, dissipation is virtually complete before significant structural load is applied to the pile. Dissipation is accompanied by an increase in the shear strength of the remoulded clay and a corresponding increase in skin friction. Thus the skin friction at the end of dissipation is normally appropriate in design.

In the case of bored piles, a thin layer of clay (of the order of 25 mm) immediately adjoining the shaft will be remoulded during boring. In addition, a gradual softening of the clay will take place adjacent to the shaft due to stress release, pore water seeping from the surrounding clay towards the shaft. Water can also be absorbed from the wet concrete when it comes into contact with the clay. Softening is accompanied by a reduction in shear strength and a reduction in skin friction. Construction of a bored pile, therefore, should be completed as quickly as possible. Limited reconsolidation of the remoulded and softened clay takes place after installation of the pile.

The relevant shear strength for the determination of the base resistance of a pile in clay is the undrained strength at base level. The ultimate bearing capacity is expressed as

$$q_b = c_u N_c \tag{8.32}$$

Based on theoretical and experimental evidence, a value of N_c of 9 is appropriate (i.e. Skempton's value for $D/B > 4$). If the clay is fissured the shear strength of a small laboratory specimen (e.g. 38 mm diameter) will be greater than the *in-situ* strength because it will be relatively less fissured than the soil mass in the vicinity of the pile base: a reduction factor should thus be applied to the laboratory strength (e.g. 0.75 has been suggested for London clay).

The skin friction can be correlated empirically with the average undrained strength ($\bar{c}_u$) of the undisturbed clay over the depth occupied by the pile, i.e.

$$q_s = \alpha \bar{c}_u \tag{8.33}$$

where α is a coefficient depending on the type of clay, the method of installation and the pile material. The appropriate value of α is obtained from the results of load tests. Values of α can range from around 0.3 to around 1.0. One difficulty with this approach is that there is usually a considerable scatter in the plot of undrained shear strength against depth and it may be difficult to define the value of $\bar{c}_u$.

An alternative approach is to express skin friction in terms of effective stress. The zone of soil disturbance around the pile is relatively thin; therefore dissipation of the positive or negative excess pore water pressure set up during installation should be virtually complete by the time the structural load is applied. In principle, therefore, an effective stress approach has more justification than one based on total stress. In terms of effective stress the skin friction can be expressed as

$$q_s = K_s \bar{\sigma}'_0 \tan \phi' \tag{8.34}$$

where K_s is the average coefficient of earth pressure and $\bar{\sigma}'_0$ the average effective overburden pressure adjacent to the pile shaft. Failure is assumed to take place in the remoulded soil close to the pile shaft and, therefore, the angle of friction between the pile and the soil is represented by the critical-state value of the angle of shearing resistance.

The product $K_s \tan \phi'$ is written as a coefficient β, thus

$$q_s = \beta \bar{\sigma}'_0 \tag{8.35}$$

Approximate values of β can be deduced by making assumptions regarding the value of K_s, especially in the case of normally consolidated clays. However, the coefficient is generally obtained empirically from the results of load tests carried out a few months after installation. For normally consolidated clays the value of β is usually within the range 0.25–0.40 but for overconsolidated clays values are significantly higher and vary within relatively wide limits.

In the case of under-reamed piles, as a result of settlement, there is a possibility that a small gap will develop between the top of the under-ream and the overlying soil, leading to a drag-down of soil on the pile shaft. Accordingly no skin friction should be taken into account below a level $2B$ above the top of the under-ream. It should be noted that in the case of under-reamed piles the reduction in pressure on the soil at base level due to the removal of soil is greater than the subsequent increase in pressure due to the weight of the pile. The left-hand side of Equation 8.26 must then be written as $(Q_f + W - \gamma D A_b)$, where W is the weight of the pile, A_b the area of the enlarged base and D the depth to base level.

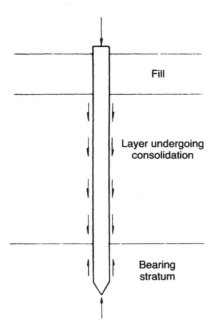

Figure 8.19 Negative skin friction.

Negative skin friction

Negative skin friction can occur on the perimeter of a pile driven through a layer of clay undergoing consolidation (e.g. due to a fill recently placed over the clay) into a firm bearing stratum (Figure 8.19). The consolidating layer exerts a downward drag on the pile and, therefore, the direction of skin friction in this layer is reversed. The force due to this downward or negative skin friction is thus carried by the pile instead of helping to support the external load on the pile. Negative skin friction increases gradually as consolidation of the clay layer proceeds, the effective overburden pressure σ'_0 gradually increasing as the excess pore water pressure dissipates. Equation 8.35 can also be used to represent negative skin friction. In normally consolidated clays, present evidence indicates that a value of β of 0.25 represents a reasonable upper limit to negative skin friction for preliminary design purposes. It should be noted that there will be a reduction in effective overburden pressure adjacent to the pile in the bearing stratum due to the transfer of part of the overlying soil weight to the pile: if the bearing stratum is sand, this will result in a reduction in bearing capacity above the critical depth.

Load tests

The loading of a test pile enables the ultimate load to be determined directly and provides a means of assessing the accuracy of predicted values. Tests may also be carried out in which loading is stopped when the proposed working load has been exceeded by a specified percentage. The results from a test on a particular pile will not necessarily reflect the performance of all other piles on the same site, and therefore an adequate number of tests are required, depending on the extent of the ground investigation. Driven piles in clays should not be tested for at least a month after installation to allow most of the increase in skin friction (a result of dissipation of the excess pore water pressure due to the driving stresses) to take place.

Two test procedures are detailed in BS 8004 [7]. In the *maintained load test* the load–settlement relationship for the test pile is obtained by loading in suitable increments, allowing sufficient time between increments for settlement to be substantially complete. The ultimate load is normally taken as that corresponding to a specified settlement, e.g. 10% of the pile diameter. Unloading stages are normally included in the test programme. In the *constant rate of penetration* (CRP) test the pile is jacked into the soil at a constant rate, the load applied in order to maintain the penetration being continuously measured. Suitable rates of penetration for tests in sands and clays are 1.5 and 0.75 mm/min, respectively. The test is continued until either shear failure of the soil takes place or the penetration is equal to 10% of the diameter of the pile base, thus defining the ultimate load. Allowance should be made for the elastic deformation of the pile under test. The settlement of a pile under maintained load cannot be estimated from the results of a CRP test. Typical load–settlement plots are shown in Figure 8.20. In Figure 8.20(b), curves A and B are typical of piles for which shaft resistance is the

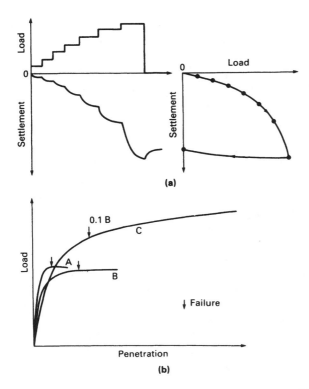

Figure 8.20 Pile loading tests: (a) maintained load test and (b) constant rate of penetration test.

dominant component of bearing capacity; curve C is typical of piles for which base resistance predominates.

In EC7 the following procedure is given for interpreting the results of pile load tests. The characteristic pile resistance is determined by dividing the average measured resistance by a correlation factor (ξ) of 1.3–1.5, depending on the number of tests. The design bearing resistance is then obtained by dividing the characteristic resistance by the appropriate partial factor as specified earlier in this section. Bearing resistance should be based either on calculations validated by load tests or on load tests alone.

Many pile load tests are not continued to the state of general shear failure, due to the cost involved and/or to the relatively large settlement required. However, a number of methods of extrapolating test data to ultimate failure have been proposed; these methods have been summarized by Fellenius [20].

Pile groups

A pile foundation may consist of a group of piles installed fairly close together (typically $2B$–$4B$ where B is the width or diameter of a single pile) and joined by a slab, known as the pile cap, cast on top of the piles. The cap is usually in contact with the soil, in which case part of the structural load is carried directly on the soil immediately below the surface. If the cap is clear of the ground surface, the piles in the group are referred to as freestanding. The principles described in this section also apply to piled rafts. In stiff clays, piles at spacings of $4B$ or greater may be installed under a raft for the prime purpose of reducing settlement. An excellent review of the design of piled rafts has been presented by Cooke [16].

In general, the ultimate load which can be supported by a group of n piles is not equal to n times the ultimate load of a single isolated pile of the same dimensions in the same soil. The ratio of the average load per pile in a group at failure to the ultimate load for a single pile is defined as the *efficiency* of the group. It is generally assumed that the distribution of load between the piles in an axially loaded group is uniform. However, experimental evidence indicates that for a group in sand the piles at the centre of the group carry greater loads than those on the perimeter; in clay, on the other hand, the piles on the perimeter of the group carry greater loads than those at the centre. It can generally be assumed that all piles in a group will settle by the same amount, due to the rigidity of the pile cap. The settlement of a pile group is always greater than the settlement of a corresponding single pile, as a result of the overlapping of the individual zones of influence of the piles in the group. The bulbs of pressure of a single pile and a pile group (with piles of the same length as the single pile) are of the form illustrated in Figure 8.21; significant stresses are thus developed over a much wider area and much greater depth in the case of a pile group than in the case of a corresponding single pile. The *settlement ratio* of a group is defined as the ratio of the settlement of the group to the settlement of a single pile when both are carrying the same proportion of their ultimate load.

The driving of a group of piles into loose or medium-dense sand causes compaction of the sand between the piles, provided that the spacing is less than about $8B$; consequently, the efficiency of the group is greater than unity. A value of 1.2 is often used in design. However, for a group of bored piles the efficiency may be as low as $\frac{2}{3}$ because the sand between the piles is not compacted during installation but the zones

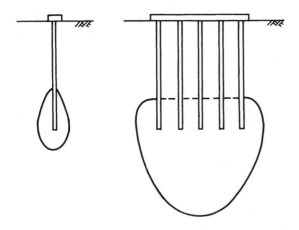

Figure 8.21 Bulbs of pressure for a single pile and a pile group.

of shear of adjacent piles will overlap. In the case of piles driven into dense sand, the group efficiency is less than unity due to loosening of the sand and the overlapping of zones of shear. If difficult driving conditions are anticipated, vibrocompaction (Section 8.6) may be a better solution.

A closely spaced group of piles in clay may fail as a unit, with shear failure taking place around the perimeter of the group and below the area covered by the piles and the enclosed soil; this is referred to as block failure. Tests by Whitaker [52] on freestanding model piles showed that for groups comprising a given number of piles of a given length there was a critical spacing of the order of $2B$ at which the mode of failure changed. For spacings above the critical value, failure occurred below individual piles. For spacings below the critical value, the group failed as a block, like an equivalent pier comprising the piles and the enclosed soil. The group efficiency at the critical spacing was between 0.6 and 0.7. However, when the pile cap was in contact with the soil, no change in the mode of failure was indicated at pile spacings above $2B$ and the efficiency exceeded unity at spacings greater than around $4B$. However, it is now considered that the length of the piles and the size and shape of the group also influence the critical spacing. It is recommended that the minimum centre-line spacing of piles in clay should not be less than the pile perimeter. The ultimate load in the case of a pile group which fails as a block is given by

$$Q_f = A_b q_b + A_s c_s \tag{8.36}$$

where A_b is equal to the base area of the group, A_s equal to the perimeter area of the group and c_s the average value of shearing resistance, per unit area, on the perimeter. The shearing resistance c_s should be taken as the undrained strength of the remoulded clay unless loading is to be delayed for at least 6 months, in which case the undrained strength of the undisturbed clay can be used. Dissipation of excess pore water pressure due to installation will take longer in the case of a pile group than in the case of a single pile and might not be complete before the early application of structural load. In design, the ultimate load should be taken as the lesser of the block failure value and the sum of the individual pile values, provided the pile cap rests on the soil. However, if the piles are freestanding the ultimate load should be the lesser of the block failure value and $\frac{2}{3}$ of the sum of the individual pile values.

The settlement of a pile group in clay can be estimated by assuming that the total load is carried by an 'equivalent raft' located at a depth of $2L/3$ where L is the length of the piles. It may be assumed, as shown in Figure 8.22(a), that the load is spread from the perimeter of the group at a slope of 1 horizontal to 4 vertical to allow for that part of the load transferred to the soil by skin friction. The vertical stress increment at any depth below the equivalent raft may be estimated by assuming in turn that the total load is spread to the underlying soil at a slope of 1 horizontal to 2 vertical. The consolidation settlement is then calculated from Equation 7.10. The immediate settlement is determined by applying Equation 5.28 to the equivalent raft.

The settlement of a pile group underlain by a depth of sand can also be estimated by means of the equivalent raft concept. In this case it may be assumed, as shown in Figure 8.22(b), that the equivalent raft is located at a depth of $2D_b/3$ in the sand stratum with a 1:4 load spread from the perimeter of the group. Again a 1:2 load spread is assumed below the equivalent raft. The settlement is determined from values of standard penetration resistance or cone penetration resistance below the equivalent raft, using the methods detailed in Section 8.4.

It is also possible to estimate the settlement due to the consolidation of a clay layer situated below a sand stratum in which a pile group is supported. The possibility of a pile group in a sand stratum punching through into an underlying layer of soft clay should also be considered in relevant cases; the vertical stress increment at the top of the clay layer should not exceed the presumed bearing value of the clay.

An alternative proposal regarding the equivalent raft is that its area should be equal to that of the pile group. In clays the equivalent raft should, as above, be located at a depth of $2L/3$ but in sands it should be located at the base of the pile group. A 1:2 load spread should be assumed below the equivalent raft in each case. The alternative proposals should be used if shaft resistance is negligible compared with base resistance.

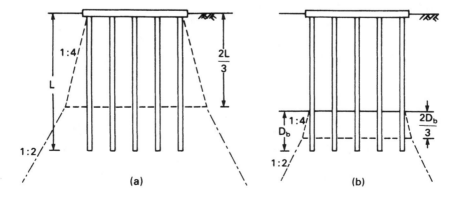

Figure 8.22 Equivalent raft concept.

A method based on elastic theory for estimating the settlement of a pile group has been developed by Poulos and Davis [36]. It should be appreciated that settlement is normally the limiting design criterion for pile groups in both sands and clays.

Pile driving formulae

A number of formulae have been proposed in which the dynamics of the pile driving operation is considered in a very idealistic way and the dynamic resistance to driving is assumed to be equal to the static bearing capacity of the pile.

Upon striking the pile, the kinetic energy of the driving hammer is assumed to be

$$Wh - \text{(energy losses)}$$

where W is the weight of the hammer and h the equivalent free fall. The energy losses may be due to friction, heat, hammer rebound, vibration and elastic compression of the pile, the packing assembly and the soil. The net kinetic energy is equated to the work done by the pile in penetrating the soil. The work done is Rs where R is the average resistance of the soil to penetration and s the set or penetration of the pile per blow. The smaller the set, the greater the resistance to penetration.

The *Engineering News* formula takes into account the energy loss due to temporary compression (c_p) resulting from elastic compression of the pile. Thus

$$R\left(s + \frac{1}{2}c_p\right) = Wh \tag{8.37}$$

from which R can be determined. In practice, empirical values are given to the term $c_p/2$ (e.g. for drop hammers $c_p/2 = 25\,\text{mm}$).

The Hiley formula takes into account the energy losses due to elastic compression of the pile, the soil and the packing assembly on top of the pile, all represented by a term c, and the energy losses due to impact, represented by an efficiency factor η. Thus

$$R\left(s + \frac{1}{2}c\right) = \eta Wh \tag{8.38}$$

The elastic compression of the pile and the soil can be obtained from the driving trace of the pile (Figure 8.23). The compression of the packing assembly must be estimated separately by assuming a value of the stress in the assembly during driving.

Driving formula should normally be used only for piles in sands and gravels and must be calibrated against the results of static load tests on similar piles in similar soil conditions.

The wave equation

The wave equation is a differential equation describing the transmission of compression waves, produced by the impact of a driving hammer, along the length of a pile. The pile is assumed to behave as a slender rod rather than as a rigid mass. A computer program can be written for the solution of the equation in finite difference form.

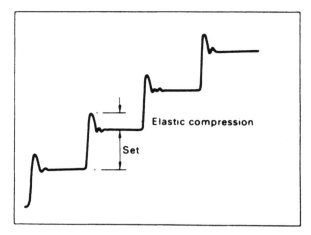

Figure 8.23 Pile driving trace.

Details of the method have been given by Smith [44]. The pile, packing assembly and driving ram are represented by a series of discrete weights and springs. The shaft and base resistances are represented by a series of springs and dashpots. Values of the parameters describing the behaviour of these elements and of the soil must be estimated.

The relationship can be obtained between the final set and the ultimate load which can be supported by the pile immediately after driving; no information can be obtained regarding the long-term behaviour of the pile. The equation also enables the stresses in the pile during driving to be determined, for use in the structural design of the pile. An assessment can also be made of the adequacy of the driving equipment to produce the desired load capacity for a given pile. Again, the validity of the analysis must be checked against the results of static load tests.

Example 8.7

A precast concrete pile 450×450 mm in section, to form part of a jetty, is to be driven into a river bed which consists of a depth of sand. The results of standard penetration tests in the sand are as follows:

Depth (m)	1.5	3.0	4.5	6.0	7.5	9.0	10.5
N		4	6	13	12	20	24	35

The pile is required to support a design compressive load of 650 kN and to withstand a design uplift load of 140 kN. Determine the depth (D_b) to which the pile must be driven (a) according to the traditional method with an overall load factor of 2.0 and (b) according to the limit state method.

(a) Required bearing resistance $= A_b q_b + A_s q_s = 2.0 \times 650 = 1300\,\text{kN}$
   Required uplift resistance $= A_s q_s = 2.0 \times 140 = 280\,\text{kN}$

where $A_b = 0.45^2$ and $A_s = 4 \times 0.45 \times D_b$

Table 8.8

D_b (m)	N	$\overline{N}$	A_sq_s (kN)	q_b (kN/m²)		A_bq_b (kN)	$A_bq_b + A_sq_s$ (kN)
				$\dfrac{40}{0.45}ND_b$	$400\,N$		
1.5	4	4	22	535		108	130
3.0	6	5	54	1600		324	378
4.5	13	8	130	5200	5 200	1053	1183
6.0	12	9	194		4 800	972	1166
7.5	20	11	297		8 000	1620	1917
9.0	24	13	421		9 600	1944	2365
10.5	35	16	605		14 000	2835	3440

Using Meyerhof's correlations,

$$q_b = 40ND_b/B \ge 400N \ (\text{kN/m}^2)$$
$$q_s = 2\overline{N} \ (\text{kN/m}^2)$$

The calculations are set out in Table 8.8. By inspection, uplift is the limiting consideration. By interpolation, the pile must be driven to at least 7.25 m.

(b) The values of q_b and q_s calculated using Meyerhof's correlations are divided by 1.50 to give the characteristic values (q_{bk} and q_{sk}). For a driven pile a partial factor of 1.30 is applied to both the characteristic bearing and uplift resistances. Thus

Required design bearing resistance, $R_{bd} = 1.30 \times 650 = 845\,\text{kN}$

Required design uplift resistance, $R_{sd} = 1.30 \times 140 = 182\,\text{kN}$

The values in Table 8.8 are re-calculated. By interpolation, the pile must be driven to at least 7.15 m, uplift again being the limiting consideration.

Example 8.8

An under-reamed bored pile is to be installed in a stiff clay. The diameters of the pile shaft and under-reamed base are 1.05 and 3.00 m, respectively. The pile is to extend from a depth of 4 m to a depth of 22 m in the clay, the top of the under-ream being at a depth of 20 m. The relationship between undrained shear strength and depth is shown in Figure 8.24 and the adhesion coefficient α is 0.4. Determine the bearing resistance of the pile (a) using the traditional method to ensure (i) an overall load factor of 2 and (ii) a load factor of 3 under the base when shaft resistance is fully mobilized, and (b) according to the limit state method.

(a) At base level (22 m) the undrained strength is 220 kN/m². Therefore

$$q_b = c_uN_c = 220 \times 9 = 1980\,\text{kN/m}^2$$

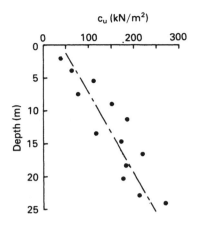

Figure 8.24 Example 8.8.

It is advisable to disregard skin friction over a length of $2B$ above the top of the under-ream, i.e. below a depth of 17.9 m. The average value of undrained strength between depths of 4 and 17.9 m is 130 kN/m². Therefore

$$q_s = \alpha \bar{c}_u = 0.4 \times 130 = 52 \, \text{kN/m}^2$$

The ultimate load is given by

$$
\begin{aligned}
Q_f &= A_b q_b + A_s q_s \\
&= \left(\frac{\pi}{4} \times 3^2 \times 1980 \right) + (\pi \times 1.05 \times 13.9 \times 52) \\
&= 13\,996 + 2384 \\
&= 16\,380 \, \text{kN}
\end{aligned}
$$

The allowable load is the lesser of:

(i) $\dfrac{Q_f}{2} = \dfrac{16\,380}{2} = 8190 \, \text{kN}$

(ii) $\dfrac{A_b q_b}{3} + A_s q_s = \dfrac{13\,996}{3} + 2384$
$$= 7049 \, \text{kN}$$

However, an allowance should be made for the difference between the pressure removed at the base of the under-ream due to boring of the shaft and the pressure subsequently applied due to the weight of the pile. Thus the allowable load may be increased by $(\gamma D A_b - W)/3$. Taking the unit weights of clay and concrete as 20 and 23.5 kN/m³, respectively, and neglecting the additional weight of the under-ream, the additional load is

$$\frac{1}{3}\left\{\left(20 \times 18 \times \frac{\pi}{4} \times 3^2\right) - \left(23.5 \times 18 \times \frac{\pi}{4} \times 1.05^2\right)\right\} = 726\,\text{kN}$$

Thus the allowable load on the pile is

$$7049 + 726 = 7775\,\text{kN}$$

(b) The characteristic undrained strength at base level is

$$c_{\text{uk}} = \frac{220}{1.50}\,\text{kN/m}^2$$

The average value of undrained strength between 4.0 and 17.9 m is

$$c_{\text{sk}} = \frac{130}{1.50}\,\text{kN/m}^2$$

Then the characteristic values of base and shaft resistances per unit area are

$$q_{\text{bk}} = \frac{9 \times 220}{1.50} = 1320\,\text{kN/m}^2 \qquad (N_c = 9)$$
$$q_{\text{sk}} = \frac{0.4 \times 130}{1.50} = 35\,\text{kN/m}^2$$

Therefore the characteristic base and shaft resistances are

$$R_{\text{bk}} = \frac{\pi}{4} \times 3^2 \times 1320 = 9330\,\text{kN}$$
$$R_{\text{sk}} = \pi \times 1.05 \times 13.9 \times 35 = 1605\,\text{kN}$$

For a bored pile the appropriate partial factors are $\gamma_b = 1.60$ and $\gamma_s = 1.30$, therefore the design bearing resistance is

$$R_{\text{cd}} = \frac{9330}{1.60} + \frac{1605}{1.30} = 5831 + 1235 = 7066\,\text{kN}$$

Example 8.9

A square group of 25 piles extends between depths of 1 and 13 m in a deposit of stiff clay 25 m thick overlying a hard stratum. The piles are 0.6 m in diameter and are spaced at 2 m centres in the group, as shown in Figure 8.25. The characteristic value of undrained shear strength of the clay at pile base level is 175 kN/m² and the average characteristic value over the length of the piles is 105 kN/m². The adhesion coefficient α is 0.45, E_u is 65 MN/m², m_v is 0.07 m²/MN and pore water pressure coefficient A is 0.24. The pile group supports a permanent load of 9000 kN and a variable load of 3000 kN. The allowable settlement is 30 mm. Are the bearing resistance and serviceability limit states satisfied according to the requirements of the limit state method?

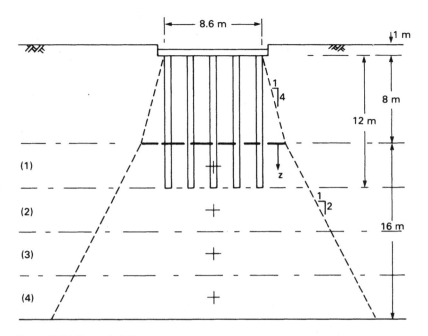

Figure 8.25 Example 8.9.

According to the limit state method, the design compressive load is

$$F_{cd} = (1.00 \times 9000) + (1.30 \times 3000) = 12\,900\,\text{kN}$$

The characteristic base and shaft resistances per unit area are

$$q_{bk} = \frac{9 \times 175}{1.50} = 1050\,\text{kN/m}^2$$
$$q_{sk} = \frac{0.45 \times 105}{1.50} = 31.5\,\text{kN/m}^2$$

Then, applying partial factors of 1.60 and 1.30 for base and shaft resistances, respectively, the design bearing resistance of a single pile is

$$R_{cd} = \frac{(\pi/4 \times 0.6^2 \times 1050)}{1.60} + \frac{(\pi \times 0.6 \times 12 \times 31.5)}{1.30}$$
$$= 186 + 548 = 734\,\text{kN}$$

The design bearing resistance of the group, assuming single pile failure and a group efficiency of 1.0 is $(25 \times 734 \times 1.0) = 18\,350\,\text{kN}$.

$R_{cd} > F_{cd}$ therefore the bearing resistance limit state is satisfied.

The width of the group is 8.6 m, therefore, assuming block failure and the full characteristic undrained strength on the perimeter, the design bearing resistance of the group is:

$$(8.6^2 \times 1050) + \left(4 \times 8.6 \times 12 \times \frac{105}{1.50}\right) = 77\,658 + 28\,896 = 106\,554\,\text{kN}.$$

Even if the remoulded strength of the clay were assumed to act on the perimeter, there would be no likelihood of block failure.

The design load for the serviceability limit state is $(9000 + 3000) = 12\,000\,\text{kN}$.

Referring to Figure 8.22(a), the equivalent raft is located 8 m ($\frac{2}{3} \times 12$ m) below the top of the piles. The width of the equivalent raft is 12.6 m. The load on the equivalent raft is assumed to be spread at a slope of 1:2 in the underlying clay. The pressure on the equivalent raft is given by

$$q = \frac{12\,000}{12.6^2} = 76\,\text{kN/m}^2$$

The immediate settlement is determined using Figure 5.15. Now

$$\frac{H}{B} = \frac{16}{12.6} = 1.3$$
$$\frac{D}{B} = \frac{9}{12.6} = 0.7$$
$$\frac{L}{B} = 1$$

Therefore $\mu_1 = 0.42$ and $\mu_0 = 0.93$; thus

$$\begin{aligned}
s_i &= \mu_0\mu_1 \frac{qB}{E_u} \\
&= 0.93 \times 0.42 \times \frac{76 \times 12.6}{65} \\
&= 6\,\text{mm}
\end{aligned}$$

To calculate the consolidation settlement the clay below the equivalent raft will be divided into four sublayers each of thickness (H) 4 m. The pressure increment ($\Delta\sigma$) at the centre of each sublayer is equal to the load of 12 000 kN divided by the spread area (Table 8.9). The settlement coefficient is obtained from Figure 7.12. The diameter of

Table 8.9

Layer	z (m)	Area (m²)	$\Delta\sigma$ (kN/m²)	$m_v\,\Delta\sigma\,H$ (mm)
1	2	14.6^2	56.3	15.8
2	6	18.6^2	34.7	9.7
3	10	22.6^2	23.5	6.6
4	14	26.6^2	17.0	4.8
				$s_{od} = 36.9$

a circle having the same area as the equivalent raft is 14.2 m, thus $H/B = 16/14.2 = 1.1$. Then from Figure 7.12, for $A = 0.24$ and $H/B = 1.1$, the value of μ is 0.52 and the consolidation settlement is

$$s_c = \mu s_{od} = 0.52 \times 36.9 = 19\,\text{mm}$$

The total settlement is

$$s = s_i + s_c = 6 + 19 = 25\,\text{mm}$$

This is less than the allowable settlement, therefore the serviceability limit state is satisfied.

Lateral loading of piles

Piles are capable of resisting lateral loading due to the resistance of the adjacent soil; lateral stresses in the soil increasing in front of the pile and decreasing behind, as load is applied. Most piles are subjected to a horizontal component of loading but if this is relatively small in relation to the vertical component, it need not be considered in design, e.g. the wind loading on a structure can normally be carried safely by the foundation piles. However, if the lateral component is relatively large the lateral resistance of the pile should be determined. Very large lateral components may require the installation of inclined (or *raked*) piles. Lateral loading on piles can also be induced by soil movement, e.g. lateral movement of the soil below an embankment behind a piled bridge abutment.

The mode of failure of a pile under lateral load depends on its length and whether or not it is restrained by a pile cap. A relatively short, rigid, unrestrained pile will rotate about a point B near the bottom, as shown in Figure 8.26(a). In the case of a relatively long flexible pile, a plastic hinge will develop at some point D along the length of the pile, as shown in Figure 8.26(b), and only above this point will there be significant displacement of the pile and soil. For piles surmounted by a pile cap which is restrained from rotation, there are three possible modes of failure. A short rigid pile will undergo translational displacement as shown in Figure 8.26(c). A pile of intermediate length will develop a plastic hinge at cap level, then rotate about a point near the bottom of the pile as illustrated in Figure 8.26(d). A long pile will develop plastic hinges at cap level and at a point along the length of the pile as indicated in Figure 8.26(e).

Close to the surface (at a depth not exceeding the width of the pile), failure in the soil is assumed to be analogous to the formation of a passive wedge in front of a retaining wall, the soil surface being pushed upwards. In a cohesionless soil the ultimate or limiting value of lateral pressure (p_1) in front of the pile can therefore be approximated to $K_p \sigma'_v$, ignoring three-dimensional effects, where K_p is the passive pressure coefficient and σ'_v the effective overburden pressure at the depth in question. At greater depth the soil in front of the pile deforms in a manner similar to that in a pressuremeter test (Section 5.1) and, based on experimental data, the limiting lateral pressure can be approximated to $K_p^2 \sigma'_v$. Behind the pile the lateral stress decreases with increasing lateral deformation of the soil. In a cohesive soil of undrained strength c_u, using the

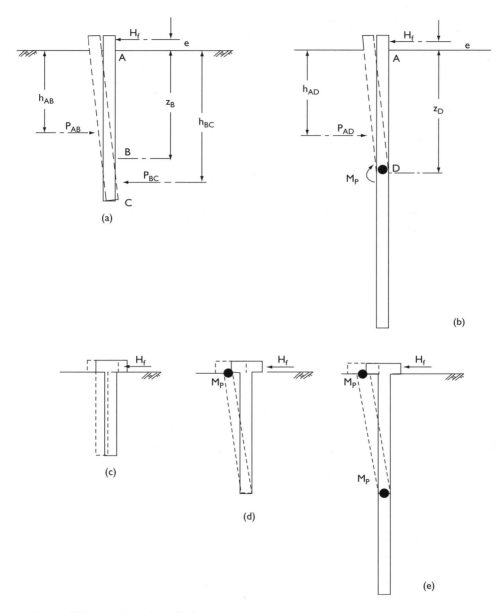

Figure 8.26 Lateral loading of piles.

pressuremeter analogy, the limiting pressure can be derived from Equation 5.5. A generalized relationship between p_1 and depth has been proposed by Fleming *et al.* [21] in which p_1 increases linearly from a value of $2c_u$ at the surface to $9c_u$ at a depth of $3B$, where B is the width or diameter of the pile, and remains at a constant value of $9c_u$ at depths below $3B$.

The horizontal force which would result in failure within the soil is denoted by H_f, acting at a distance e above the surface. For a short unrestrained pile the depth of the point of rotation is written as z_B. The limit forces on the front of the pile above the point of rotation and on the back of the pile below the point of rotation are denoted by P_{AB} and P_{BC}, respectively, calculated using the above values of limit pressure. These forces act at depths of h_{AB} and h_{BC}, respectively. Then for horizontal equilibrium

$$H_f = P_{AB} - P_{BC}$$

and for moment equilibrium

$$H_f(e + z_B) = P_{AB}(z_B - h_{AB}) + P_{BC}(h_{BC} - z_B)$$

The unknown quantities are H_f and z_B.

For a long unrestrained pile a plastic hinge develops at depth z_D and at this point the bending moment will be a maximum, of value M_p, and the shear force will be zero. Only the forces above the hinge, i.e. over the length AD, need be considered. Then for horizontal equilibrium

$$H_f = P_{AD}$$

and for moment equilibrium

$$\begin{aligned} M_p &= H_f[e + z_D - (z_D - h_{AD})] \\ &= H_f(e + h_{AD}) \end{aligned}$$

In the case of piles restrained by a pile cap a plastic hinge forms at the base of the cap and the additional moment at this point must be introduced into the above equations.

Fleming *et al.* [21] produced design charts, in dimensionless form, based on the above equations for short and long piles, unrestrained and restrained, in uniform soil deposits. An appropriate load factor should be applied to the value of H_f in traditional design. In limit state design a partial factor should be applied to the shear strength parameter in question.

8.6 GROUND IMPROVEMENT TECHNIQUES

An alternative to the use of deep foundations is the improvement of the soil properties near the surface; shallow foundations are then a possibility. The improvement techniques described below all require the services of a specialist contractor.

Vibrocompaction

The density index of loose to medium-dense sand deposits can be increased by the process of vibrocompaction. The technique employs a depth vibrator suspended from the jib of a crane or carried on special mountings, typical vibrators having lengths of 3–5 m and diameters of 300–450 mm. The vibrator section is located at the lower end of, and isolated from, the main body of the unit. The vibrator, which can be hydraulically or electrically powered, operates with a gyratory motion in a horizontal plane, produced by the rotation of eccentric masses. The unit penetrates the soil under its own weight, usually assisted by jets of water emitted from the conical point of the vibrator. The combined effects of vibration and water jetting induce local liquefaction of the adjacent soil, enabling the unit to penetrate readily under its own weight. After reaching the required depth, jetting is halted or reduced, the vibrator is gradually withdrawn and the surrounding soil is compacted by the horizontal vibratory action. The process creates a conical depression at the surface which is continuously filled with granular material, either from the site or imported, as the vibrator is withdrawn. Significant compaction of the soil can usually be achieved to a radius of up to 2.5 m from the axis of the vibrator, depending on the particle size distribution and initial density of the soil and the characteristics of the equipment. The soil should be compacted to at least the significant depth of the foundations in question, depths up to 12 m having been treated. The process is repeated at suitable spacings over the area in question creating a soil mass of increased bearing capacity. Vibrocompaction cannot be used in fine soils, especially saturated clays, because the vibrations would be damped within a relatively small radius. The process may be less efficient if the soil has a significant content of fine sand and non-plastic silt particles.

The effectiveness of the vibrocompaction process can be assessed by performing either standard penetration or cone penetration tests before and after the event. In the larger contracts it is usual to commission a test programme to determine the optimum spacing between insertions of the vibrator.

Vibroreplacement

Vibroreplacement involves the reinforcement of fine soil deposits with 'stone columns' to provide adequate support for relatively light structures. The columns do not transfer load to greater depth, i.e. they do not function in the same way as piles – they rely largely on the lateral resistance of the surrounding soil – therefore are not adequate to support relatively heavy loading. Stone columns also fulfil a similar function to vertical sand drains in accelerating the rate of consolidation of the surrounding soil.

A depth vibrator is again used to penetrate the soil. Water or air jets may be used to facilitate the process. The soil is displaced radially by the vibrator resulting in the formation of a cylindrical cavity. The vibrator is then withdrawn, compressed air being introduced to break the suction and the cavity is filled in stages with layers of 50–75 mm angular aggregate, each layer being compacted by re-inserting the vibrator. The aggregate is displaced both laterally and downwards with further displacement of the adjacent soil. A stone column is thus formed which interlocks with the surrounding soil. Stone columns may be installed either in a grid configuration over the area in question, forming a composite soil mass of enhanced bearing capacity, or in positions where structural columns are to be located. The strength and stiffness of stone columns depends on their degree of lateral confinement within the surrounding soil. Again, field trials are usually performed to confirm optimum design details. Some design charts for stone columns have been published by Moseley [30].

In soft clays, material may be removed by means of water emitted under pressure through the jet holes at the point of the vibrator, i.e. the soil is then not displaced. It is uncertain if adequate support can be relied upon in soft clays if the rate of load application is slow. Soft clay may be gradually squeezed into the voids within the column in which case there will be reduced lateral resistance and reduced efficiency as a drain.

Dynamic deep compaction

This process involves the use of high-energy tamping to improve the engineering properties of relatively weak soil, improvement to depths of around 10 m being possible. The technique consists of dropping a heavy mass, usually within the range 6–20 tonnes, onto the ground surface from a height of 5–20 m (although greater masses and heights have been used), the *drop energy* per blow being the mass multiplied by the drop height. A crawler crane or lifting frame is used to raise the tamper, then release it in free fall. The impact of the tamper creates a hole, known as the imprint, in the ground surface and causes shock waves to be transmitted through the soil to a considerable depth. Typically the tamper is dropped 5–10 times at each position and the process is repeated at 5–15 m centres in a square grid over the area being treated. In practice different values of drop energy may be used to treat different depth ranges within the soil. The full available energy at relatively wide drop spacing may be employed to treat the deepest levels of soil and reduced energy at closer drop spacing to treat levels nearer the surface. The average reduction in the level of the area under treatment is referred to as the induced settlement and is dependent on the total energy applied and details of the application sequence. Coarse and fine soils behave in different ways under the process.

In coarse soils above the water table the shock waves produced by impact cause the particles to be packed closer together, resulting in a higher density index and consequently an increase in the bearing capacity of the soil. Below the water table, excess pore water pressure is developed by the impact stresses and, depending on the grading and density index of the soil, liquefaction may be initiated. The density index will subsequently increase as the excess pore water pressure dissipates. Liquefaction may be avoided by the use of lower drop energy.

In fine soils the improvements in properties achieved are normally less pronounced, and require a larger number of blows, than in coarse soils. Reasonable results can be achieved in soils above the water table but very little improvement can be expected in thick layers of saturated clay. Impact by the tamper induces local excess pore water pressure which dissipates by outward drainage into the surrounding soil. In addition, drainage channels can be formed by hydraulic fracture and the development of shear planes within the soil below the impact area, producing a temporary increase in mass permeability and resulting in an acceleration in the rate of consolidation. In fine soils the process has been referred to as dynamic consolidation.

The effectiveness of the process can be monitored by means of load tests and/or *in-situ* penetration tests. Induced settlement over the site would also be measured as operations proceed. Initial trials are normally commissioned to determine the optimum drop energy, number of drops and imprint spacing. In fine soils, piezometers might be installed to determine the time necessary for dissipation of excess pore water pressure. If necessary, vibration gauges would be used to ensure that the operations did not cause adverse effects on structures in the vicinity.

Lime stabilization

The bearing properties of soft clay and silt can be enhanced by the formation of a group of *lime columns* within the soil. The technique uses a special mixing tool mounted on a long vertical shaft of hollow section which passes through the rotary drive unit of the rig. The mixer is rotated into the soil, penetrating it to the required depth. The mixer is then gradually withdrawn and at the same time unslaked lime (CaO) is introduced through holes immediately above the mixing blades, the lime being forced down the inside of the shaft by compressed air. The orientation of the mixer blades is such that the soil–lime mixture is compacted as the tool is withdrawn. The proportion of lime is normally within the range of 3–10% of the dry weight of the soil. Lime columns are usually 300–600 mm in diameter, typical spacing is at 1–2 m centres and they can be up to 15 m in length. Lime stabilization results in higher bearing capacity and lower compressibility of the treated soil mass.

The added lime reacts with the pore water, resulting in chemical bonding between soil particles, a reduction in water content and, in turn, an increase in undrained shear strength. The heat of hydration of unslaked lime contributes to the reduction in water content. Base exchange takes place, cations adsorbed on the particle surfaces being replaced by calcium, resulting in the coagulation of particles. The process depends, therefore, on there being a relatively high clay mineral content in the soil and is most successful in soils with a clay content of at least 30–40%. Fly ash can be added to the soil–lime mix to compensate for a lower clay content. The addition of gypsum accelerates the chemical reaction and increases the strength of the treated soil. Although the shear strength is initially decreased by the rotation of the mixing tool, it subsequently increases rapidly and becomes higher than the original value shortly after completion of the column. The strength continues to increase for many months due to the pozzolanic reaction of the lime with the silicates and aluminates of the clay. The stabilized soil is likely to be more friable than the untreated mass because strength will be lost if particle bonds are broken.

Lime columns can provide an economic foundation for relatively light structures compared to piling especially if the allowable bearing capacity of the latter is not fully utilized. The columns are installed primarily to reduce settlement, the shear strength of the untreated soil usually being adequate to support the imposed loading. The permeability of a lime column can be much higher than that of the surrounding soil, the column then acting as a vertical drain, increasing the rate of settlement. The bearing capacity of a column depends on both the undrained shear strength of the stabilized soil and the lateral confining pressure of the surrounding soil, which is normally greater than the total overburden pressure. The column material is subject to creep, i.e. slow continuous deformation will occur at constant load. The load at which creep is initiated is normally 65–80% of the (short-term) ultimate load.

There are alternative construction techniques and other materials such as cement and slaked (hydrated) lime can be used. Full details of the process, its uses and methods of design have been given by Broms [30].

8.7 EXCAVATIONS

Foundation works may require a relatively deep excavation with vertical sides. The sides may be supported by soldier piles with timber sheeting, sheet pile walls or diaphragm walls; these structures can be braced by means of horizontal or inclined struts or by tie-backs. In addition to the design of the supporting structure, consideration must be given to the ground movements which will occur around the excavation, especially if the excavation is close to existing structures. The following movements (Figure 8.27) should be considered:

1   settlement of the ground surface adjacent to the excavation,
2   lateral movement of the vertical supports, and
3   heave of the base of the excavation.

To a large extent the above movements are interdependent because they are a result of strains in the soil mass due to stress relief when excavation takes place. The magnitude and distribution of the ground movements depend on the type of soil, the dimensions of the excavation, details of the construction procedure and the standard of workmanship. Ground movements should be monitored during excavation so that advance warning of excessive movement or possible instability can be obtained.

Assuming comparable construction techniques and workmanship, the magnitude of settlement adjacent to an excavation is likely to be relatively small in dense cohesionless soils but can be excessive in soft plastic clays. Envelopes of the upper limits of observed settlements in various types of soil have been produced by Peck [33], settlement being given in relation to maximum depth of excavation and distance from the edge of the excavation. These envelopes, shown in Figure 8.28, are applicable to excavations supported by sheet piling or soldier piles, with bracing or tie-backs, and relate to average workmanship. For excavations supported by diaphragm walls the settlements are likely to be significantly lower than those indicated by Peck's envelopes.

Settlement can be reduced by adopting construction procedures which decrease lateral movement and base heave. For a given type of soil, therefore, settlement can be kept to a minimum by installing the struts or tie-backs as soon as possible and before excavation proceeds significantly below the point of support. Care should also be taken to ensure that no voids are left between the supporting structure and the soil. In cohesionless soils it is vital that groundwater flow is controlled; otherwise erratic settlement may be caused by a loss of soil into the excavation. It should be realized that

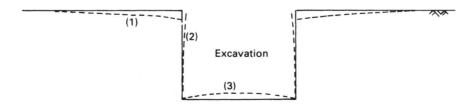

Figure 8.27 Ground movements associated with deep excavation.

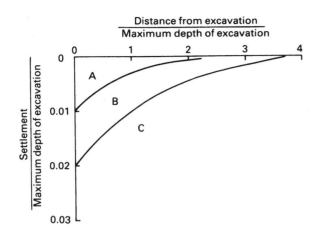

Figure 8.28 Envelopes of settlement adjacent to excavation: (A) sand and firm to stiff clay, (B) very soft to soft clay of limited depth and (C) very soft to soft clay of considerable depth. (Reproduced by permission of the Mexican Society SMFE.)

for a given method of construction and the best possible standard of workmanship, the settlement at a given point cannot be reduced below a minimum value which depends on the type of soil and the depth of excavation.

The magnitude and distribution of lateral movement depends, to a large extent, on the mode of deformation of the supporting structure (e.g. whether the structure is allowed to deflect as a cantilever or whether it is braced near the surface with the maximum deflection then taking place at greater depth). Lateral movement thus depends on the spacing and timing of installation of the struts or tie-backs. As in the case of settlement, excessive movements can occur if excavation is allowed to proceed too far before the first strut or tie-back is installed. The other main factor is the type of soil. Under comparable conditions, lateral movements in soft to medium clays are substantially greater than those in dense cohesionless soils.

Most problems concerning braced excavations are the result of excessive ground movements and the control of such movements should be considered at the beginning of the design process. The design of the support system should be based on the requirements of movement control, i.e. a serviceability limit state. There are three general approaches to the estimation of ground movements, namely empirical correlations based on *in-situ* measurements (such as Peck's envelopes), the use of analytical procedures such as the finite element method and semi-empirical procedures which combine *in-situ* observations with an analytical framework.

A semi-empirical method for estimating the maximum lateral movement of a braced wall in clay was proposed by Mana and Clough [24]. Maximum lateral movement, as a percentage of excavation depth, was correlated with the factor of safety against base heave (Equation 8.11), using both *in-situ* measurements and results from a finite element analysis. Factor of safety against base heave was used in the correlation because it takes into account the effects of variables such as shear strength and excavation geometry. Modification factors were also presented to account for the effects of wall stiffness, strut spacing, stiffness and preload, depth to a firm stratum, excavation width and soil modulus. Subsequently the chart shown in Figure 8.29 was presented by Clough *et al.* [15] in which the effects of wall stiffness per horizontal unit length (EI) and average vertical strut spacing (h) were incorporated in the primary correlation. The chart is based on 'average conditions' and good workmanship, and on the assumption that the struts are placed before significant movement takes place. The aim of the method is to ensure that the value of maximum lateral movement obtained from the chart is unlikely to be exceeded, i.e. the approach is analogous to that of the Terzaghi and Peck method (Section 8.4) which aims to limit the settlement of shallow foundations on sand to 25 mm.

Maximum surface settlement is generally less than maximum lateral movement. *In-situ* observations indicate that maximum settlement is within the range of 1.0–0.5 of maximum lateral movement, while finite element analysis indicates a range of 0.8–0.4. Conservatively, maximum settlement could be taken to be equal to maximum lateral movement in design. The chart shown in Figure 8.29 enables an estimation to be made of the variation of settlement with distance from the wall.

Base heave is generally a problem only in cohesive soils. The soil outside the excavation acts as a surcharge with respect to that below the base of the excavation,

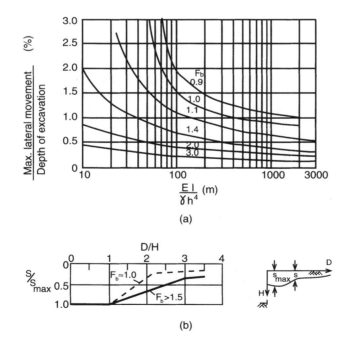

Figure 8.29 Braced excavation: (a) lateral movement and (b) surface settlement. (Reproduced from G.W. Clough *et al.* (1989) Movement control of excavation support systems by iterative design, in *Proceedings of the ASCE Congress on Foundation Engineering – Current Principles and Practices*, by permission of the American Society of Civil Engineers.)

and therefore upward deformation, and in extreme cases shear failure (Section 8.2), will occur. Short-term heave will be mainly elastic, unless the factor of safety against base failure is low, but additional heave will occur due to swelling if the base remains unloaded for any length of time. In heavily overconsolidated clays, heave can be associated with the relief of the high lateral stresses existing in the clay prior to excavation.

8.8 GROUND ANCHORS

A ground anchor normally consists of a high-tensile steel cable or bar, called the tendon, one end of which is held securely in the soil by a mass of cement grout or grouted soil; the other end of the tendon is anchored against a bearing plate on the structural unit to be supported. The main application of ground anchors is in the construction of tie-backs for diaphragm or sheet pile walls. Other applications are in the anchoring of structures subjected to overturning, sliding or buoyancy, in the provision of reaction for *in-situ* load tests and in preloading to reduce settlement. Ground anchors can be constructed in sands (including gravelly sands and silty sands) and stiff clays, and they can be used in situations where either temporary or permanent support is required.

The grouted length of tendon, through which force is transmitted to the surrounding soil, is called the fixed anchor length. The length of tendon between the fixed anchor and the bearing plate is called the free anchor length: no force is transmitted to the soil over this length. For temporary anchors the tendon is normally greased and covered with plastic tape over the free anchor length. This allows for free movement of the tendon and gives protection against corrosion. For permanent anchors the tendon is normally greased and sheathed with polythene under factory conditions; on site the tendon is stripped and degreased over what will be the fixed anchor length.

The ultimate load which can be carried by an anchor depends on the soil resistance (principally skin friction) mobilized adjacent to the fixed anchor length. (This, of course, assumes that there will be no prior failure at the grout–tendon interface or of the tendon itself.) Anchors are usually prestressed in order to reduce the lateral displacement required to mobilize soil resistance and to minimize ground movements in general. Each anchor is subjected to a test loading after installation, temporary anchors usually being tested to 1.2 times the working load and permanent anchors to 1.5 times the working load. Finally, prestressing of the anchor takes place. Creep displacements under constant load will occur in ground anchors. A creep coefficient, defined as the displacement per unit log time, can be determined by means of a load test. It has been suggested that this coefficient should not exceed 1 mm for 1.5 times the working load.

A comprehensive ground investigation is essential in any location where ground anchors are to be employed. The soil profile must be determined accurately, any variations in the level and thickness of strata being particularly important. In the case of sands the particle size distribution should be determined, in order that permeability and grout acceptability can be estimated. The density index of sands is also required to allow an estimate of ϕ' to be made. In the case of stiff clays the undrained shear strength should be determined.

Full details concerning the design, construction and testing of ground anchors are given in BS 8081: 1989 [8], the UK code of practice for ground anchorages.

Anchors in sands

In general the sequence of construction is as follows. A cased borehole (diameter usually within the range 75–125 mm) is advanced through the soil to the required depth. The tendon is then positioned in the hole and cement grout is injected under pressure over the fixed anchor length as the casing is withdrawn. The grout penetrates the soil around the borehole, to an extent depending on the permeability of the soil and on the injection pressure, forming a zone of grouted soil, the diameter of which can be up to four times that of the borehole (Figure 8.30(a)). Care must be taken to ensure that the injection pressure does not exceed the overburden pressure of the soil above the anchor, otherwise heaving or fissuring may result. When the grout has achieved adequate strength, the other end of the tendon is anchored against the bearing plate. The space between the sheathed tendon and the sides of the borehole, over the free anchor length, is normally filled with grout (under low pressure): this grout gives additional corrosion protection to the tendon.

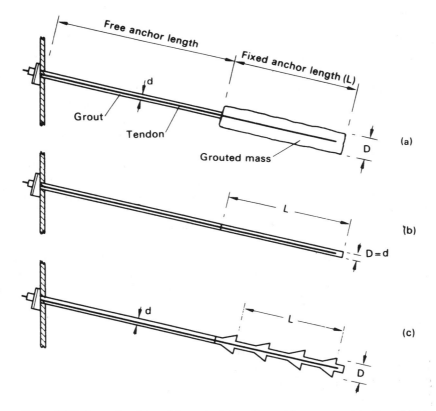

Figure 8.30 Ground anchors: (a) grouted mass formed by pressure injection, (b) grout cylinder and (c) multiple under-reamed anchor.

The ultimate resistance of an anchor to pull-out is equal to the sum of the side resistance and the end resistance of the grouted mass. The following theoretical expression has been proposed:

$$T_f = A\sigma'_v \pi DL \tan \phi' + B\gamma'h\frac{\pi}{4}(D^2 - d^2) \qquad (8.39)$$

where T_f = ultimate load capacity of anchor, A = ratio of normal pressure at interface to effective overburden pressure, σ'_v = effective overburden pressure adjacent to the fixed anchor and B = bearing capacity factor.

It was suggested that the value of A is normally within the range 1–2. The factor B is analogous to the bearing capacity factor N_q in the case of piles and it was suggested that the ratio N_q/B is within the range 1.3–1.4, using the N_q values of Berezantzev, Khristoforov and Golubkov. However, the above expression is unlikely to represent all the relevant factors in a complex problem. The ultimate resistance also depends on details of the installation technique and a number of semi-empirical formulae have been proposed by specialist contractors, suitable for use with their particular technique. An example of such a formula is

$$T_f = Ln \tan \phi' \qquad (8.40)$$

The value of the empirical factor n is normally within the range 400–600 kN/m for coarse sands and gravels, and within the range 130–165 kN/m for fine to medium sands.

Anchors in stiff clays

The simplest construction technique for anchors in stiff clays is to auger a hole to the required depth, position the tendon and grout the fixed anchor length using a tremie pipe (Figure 8.30(b)). However, such a technique would produce an anchor of relatively low capacity because the skin friction at the grout–clay interface would be unlikely to exceed $0.3c_u$ (i.e. $\alpha = 0.3$).

Anchor capacity can be increased by the technique of gravel injection. The augered hole is filled with pea gravel over the fixed anchor length, then a casing, fitted with a pointed shoe, is driven into the gravel, forcing it into the surrounding clay. The tendon is then positioned and grout is injected into the gravel as the casing is withdrawn (leaving the shoe behind). This technique results in an increase in the effective diameter of the fixed anchor (of the order of 50%) and an increase in side resistance: a value of α of around 0.6 can be expected. In addition there will be some end resistance. The borehole is again filled with grout over the free anchor length.

Another technique employs an expanding cutter to form a series of enlargements (or under-reams) of the augered hole at close intervals over the fixed anchor length (Figure 8.30(c)): the cuttings are generally removed by flushing with water. The cable is then positioned and grouting takes place. A value of α of unity can normally be assumed along the cylindrical surface through the extremities of the enlargements.

The following design formula can be used for anchors in stiff clays:

$$T_f = \pi DL\alpha c_u + \frac{\pi}{4}(D^2 - d^2)c_u N_c \qquad (8.41)$$

where T_f = ultimate load capacity of anchor, L = fixed anchor length, D = diameter of fixed anchor, d = diameter of borehole, α = skin friction coefficient and N_c = bearing capacity factor (generally assumed to be 9). Resistance at the grout–clay interface along the free anchor length may also be taken into account.

PROBLEMS

8.1   A load of 425 kN/m is carried on a strip footing 2 m wide at a depth of 1 m in a stiff clay of saturated unit weight 21 kN/m³, the water table being at ground level. Determine the factor of safety with respect to shear failure (a) when $c_u = 105$ kN/m² and $\phi_u = 0$ and (b) when $c' = 10$ kN/m² and $\phi' = 28°$.

8.2   A strip footing 1.5 m wide is located at a depth of 0.75 m in a sand of unit weight 18 kN/m³, the water table being well below foundation level. The characteristic shear strength parameters are $c' = 0$ and $\phi' = 38°$. The footing supports a design load of 500 kN/m. Determine the factor of safety with respect to shear failure. Using the limit state method, determine if the bearing resistance limit state is satisfied.

8.3   Determine the design load on a footing 4.50×2.25 m at a depth of 3.50 m in a stiff clay if a factor of safety of 3 with respect to shear failure is specified. The saturated unit weight of the clay is 20 kN/m³ and the characteristic shear strength parameters are $c_u = 135$ kN/m² and $\phi_u = 0$. Is the bearing resistance limit state then satisfied in accordance with the limit state method?

8.4   A footing 2.5×2.5 m carries a pressure of 400 kN/m² at a depth of 1 m in a sand. The saturated unit weight of the sand is 20 kN/m³ and the unit weight above the water table is 17 kN/m³. The design shear strength parameters are $c' = 0$ and $\phi' = 40°$. Determine the factor of safety with respect to shear failure for the following cases:
(a)   the water table is 5 m below ground level,
(b)   the water table is 1 m below ground level,
(c)   the water table is at ground level and there is seepage vertically upwards under a hydraulic gradient of 0.2.

8.5   A foundation 3.0×3.0 m supports a permanent load of 4000 kN and a variable load of 1500 kN at a depth of 1.5 m in sand. The water table is at the surface, the saturated unit weight of the sand being 20 kN/m³. Characteristic values of the shear strength parameters are $c' = 0$ and $\phi' = 39°$. Is the bearing resistance limit state satisfied in accordance with the limit state method?

8.6   A foundation 4×4 m is located at a depth of 1 m in a layer of saturated clay 13 m thick. Characteristic parameters for the clay are $c_u = 100$ kN/m², $\phi_u = 0$, $c' = 0$, $\phi' = 32°$, $m_v = 0.065$ m²/MN, $A = 0.42$, $\gamma_{sat} = 21$ kN/m³. Determine the design load of the foundation to ensure (a) a factor of safety with respect to shear failure of 3 using the traditional method, (b) the bearing resistance limit state is satisfied using the limit state recommendations and (c) consolidation settlement does not exceed 30 mm.

8.7   A long braced excavation in soft clay is 4 m wide and 8 m deep. The saturated unit weight of the clay is 20 kN/m³ and the undrained shear strength adjacent to the bottom of the excavation is given by $c_u = 40$ kN/m²($\phi_u = 0$). Determine the factor of safety against base failure of the excavation.

8.8   A permanent load of 2500 kN and an imposed load of 1250 kN are to be supported on a foundation 2.50×2.50 m at a depth of 1.0 m in a deposit of gravelly sand extending from the surface to a depth of 6.0 m. A layer of clay 2.0 m thick lies immediately below the sand. The water table may rise to foundation level. The unit weight of the sand above the water table is $17 \, \text{kN/m}^3$ and below the water table the saturated unit weight is $20 \, \text{kN/m}^3$. Characteristic values of the shear strength parameters for the sand are $c' = 0$ and $\phi' = 38°$. The coefficient of volume compressibility for the clay is $0.15 \, \text{m}^2/\text{MN}$. It is specified that the long-term settlement of the foundation due to consolidation of the clay should not exceed 20 mm. In accordance with the limit state recommendations, are the bearing resistance and serviceability (settlement) limit states satisfied?

8.9   A foundation 3.5×3.5 m is to be constructed at a depth of 1.2 m in a deep sand deposit, the water table being 3.0 m below the surface. The following values of standard penetration resistance were determined at the location:

Depth (m)	0.70	1.35	2.20	2.95	3.65	4.40	5.15	6.00
N	6	9	10	8	12	13	17	23

If the settlement is not to exceed 25 mm, determine the allowable bearing capacity according to the following design procedures: (a) Terzaghi and Peck, (b) Meyerhof and (c) Burland and Burbidge.

8.10   A footing 3.0×3.0 m carries a net foundation pressure of $130 \, \text{kN/m}^2$ at a depth of 1.2 m in a deep deposit of sand of unit weight $16 \, \text{kN/m}^3$, the water table being well below the surface. The variation of cone penetration resistance (q_c) with depth (z) is as follows:

z (m)	1.2	1.6	2.0	2.4	2.6	3.0	3.4	3.8	4.2
q_c (MN/m^2)	3.2	2.1	2.8	2.3	6.1	5.0	3.6	4.5	3.5

z (m)	4.6	5.0	5.4	5.8	6.2	6.6	7.0	7.4	8.0
q_c (MN/m^2)	4.0	8.1	6.4	7.6	6.9	13.2	11.7	12.9	14.8

Determine the settlement of the footing using Schmertmann's method.

8.11   A bored pile with an enlarged base is to be installed in a stiff clay, the characteristic undrained strength at base level being $220 \, \text{kN/m}^2$. The saturated unit weight of the clay is $21 \, \text{kN/m}^3$. The diameters of the pile shaft and base are 1.05 and 3.00 m, respectively. The pile extends from a depth of 4 m to a depth of 22 m, the top of the under-ream being at a depth of 20 m. Past experience indicates that a skin friction coefficient β of 0.70 is appropriate for the clay. Determine the design load of the pile (a) according to the traditional method, ensuring (i) an overall load factor of 2 and (ii) a load factor of 3 under the base when shaft resistance is fully mobilized, and (b) according to the limit state method.

8.12   Thirty six piles, 0.60 m in diameter, are spaced at 2.40 m centres in a 6×6 group. The piles extend between depths of 3 and 18 m in a deposit of stiff clay 28 m thick overlying rock. The characteristic undrained strength of the clay at a depth of 18 m is $145 \, \text{kN/m}^2$ and the average characteristic value over the pile length is $105 \, \text{kN/m}^2$. The following parameters have also been determined for the clay: $\alpha = 0.40$, $A = 0.28$, $E_u = 65\,MN/m^2$, $m_v = 0.08\,m^2/MN$. If the pile group supports a permanent load of 15 MN and an imposed load of 6 MN, (a) determine the overall load factor according to the traditional method, (b) check that the bearing resistance limit state is satisfied according to the limit state recommendations and (c) determine the total settlement of the group.

8.13 At a particular site the soil profile consists of a layer of soft clay underlain by a depth of sand. The values of standard penetration resistance at depths of 0.75, 1.50, 2.25, 3.00 and 3.75 m in the sand are 18, 24, 26, 34 and 32, respectively. Nine precast concrete piles, in a square group, are driven through the clay and 2 m into the sand. The piles are 0.25×0.25 m in section and are spaced at 0.75 m centres. The pile group supports a permanent load of 2000 kN and an imposed load of 1000 kN. Settlement should not exceed 20 mm. Neglecting skin friction in the clay, (a) determine the load factor using the traditional method, (b) check that the bearing resistance limit state is satisfied using the limit state method and (c) check that the serviceability limit state is satisfied.

8.14 A ground anchor in a stiff clay, formed by the gravel injection technique, has a fixed anchor length of 5 m and an effective fixed anchor diameter of 200 mm: the diameter of the borehole is 100 mm. The relevant shear strength parameters for the clay are $c_u = 110\,kN/m^2$ and $\phi_u = 0$. What would be the expected ultimate load capacity of the anchor, assuming a skin friction coefficient of 0.6?

REFERENCES

1 Atkinson, J.H. (1981) *Foundations and Slopes: An Introduction to Applications of Critical State Soil Mechanics*, McGraw-Hill, London.

2 Berezantzev, V.G., Khristoforov, V.S. and Golubkov, V.N. (1961) Load bearing capacity and deformation of piled foundations, in *Proceedings 5th International Conference SMFE, Paris*, Vol. 2, pp. 11–15.

3 Bjerrum, L. (1963) Discussion, in *Proceedings European Conference SMFE, Wiesbaden*, Vol. 3, pp. 135–7.

4 Bjerrum, L. and Eggestad, A. (1963) Interpretation of loading tests on sand, in *Proceedings European Conference SMFE, Wiesbaden*, Vol. 1, pp. 199–203.

5 Bjerrum, L. and Eide, O. (1956) Stability of strutted excavations in clay, *Geotechnique*, **6**, 32–47.

6 British Standard 1377 (1990) *Methods of Test for Soils for Civil Engineering Purposes*, British Standards Institution, London.

7 British Standard 8004 (1986) *Code of Practice for Foundations*, British Standards Institution, London.

8 British Standard 8081 (1989) *Code of Practice for Ground Anchorages*, British Standards Institution, London.

9 Burland, J.B. (1973) Shaft friction of piles in clay, *Ground Engineering*, **6** (3), 30–42.

10 Burland, J.B., Broms, B.B. and De Mello, V.F.B. (1977) Behaviour of foundations and structures, in *Proc. 9th International Conference SMFE, Tokyo*, Vol. 2, Japanese Society SMFE, Tokyo, pp. 495–538.

11 Burland, J.B. and Burbidge, M.C. (1985) Settlement of foundations on sand and gravel, *Proc. Institution of Civil Engineers*, Part 1, **78**, December, pp. 1325–81.

12 Burland, J.B. and Cooke, R.W. (1974) The design of bored piles in stiff clays, *Ground Engineering*, **7** (4), 28–35.

13 Burland, J.B. and Wroth, C.P. (1975) Settlement of buildings and associated damage, in *Proceedings of Conference on Settlement of Structures* (British Geotechnical Society), Pentech Press, London, pp. 611–53.

14 Clayton, C.R.I. (1995) The standard penetration test (SPT): methods and use, *CIRIA Report 143*, London.

15 Clough, G.W., Smith, E.M. and Sweeney, B.P. (1989) Movement control of excavation support systems by iterative design, *Proc. ASCE Congress on Foundation Engineering – Current Principles and Practices*, Evanston, IL.

16 Cooke, R.W. (1986) Piled raft foundations on stiff clays – a contribution to design philosophy, *Geotechnique*, **36**, 169–203.

17 DeBeer, E.E. (1970) Experimental determination of the shape factors and the bearing capacity factors of sand, *Geotechnique*, **20**, 387–411.

18 Durgunoglu, H.T. and Mitchell, J.K. (1975) Static penetration resistance of soils, *Proceedings ASCE Conference on In-situ Measurement of Soil Properties*, Raleigh, NC.

19 European Committee for Standardization (1994) *Eurocode 7: Geotechnical Design – Part 1: General Rules*, ENV 1997-1, Brussels.

20 Fellenius, B.H. (1980) The analysis of results from routine pile load tests, *Ground Engineering*, **13** (6), 19–31.

21 Fleming, W.G.K., Weltman, A.J., Randolph, M.F. and Elson, W.K. (1992) *Piling Engineering*, 2nd edn, Blackie, Glasgow.

22 Gibbs, H.J. and Holtz, W.G. (1957) Research on determining the density of sands by spoon penetration testing, in *Proceedings 4th International Conference SMFE, London*, Vol. 1, Butterworths, London, pp. 35–9.

23 Hansen, J.B. (1968) A revised extended formula for bearing capacity, *Danish Geotechnical Institute Bulletin*, No. 28.

24 Mana, A.I. and Clough, G.W. (1981) Prediction of movements for braced cuts in clay, *Journal ASCE*, **107** (GT6), 759–77.

25 Meigh, A.C. (1987) *Cone Penetration Testing: Methods and Interpretation*, CIRIA/ Butterworths, London.

26 Meyerhof, G.G. (1956) Penetration tests and bearing capacity of cohesionless soils, *Proceedings ASCE*, **82**, No. SM1, Paper 866, pp. 1–19.

27 Meyerhof, G.G. (1963) Some recent research on the bearing capacity of foundations, *Canadian Geotechnical Journal*, **1** (1), 16–26.

28 Meyerhof, G.G. (1965) Shallow foundations, *Proceedings ASCE*, **91** (SM2), 21–31.

29 Meyerhof, G.G. (1976) Bearing capacity and settlement of pile foundations, *Proceedings ASCE*, **102** (GT3), 195–228.

30 Moseley, M.P. (ed.) (1993) *Ground Improvement*, Blackie Academic & Professional, Glasgow.

31 Orr, T.L.L. and Farrell, E.R. (1999) *Geotechnical Design to Eurocode 7*, Springer-Verlag, London.

32 Parry, R.H.G. (1995) *Mohr Circles, Stress Paths and Geotechnics*, E. & F.N. Spon, London.

33 Peck, R.B. (1969) Deep excavations and tunnelling in soft ground, *7th International Conference SMFE, Mexico*, State of the Art Volume, pp. 225–90.

34 Peck, R.B., Hanson, W.E. and Thornburn, T.H. (1974) *Foundation Engineering*, John Wiley & Sons, New York.

35 Polshin, D.E. and Tokar, R.A. (1957) Maximum allowable non-uniform settlement of structures, in *Proceedings 4th International Conference SMFE, London*, Vol. 1, Butterworths, London, pp. 402–5.

36 Poulos, H.G. and Davis, E.H. (1980) *Pile Foundation Analysis and Design*, John Wiley & Sons, New York.

37   Schmertmann, J.H. (1970) Static cone to compute static settlement over sand, *Proceedings ASCE*, **96** (SM3), 1011–43.

38   Schmertmann, J.H. (1975) Measurement of *in-situ* shear strength, in *Proceedings of Conference on In-Situ Measurement of Soil Properties*, ASCE, New York, Vol. II, pp. 57–138.

39   Schmertmann, J.H., Hartman, J.P. and Brown, P.R. (1978) Improved strain influence factor diagrams, *Proceedings ASCE*, **104** (GT8), 1131–35.

40   Skempton, A.W. (1951) The bearing capacity of clays, *Proceedings Building Research Congress*, **1**, 180–9.

41   Skempton, A.W. (1959) Cast-*in-situ* bored piles in London clay, *Geotechnique*, **9**, 153–73.

42   Skempton, A.W. (1986) Standard penetration test procedures and the effects in sands of overburden pressure, relative density, particle size, ageing and overconsolidation, *Geotechnique*, **36**, 425–47.

43   Skempton, A.W. and MacDonald, D.H. (1956) Allowable settlement of buildings, *Proceedings ICE*, **5** (Part 3), 727–68.

44   Smith, E.A.L. (1960) Pile driving analysis by the wave equation, *Proceedings ASCE*, **86** (SM4), 35–61.

45   Terzaghi, K. (1943) *Theoretical Soil Mechanics*, John Wiley & Sons, New York.

46   Terzaghi, K., Peck, R.B. and Mesri, G. (1996) *Soil Mechanics in Engineering Practice*, 3rd edn, John Wiley & Sons, New York.

47   Thorburn, S. (1963) Tentative correction chart for the standard penetration test in non-cohesive soils, *Civil Engineering and Public Works Review*, **58**, 752–3.

48   Thorburn, S. (1975) Building structures supported by stabilised ground, *Geotechnique*, **25**, 83–94.

49   Tomlinson, M.J. (1994) *Pile Design and Construction Practice*, 4th edn, E. & F.N. Spon, London.

50   Tomlinson, M.J. (2001) *Foundation Design and Construction*, 7th edn, Longman Scientific & Technical, Harlow, Essex.

51   Vesic, A.S. (1973) Analysis of ultimate loads of shallow foundations, *Journal ASCE*, **99** (SM1), 45–73.

52   Whitaker, T. (1957) Experiments with models piles in groups, *Geotechnique*, **7**, 147–67.

53   Whitaker, T. (1976) *The Design of Piled Foundations*, Pergamon Press, Oxford.

Chapter 9

Stability of slopes

9.1 INTRODUCTION

Gravitational and seepage forces tend to cause instability in natural slopes, in slopes formed by excavation and in the slopes of embankments. The most important types of slope failure are illustrated in Figure 9.1. In *rotational* slips the shape of the failure surface in section may be a circular arc or a non-circular curve. In general, circular slips are associated with homogeneous, isotropic soil conditions and non-circular slips with non-homogeneous conditions. *Translational* and *compound* slips occur where the form of the failure surface is influenced by the presence of an adjacent stratum of significantly different strength, most of the failure surface being likely to pass through the stratum of lower shear strength. The form of the surface would also be influenced by the presence of discontinuities such as fissures and pre-existing slips. Translational slips tend to occur where the adjacent stratum is at a relatively shallow depth below the surface of the slope, the failure surface tending to be plane and roughly parallel to the slope. Compound slips usually occur where the adjacent stratum is at greater depth, the failure surface consisting of curved and plane sections. In most cases, slope stability can be considered as a two-dimensional problem, conditions of plane strain being assumed.

Design is based on the requirement to maintain stability rather than on the need to minimize deformation. If deformation were such that the strain in an element of soil exceeded the value corresponding to peak strength, then the strength would fall towards the ultimate value. Thus it is appropriate to use the critical-state strength in analysing stability. However, if a pre-existing slip surface were to be present within the soil, use of the residual strength would be appropriate. Limiting equilibrium methods are normally used in the analysis of slope stability in which it is considered that failure is on the point of occurring along an assumed or a known failure surface. In the traditional approach the shear strength required to maintain a condition of limiting equilibrium is compared with the available shear strength of the soil, giving the average (lumped) factor of safety along the failure surface.

Alternatively, the limit state method can be used in which partial factors are applied to the shear strength parameters. Case C (Section 8.1) applies to slope problems, the greatest uncertainties being the soil properties. The ultimate limit state of overall stability is then satisfied if, depending on the method of analysis, either the design disturbing force (S_d) is less than or equal to the design resisting force (R_d) along the potential failure surface or the design disturbing moment is less than or equal to the

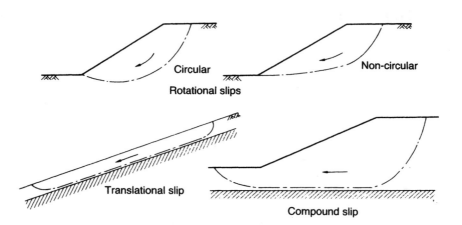

Figure 9.1 Types of slope failure.

design resisting moment. Characteristic values of shear strength parameters c' and $\tan \phi'$ should be divided by factors 1.60 and 1.25, respectively. (However, the value of c' is zero if the critical-state strength is used.) The characteristic value of parameter c_u should be divided by 1.40. A factor of unity is appropriate for the self-weight of the soil and for pore water pressures. However, variable loads on the soil surface adjacent to the slope should be multiplied by a factor of 1.30.

The following limit states should be considered as appropriate:

1   Loss of overall stability due to slip failure.
2   Bearing resistance failure below embankments.
3   Internal erosion due to high hydraulic gradients and/or poor compaction.
4   Failure as a result of surface erosion.
5   Failure due to hydraulic uplift.
6   Excessive soil deformation resulting in structural damage to, or loss of service-ability of, adjacent structures, highways or services.

9.2   ANALYSIS FOR THE CASE OF $\phi_u = 0$

This analysis, in terms of total stress, covers the case of a fully saturated clay under undrained conditions, i.e. for the condition immediately after construction. Only moment equilibrium is considered in the analysis. In section, the potential failure surface is assumed to be a circular arc. A trial failure surface (centre O, radius r and length L_a) is shown in Figure 9.2. Potential instability is due to the total weight of the soil mass (W per unit length) above the failure surface. For equilibrium the shear strength which must be mobilized along the failure surface is expressed as

$$\tau_m = \frac{\tau_f}{F} = \frac{c_u}{F}$$

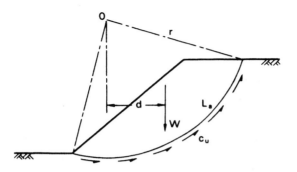

Figure 9.2 The $\phi_u = 0$ analysis.

where F is the factor of safety with respect to shear strength. Equating moments about O:

$$Wd = \frac{c_u}{F} L_a r$$

Therefore

$$F = \frac{c_u L_a r}{Wd} \tag{9.1}$$

The moments of any additional forces must be taken into account. In the event of a tension crack developing, the arc length L_a is shortened and a hydrostatic force will act normal to the crack if it fills with water. It is necessary to analyse the slope for a number of trial failure surfaces in order that the minimum factor of safety can be determined.

Based on the principle of geometric similarity, Taylor [19] published *stability coefficients* for the analysis of homogeneous slopes in terms of total stress. For a slope of height H the stability coefficient (N_s) for the failure surface along which the factor of safety is a minimum is

$$N_s = \frac{c_u}{F\gamma H} \tag{9.2}$$

For the case of $\phi_u = 0$, values of N_s can be obtained from Figure 9.3. The coefficient N_s depends on the slope angle β and the depth factor D, where DH is the depth to a firm stratum.

Gibson and Morgenstern [8] published stability coefficients for slopes in normally consolidated clays in which the undrained strength $c_u(\phi_u = 0)$ varies linearly with depth.

In limit state design the characteristic value of undrained strength (c_{uk}) is divided by the appropriate partial factor to obtain the design value (c_{ud}). The limit state of overall stability is satisfied if the design disturbing moment (Wd) is less than or equal to the design resisting moment ($c_{ud} L_a r$).

A three-dimensional analysis for slopes in clay under undrained conditions has been presented by Gens *et al.* [7].

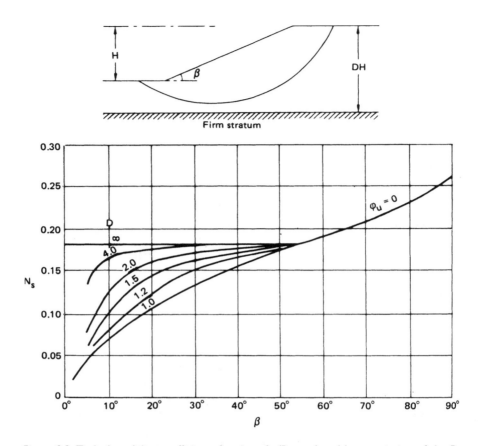

Figure 9.3 Taylor's stability coefficients for $\phi_u = 0$. (Reproduced by permission of the Boston Society of Civil Engineers.)

Example 9.1

A 45° slope is excavated to a depth of 8 m in a deep layer of saturated clay of unit weight 19 kN/m³: the relevant shear strength parameters are $c_u = 65\,\text{kN/m}^2$ and $\phi_u = 0$. Determine the factor of safety for the trial failure surface specified in Figure 9.4. Check that no loss of overall stability will occur according to the limit state approach.

In Figure 9.4, the cross-sectional area ABCD is 70 m².

$$\text{Weight of soil mass} = 70 \times 19 = 1330\,\text{kN/m}$$

The centroid of ABCD is 4.5 m from O. The angle AOC is $89\frac{1}{2}°$ and radius OC is 12.1 m. The arc length ABC is calculated as 18.9 m. The factor of safety is given by

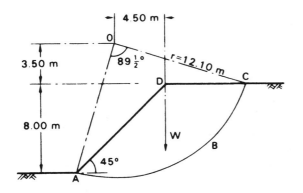

Figure 9.4 Example 9.1.

$$F = \frac{c_u L_a r}{Wd}$$
$$= \frac{65 \times 18.9 \times 12.1}{1330 \times 4.5} = 2.48$$

This is the factor of safety for the trial failure surface selected and is not necessarily the minimum factor of safety.

The minimum factor of safety can be estimated by using Equation 9.2. From Figure 9.3, $\beta = 45°$ and assuming that D is large, the value of N_s is 0.18. Then

$$F = \frac{c_u}{N_s \gamma H}$$
$$= \frac{65}{0.18 \times 19 \times 8}$$
$$= 2.37$$

Using the limit state method the characteristic value of undrained strength (c_{uk}) is divided by a partial factor of 1.40. Thus the design value of the parameter (c_{ud}) is $65/1.40$ i.e. $46\,\text{kN/m}^2$, hence

design disturbing moment per m $= Wd = 1330 \times 4.5 = 5985\,\text{kNm}$

design resisting moment per m $= c_{ud} L_a r = 46 \times 18.9 \times 12.1 = 10\,520\,\text{kN m}$

The design disturbing moment is less than the design resisting moment, therefore the overall stability limit state is satisfied.

9.3   THE METHOD OF SLICES

In this method the potential failure surface, in section, is again assumed to be a circular arc with centre O and radius r. The soil mass (ABCD) above a trial failure surface

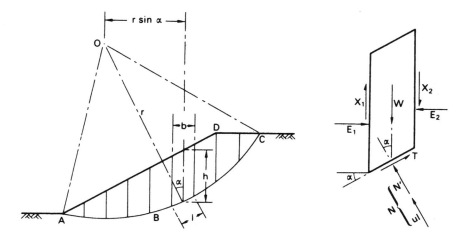

Figure 9.5 The method of slices.

(AC) is divided by vertical planes into a series of slices of width b, as shown in Figure 9.5. The base of each slice is assumed to be a straight line. For any slice the inclination of the base to the horizontal is α and the height, measured on the centre-line, is h. The analysis is based on the use of a lumped factor of safety (F), defined as the ratio of the available shear strength (τ_f) to the shear strength (τ_m) which must be mobilized to maintain a condition of limiting equilibrium, i.e.

$$F = \frac{\tau_f}{\tau_m}$$

The factor of safety is taken to be the same for each slice, implying that there must be mutual support between slices, i.e. forces must act between the slices.

The forces (per unit dimension normal to the section) acting on a slice are:

1   The total weight of the slice, $W = \gamma bh$ (γ_{sat} where appropriate).
2   The total normal force on the base, N (equal to σl). In general this force has two components, the effective normal force N' (equal to $\sigma' l$) and the boundary water force U (equal to ul), where u is the pore water pressure at the centre of the base and l the length of the base.
3   The shear force on the base, $T = \tau_m l$.
4   The total normal forces on the sides, E_1 and E_2.
5   The shear forces on the sides, X_1 and X_2.

Any external forces must also be included in the analysis.

The problem is statically indeterminate and in order to obtain a solution assumptions must be made regarding the interslice forces E and X; in general the resulting solution for factor of safety is not exact.

Considering moments about O, the sum of the moments of the shear forces T on the failure arc AC must equal the moment of the weight of the soil mass ABCD. For any slice the lever arm of W is $r \sin \alpha$, therefore

$$\Sigma Tr = \Sigma Wr \sin \alpha$$

Now

$$T = \tau_m l = \frac{\tau_f}{F} l$$

$$\therefore \Sigma \frac{\tau_f}{F} l = \Sigma W \sin \alpha$$

$$\therefore F = \frac{\Sigma \tau_f l}{\Sigma W \sin \alpha}$$

For an effective stress analysis (in terms of tangent parameters c' and ϕ'):

$$F = \frac{\Sigma(c' + \sigma' \tan \phi')l}{\Sigma W \sin \alpha}$$

or

$$F = \frac{c' L_a + \tan \phi' \Sigma N'}{\Sigma W \sin \alpha} \tag{9.3a}$$

where L_a is the arc length AC. Equation 9.3(a) is exact but approximations are introduced in determining the forces N'. For a given failure arc the value of F will depend on the way in which the forces N' are estimated.

However, the critical-state strength is normally appropriate in the analysis of slope stability, i.e. $\phi' = \phi'_{cv}$ and $c' = 0$, therefore the factor of safety is given by

$$F = \frac{\tan \phi'_{cv} \Sigma N'}{\Sigma W \sin \alpha} \tag{9.3b}$$

The Fellenius (or Swedish) solution

In this solution it is assumed that for each slice the resultant of the interslice forces is zero. The solution involves resolving the forces on each slice normal to the base, i.e.

$$N' = W \cos \alpha - ul$$

Hence the factor of safety in terms of effective stress (Equation 9.3(a)) is given by

$$F = \frac{c' L_a + \tan \phi' \Sigma(W \cos \alpha - ul)}{\Sigma W \sin \alpha} \tag{9.4}$$

The components $W \cos \alpha$ and $W \sin \alpha$ can be determined graphically for each slice. Alternatively, the value of α can be measured or calculated. Again, a series of trial failure surfaces must be chosen in order to obtain the minimum factor of safety. This solution underestimates the factor of safety: the error, compared with more accurate methods of analysis, is usually within the range 5–20%.

For an analysis in terms of total stress the parameter c_u is used in Equation 9.3(a) (with $\phi_u = 0$) and the value of u is zero. The factor of safety then becomes

$$F = \frac{c_u L_a}{\Sigma W \sin \alpha} \tag{9.5}$$

As N' does not appear in Equation 9.5, an exact value of F is obtained.

Use of the Fellenius method is not now recommended in practice.

The Bishop routine solution

In this solution it is assumed that the resultant forces on the sides of the slices are horizontal, i.e.

$$X_1 - X_2 = 0$$

For equilibrium the shear force on the base of any slice is

$$T = \frac{1}{F}(c'l + N' \tan \phi')$$

Resolving forces in the vertical direction:

$$W = N' \cos \alpha + ul \cos \alpha + \frac{c'l}{F} \sin \alpha + \frac{N'}{F} \tan \phi' \sin \alpha$$

$$\therefore \ N' = \frac{[W - (c'l/F) \sin \alpha - ul \cos \alpha]}{[\cos \alpha + (\tan \phi' \sin \alpha)/F]} \tag{9.6}$$

It is convenient to substitute

$$l = b \sec \alpha$$

From Equation 9.3(a), after some rearrangement,

$$F = \frac{1}{\Sigma W \sin \alpha} \sum \left[\{c'b + (W - ub) \tan \phi'\} \frac{\sec \alpha}{1 + (\tan \alpha \tan \phi'/F)} \right] \tag{9.7}$$

Bishop [2] also showed how non-zero values of the resultant forces $(X_1 - X_2)$ could be introduced into the analysis but this refinement has only a marginal effect on the factor of safety.

The pore water pressure can be related to the total 'fill pressure' at any point by means of the dimensionless *pore pressure ratio*, defined as

$$r_u = \frac{u}{\gamma h} \tag{9.8}$$

(γ_{sat} where appropriate). For any slice,

$$r_u = \frac{u}{W/b}$$

Hence Equation 9.7 can be written as

$$F = \frac{1}{\Sigma W \sin \alpha} \Sigma \left[\{c'b + W(1 - r_u)\tan\phi'\} \frac{\sec\alpha}{1 + (\tan\alpha\tan\phi'/F)} \right] \qquad (9.9)$$

As the factor of safety occurs on both sides of Equation 9.9, a process of successive approximation must be used to obtain a solution but convergence is rapid.

Due to the repetitive nature of the calculations and the need to select an adequate number of trial failure surfaces, the method of slices is particularly suitable for solution by computer. More complex slope geometry and different soil strata can be introduced.

In most problems the value of the pore pressure ratio r_u is not constant over the whole failure surface but, unless there are isolated regions of high pore pressure, an average value (weighted on an area basis) is normally used in design. Again, the factor of safety determined by this method is an underestimate but the error is unlikely to exceed 7% and in most cases is less than 2%.

Spencer [18] proposed a method of analysis in which the resultant interslice forces are parallel and in which both force and moment equilibrium are satisfied. Spencer showed that the accuracy of the Bishop routine method, in which only moment equilibrium is satisfied, is due to the insensitivity of the moment equation to the slope of the interslice forces.

Dimensionless stability coefficients for homogeneous slopes, based on Equation 9.9, have been published by Bishop and Morgenstern [4]. It can be shown that for a given slope angle and given soil properties the factor of safety varies linearly with r_u and can thus be expressed as

$$F = m - nr_u \qquad (9.10)$$

where m and n are the stability coefficients. The coefficients m and n are functions of β, ϕ', depth factor D and the dimensionless factor $c'/\gamma H$ (which is zero if the critical-state strength is used).

Example 9.2

Using the Fellenius method of slices, determine the factor of safety, in terms of effective stress, of the slope shown in Figure 9.6 for the given failure surface (a) using peak strength parameters $c' = 10\,\text{kN/m}^2$ and $\phi' = 29°$ and (b) using critical-state parameter $\phi'_{cv} = 31°$. The unit weight of the soil both above and below the water table is $20\,\text{kN/m}^3$.

(a) The factor of safety is given by Equation 9.4. The soil mass is divided into slices 1.5 m wide. The weight (W) of each slice is given by

$$W = \gamma bh = 20 \times 1.5 \times h = 30h\,\text{kN/m}$$

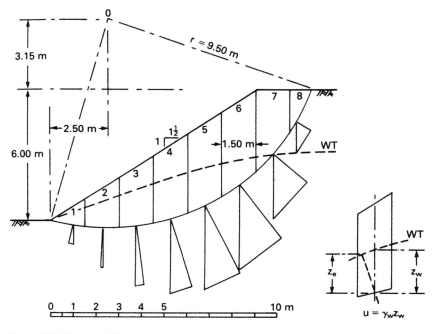

Figure 9.6 Example 9.2.

The height h for each slice is set off below the centre of the base, and the normal and tangential components $h \cos \alpha$ and $h \sin \alpha$, respectively, are determined graphically, as shown in Figure 9.6. Then

$$W \cos \alpha = 30 h \cos \alpha$$

$$W \sin \alpha = 30 h \sin \alpha$$

The pore water pressure at the centre of the base of each slice is taken to be $\gamma_w z_w$, where z_w is the vertical distance of the centre point below the water table (as shown in the figure). This procedure slightly overestimates the pore water pressure which strictly should be $\gamma_w z_e$, where z_e is the vertical distance below the point of intersection of the water table and the equipotential through the centre of the slice base. The error involved is on the safe side.

The arc length (L_a) is calculated as 14.35 m. The results are given in Table 9.1.

$$\Sigma W \cos \alpha = 30 \times 17.50 = 525 \, \text{kN/m}$$

$$\Sigma W \sin \alpha = 30 \times 8.45 = 254 \, \text{kN/m}$$

$$\Sigma (W \cos \alpha - ul) = 525 - 132 = 393 \, \text{kN/m}$$

Table 9.1

Slice No.	$h \cos \alpha$ (m)	$h \sin \alpha$ (m)	u (kN/m²)	l (m)	ul (kN/m)
1	0.75	−0.15	5.9	1.55	9.1
2	1.80	−0.10	11.8	1.50	17.7
3	2.70	0.40	16.2	1.55	25.1
4	3.25	1.00	18.1	1.60	29.0
5	3.45	1.75	17.1	1.70	29.1
6	3.10	2.35	11.3	1.95	22.0
7	1.90	2.25	0	2.35	0
8	0.55	0.95	0	2.15	0
	17.50	8.45		14.35	132.0

$$F = \frac{c'L_a + \tan \phi' \Sigma (W \cos \alpha - ul)}{\Sigma W \sin \alpha}$$

$$= \frac{(10 \times 14.35) + (0.554 \times 393)}{254}$$

$$= \frac{143.5 + 218}{254} = 1.42$$

(b) In terms of critical-state strength

$$F = \frac{\tan 31° \times 393}{254} = 0.93$$

Deformation is likely to result in strains along a potential failure surface exceeding the value corresponding to peak strength. Therefore the strength mobilized for stability is likely to fall below the peak value and to approach the critical-state value. Therefore the slope is unsafe. It should be noted that in case (a), the proportion of shear strength represented by c', generally a parameter of uncertain value, is 40%.

9.4 ANALYSIS OF A PLANE TRANSLATIONAL SLIP

It is assumed that the potential failure surface is parallel to the surface of the slope and is at a depth that is small compared with the length of the slope. The slope can then be considered as being of infinite length, with end effects being ignored. The slope is inclined at angle β to the horizontal and the depth of the failure plane is z, as shown in section in Figure 9.7. The water table is taken to be parallel to the slope at a height of $mz(0 < m < 1)$ above the failure plane. Steady seepage is assumed to be taking place in a direction parallel to the slope. The forces on the sides of any vertical slice are equal and opposite, and the stress conditions are the same at every point on the failure plane.

In terms of effective stress, the shear strength of the soil along the failure plane (using the critical-state strength) is

$$\tau_f = (\sigma - u) \tan \phi'_{cv}$$

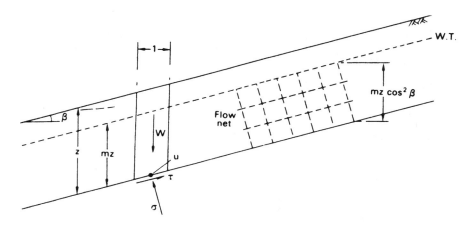

Figure 9.7 Plane translational slip.

and the factor of safety is

$$F = \frac{\tau_f}{\tau}$$

The expressions for σ, τ and u are

$$\sigma = \{(1 - m)\gamma + m\gamma_{sat}\}z \cos^2 \beta$$
$$\tau = \{(1 - m)\gamma + m\gamma_{sat}\}z \sin \beta \cos \beta$$
$$u = mz\gamma_w \cos^2 \beta$$

If the soil between the surface and the failure plane is not fully saturated (i.e. $m = 0$) then

$$F = \frac{\tan \phi'_{cv}}{\tan \beta} \tag{9.11}$$

If the water table coincides with the surface of the slope (i.e. $m = 1$) then

$$F = \frac{\gamma' \tan \phi'_{cv}}{\gamma_{sat} \tan \beta} \tag{9.12}$$

For a total stress analysis the shear strength parameter c_u is used (with $\phi_u = 0$) and the value of u is zero.

Example 9.3

A long natural slope in an overconsolidated fissured clay of saturated unit weight $20 \, \text{kN/m}^3$ is inclined at $12°$ to the horizontal. The water table is at the surface and seepage is roughly parallel to the slope. A slip has developed on a plane parallel to the surface at a depth of 5 m. (1) Determine the factor of safety along the slip plane using (a) the critical-state parameter $\phi'_{cv} = 28°$ and (b) the residual strength parameter $\phi'_r = 20°$. (2) Analyse the stability of the slope by the limit state method.

1   Equation 9.12 applies in both cases.

(a) In terms of critical-state strength

$$F = \frac{10.2 \tan 28°}{20 \tan 12°} = 1.28$$

(b) In terms of residual strength

$$F = \frac{10.2 \tan 20°}{20 \tan 12°} = 0.87$$

2   In the limit state method the characteristic values of the ϕ' parameters are divided by the partial factor 1.25. Thus the design values are

$$\phi'_{cv} = \tan^{-1}\left(\frac{\tan 28°}{1.25}\right) = 23°$$

$$\phi'_r = \tan^{-1}\left(\frac{\tan 20°}{1.25}\right) = 16°$$

With the water table at the surface the value of $m = 1$

(a) The design disturbing force per m^2 is

$$S_d = \gamma_{sat} z \sin \beta \cos \beta$$
$$= 20 \times 5 \times \sin 12° \cos 12°$$
$$= 20.3 \, \text{kN}$$

The design resisting force per m^2 is

$$R_d = (\sigma - u) \tan \phi'_{cv}$$
$$= (\gamma_{sat} - \gamma_w) z \cos^2 \beta \tan 23°$$
$$= 10.2 \times 5 \times \cos^2 12° \tan 23°$$
$$= 20.7 \, \text{kN}$$

The design disturbing force is less than the design resisting force; therefore, in terms of critical-state strength, the limit state for overall stability is satisfied.

(b) With $\phi'_r = 16°$ the design resisting force becomes 14.0 kN. The design disturbing force remains 20.3 kN; therefore, in terms of residual strength, the limit state is not satisfied.

Both methods (1) and (2) show that failure of the slope has taken place as shear strength has decreased below the critical-state value towards the residual value.

9.5   GENERAL METHODS OF ANALYSIS

Morgenstern and Price [12] developed a general analysis in which all boundary and equilibrium conditions are satisfied and in which the failure surface may be any shape, circular, non-circular or compound. The ground surface is represented by a function $y = z(x)$ and the trial failure surface by $y = y(x)$ as shown in Figure 9.8. The forces acting on an infinitesimal slice of width dx are also shown in the figure. The forces are denoted as follows:

$E' = $ effective normal force on a side of the slice,
$X = $ shear force on a side,
$P_w = $ boundary water force on a side,
$dN' = $ effective normal force on the base of the slice,
$dS = $ shear force on the base,
$dP_b = $ boundary water force on the base,
$dW = $ total weight of the slice.

The line of thrust of the effective normal forces (E') is represented by a function $y = y'_t(x)$ and that of the internal water forces (P_w) by $y = h(x)$. Two governing differential equations are obtained by equating moments about the mid-point of the

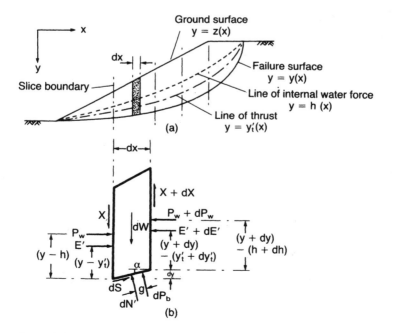

Figure 9.8 The Morgenstern–Price method.

base, and forces perpendicular and parallel to the base, to zero. The equations are simplified by working in terms of the total normal force (E), where

$$E = E' + P_w$$

The position of force E on a side of the slice is obtained from the expression

$$Ey_t = E'y_t' + P_w h$$

The problem is rendered statically determinate by assuming a relationship between the forces E and X of the form

$$X = \lambda f(x)E \qquad (9.13)$$

where $f(x)$ is a function chosen to represent the pattern of variation of the ratio X/E across the failure mass and λ is a scale factor. The value of λ is obtained as part of the solution along with the factor of safety F.

To obtain a solution the soil mass above a trial failure surface is divided into a series of slices of finite width such that the failure surface within each slice can be assumed to be linear. The boundary conditions at each end of the failure surface are in terms of the force E and a moment M which is given by the integral of an expression containing both E and X: normally both E and M are zero at each end of the failure surface. The method of solution involves choosing trial values of λ and F, setting the force E to zero at the beginning of the failure surface and integrating across each slice in turn, obtaining values of E, X and y_t: the resulting values of E and M at the end of the failure surface will in general not be zero. A systematic iteration technique, based on the Newton–Raphson method and described by Morgenstern and Price [13], is used to modify the values of λ and F until the resulting values of both E and M at the end of the failure surface are zero. The factor of safety is not significantly affected by the choice of the function $f(x)$ and as a consequence $f(x) = 1$ is a common assumption.

For any assumed failure surface it is necessary to examine the solution to ensure that it is valid in respect of the implied state of stress within the soil mass above that surface. Accordingly, a check is made to ensure that neither shear failure nor a state of tension is implied within the mass. The first condition is satisfied if the available shearing resistance on each vertical interface is greater than the corresponding value of the force X: the ratio of these two forces represents the local factor of safety against shear failure along the inter-face. The requirement that no tension should be developed is satisfied if the line of thrust of the E forces, as given by the computed values of y_t, lies wholly above the failure surface.

Computer software for the Morgenstern–Price analysis is readily available. The method can be fully exploited if an interactive approach, using computer graphics, is adopted.

Bell [1] proposed a method of analysis in which all the conditions of equilibrium are satisfied and the assumed failure surface may be of any shape. The soil mass is divided into a number of vertical slices and statical determinacy is obtained by means of an assumed distribution of normal stress along the failure surface.

Sarma [15] developed a method, based on the method of slices, in which the critical earthquake acceleration required to produce a condition of limiting equilibrium is determined. An assumed distribution of vertical interslice forces is used in the analysis. Again, all the conditions of equilibrium are satisfied and the assumed failure surface may be of any shape. The static factor of safety is the factor by which the shear strength of the soil must be reduced such that the critical acceleration is zero.

The use of a computer is also essential for the Bell and Sarma methods and all solutions must be checked to ensure that they are physically acceptable.

9.6   END-OF-CONSTRUCTION AND LONG-TERM STABILITY

Excavated slopes

When a slope is formed by excavation, the decreases in total stress result in changes in pore water pressure in the vicinity of the slope and, in particular, along a potential failure surface. For the case illustrated in Figure 9.9, the initial pore water pressure (u_0) depends on the depth of the point in question below the initial (static) water table (i.e. $u_0 = u_s$). The change in pore water pressure (Δu) due to excavation is given theoretically by Equation 4.25 or 4.26. After excavation, pore water will flow towards the slope and drawdown of the water table will occur: a steady seepage condition will become established for which a flow net can be drawn. The final pore water pressure (u_f), after dissipation of excess pore water pressure is complete, will be the steady seepage value determined from the flow net (i.e. $u_f = u_{ss}$).

If the permeability of the soil is low, a considerable time will elapse before any significant dissipation of excess pore water pressure will have taken place. At the end of construction the soil will be virtually in the undrained condition and a total stress analysis will be relevant. In principle an effective stress analysis is also possible for the end-of-construction condition using the appropriate value of pore water pressure ($u_0 + \Delta u$) for this condition. However, because of its greater simplicity, a total stress analysis is generally used. It should be realized that the same factor of safety will not generally be obtained from a total stress and an effective stress analysis of the

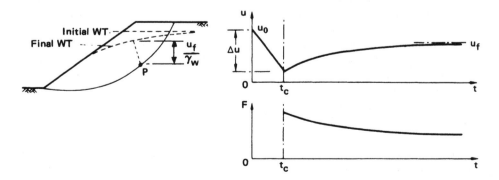

Figure 9.9 Pore pressure dissipation and factor of safety after excavation. (Reproduced from A.W. Bishop and L. Bjerrum (1960) *Proceedings ASCE Research Conference on Shear Strength of Cohesive Soils*, Boulder, Colorado, by permission of the American Society of Civil Engineers.)

end-of-construction condition. In a total stress analysis it is implied that the pore water pressures are those for a failure condition (being the equivalent of the pore water pressure at failure in an undrained triaxial test): in an effective stress analysis the pore water pressures used are those predicted for a non-failure condition. In the long term, the fully drained condition will be reached and only an effective stress analysis will be appropriate.

On the other hand, if the permeability of the soil is high, dissipation of excess pore water pressure will be largely complete by the end of construction. An effective stress analysis is relevant for all conditions with values of pore water pressure being obtained from the static water table level or the steady seepage flow net.

It is important to identify the most dangerous condition in any practical problem in order that the appropriate shear strength parameters are used in design.

Equation 4.25, with $B = 1$ for a fully saturated clay, can be rearranged as follows:

$$\Delta u = \frac{1}{2}(\Delta\sigma_1 + \Delta\sigma_3) + \left(A - \frac{1}{2}\right)(\Delta\sigma_1 - \Delta\sigma_3) \tag{9.14}$$

For a typical point P on a potential failure surface (Figure 9.9) the first term in Equation 9.14 is negative and the second term will also be negative if the value of A is less than 0.5. Overall, the pore water pressure change Δu is negative. The effect of the rotation of the principal stress directions is neglected. As dissipation proceeds the pore water pressure increases to the steady seepage value as shown in Figure 9.9. The factor of safety will therefore have a lower value in the long term, when dissipation is complete, than at the end of construction.

Slopes in overconsolidated fissured clays require special consideration. A number of cases are on record in which failures in this type of clay have occurred long after dissipation of excess pore water pressure had been completed. Analysis of these failures showed that the average shear strength at failure was well below the peak value. It is probable that large strains occur locally due to the presence of fissures, resulting in the peak strength being reached, followed by a gradual decrease towards the critical-state value. The development of large local strains can lead eventually to a progressive slope failure. However, fissures may not be the only cause of progressive failure; there is considerable non-uniformity of shear stress along a potential failure surface and local overstressing may initiate progressive failure. It is also possible that there could be a pre-existing slip surface in this type of clay and that it could be reactivated by excavation. In such cases a considerable slip movement could have taken place previously, sufficiently large for the shear strength to fall below the critical-state value and towards the residual value.

Thus for an initial failure (a 'first time' slip) in overconsolidated fissured clay the relevant strength for the analysis of long-term stability is the critical-state value. However, for failure along a pre-existing slip surface the relevant strength is the residual value. Clearly it is vital that the presence of a pre-existing slip surface in the vicinity of a projected excavation should be detected during the ground investigation.

The strength of an overconsolidated clay at the critical state, for use in the analysis of a potential first time slip, is difficult to determine accurately. Skempton [17] has suggested that the maximum strength of the remoulded clay in the normally consolidated condition can be taken as a practical approximation to the strength of the

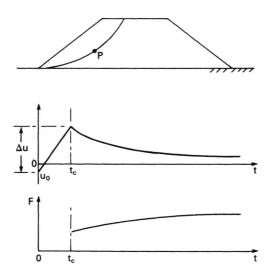

Figure 9.10 Pore pressure dissipation and factor of safety in an embankment.

overconsolidated clay at the critical state, i.e. when it has fully softened adjacent to the slip plane as the result of expansion during shear.

Embankments

The construction of an embankment results in increases in total stress, both within the embankment itself as successive layers of fill are placed and in the foundation soil. The initial pore water pressure (u_0) depends primarily on the placement water content of the fill. The construction period of a typical embankment is relatively short and, if the permeability of the compacted fill is low, no significant dissipation is likely during construction. Dissipation proceeds after the end of construction with the pore water pressure decreasing to the final value in the long term, as shown in Figure 9.10. The factor of safety of an embankment at the end of construction is therefore lower than in the long term. Shear strength parameters for the fill material should be determined from tests on specimens compacted to the values of dry density and water content to be specified for the embankment.

The stability of an embankment may also depend on the shear strength of the foundation soil. The possibility of failure along a surface such as that illustrated in Figure 9.11 should be considered in appropriate cases.

9.7   EMBANKMENT DAMS

An embankment dam would normally be used where the foundation and abutment conditions were unsuitable for a concrete dam and where suitable materials for the embankment were present at or close to the site. An extensive ground investigation is essential, general at first but becoming more detailed as design studies proceed, to

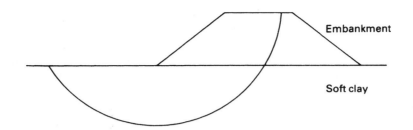

Figure 9.11 Failure below an embankment.

determine foundation and abutment conditions and to identify suitable borrow areas. It is important to determine both the quantity and quality of available material. The natural water content of fine soils should be determined for comparison with the optimum water content for compaction.

Most embankment dams are not homogeneous but are of zoned construction, the detailed section depending on the availability of soil types. Typically a dam will consist of a core of low-permeability soil with shoulders of other suitable material on each side. The upstream slope is usually covered by a thin layer of rockfill (known as rip-rap) to protect it from erosion by wave action. The downstream slope is usually grassed. An internal drainage system, to alleviate the detrimental effects of seeping water, would normally be incorporated. Depending on the materials used, horizontal drainage layers may also be incorporated to accelerate the dissipation of excess pore water pressure. Slope angles should be such that stability is ensured but overconservative design must be avoided: a decrease in slope angle of as little as 2–3° (to the horizontal) would mean a significant increase in the volume of fill for a large dam.

Failure of an embankment dam could result from the following causes: (1) instability of either the upstream or downstream slope, (2) internal erosion and (3) erosion of the crest and downstream slope by overtopping. (The third cause arises basically from errors in the hydrological predictions.)

The factor of safety for both slopes must be determined as accurately as possible for the most critical stages in the life of the dam. The potential failure surface may lie entirely within the embankment or may pass through the embankment and the foundation soil. In the case of the upstream slope the most critical stages are at the end of construction and during rapid drawdown of the reservoir level. The critical stages for the downstream slope are at the end of construction and during steady seepage when the reservoir is full. The pore water pressure distribution at any stage has a dominant influence on the factor of safety of the slopes and it is common practice to install a piezometer system so that the actual pore water pressures can be measured and compared with the predicted values used in design (provided an effective stress analysis has been used). Remedial action could then be taken if the factor of safety, based on the measured values, was considered to be too low.

The Morgenstern–Price analysis is the most appropriate because of its inherent accuracy and because it can deal with non-circular failure surfaces. Values of the parameters c', ϕ', r_u and γ are required for each soil zone. However, it must be recognized that although the analysis itself is accurate, the values of factor of safety produced depend on the correctness of the parameters used.

If a potential failure surface were to pass through foundation material containing fissures, joints or pre-existing slip surfaces, then progressive failure (Section 9.6) would be a possibility. The different stress–strain characteristics of various zone materials through which a potential failure surface passes, together with non-uniformity of shear stress, could also lead to progressive failure.

Another problem is the danger of cracking due to differential movements between soil zones and between the dam and the abutments. The possibility of hydraulic fracturing, particularly within the clay core, should also be considered. Hydraulic fracturing occurs on a plane where the total normal stress is less than the local value of pore water pressure. Following the completion of construction the clay core tends to settle relative to the rest of the embankment due to long-term consolidation: consequently the core will be partially supported by the rest of the embankment. Thus vertical stress in the core will be reduced and the chances of hydraulic fracture increased. The transfer of stress from the core to the shoulders of the embankment is another example of the arching phenomenon (Section 6.7). Following fracture or cracking the resulting leakage could lead to serious internal erosion and impair stability. The finite element method has been used to predict the stresses and deformations within embankment dams, enabling potential areas of cracking and fracture to be predicted.

End of construction

Most slope failures in embankment dams occur either during construction or at the end of construction. Pore water pressures depend on the placement water content of the fill and on the rate of construction. A commitment to achieve rapid completion will result in the maximization of pore water pressure at the end of construction. However, the construction period of an embankment dam is likely to be long enough to allow partial dissipation of excess pore water pressure, especially for a dam with internal drainage. A total stress analysis, therefore, would result in an overconservative design. An effective stress analysis is preferable, using predicted values of r_u. An upper bound value of r_u can be deduced.

The pore pressure (u) at any point can be written as

$$u = u_0 + \Delta u$$

where u_0 is the initial value and Δu the change in pore water pressure under undrained conditions. In terms of the change in total major principal stress

$$u = u_0 + \overline{B}\Delta\sigma_1$$

Then

$$r_u = \frac{u_0}{\gamma h} + \overline{B}\frac{\Delta\sigma_1}{\gamma h}$$

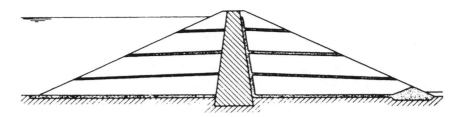

Figure 9.12 Horizontal drainage layers.

If it is assumed that the increase in total major principal stress is approximately equal to the fill pressure along a potential failure surface

$$r_u = \frac{u_0}{\gamma h} + \overline{B} \tag{9.15}$$

The soil is partially saturated when compacted and therefore the initial pore water pressure (u_0) is negative. The actual value of u_0 depends on the placement water content; the higher the water content, the closer the value of u_0 to zero. The value of $\overline{B}$ also depends on the placement water content; the higher the water content, the higher the value of $\overline{B}$. Thus, for an upper bound

$$r_u = \overline{B} \tag{9.16}$$

The value of $\overline{B}$ must correspond to the stress conditions in the dam. Equations 9.15 and 9.16 assume no dissipation during construction. A factor of safety as low as 1.3 may be acceptable at the end of construction provided there is reasonable confidence in the design data.

If high values of r_u are anticipated, dissipation of excess pore water pressure can be accelerated by means of horizontal drainage layers incorporated in the dam, drainage taking place vertically towards the layers: a typical dam section is shown in Figure 9.12. The efficiency of drainage layers has been examined theoretically by Gibson and Shefford [9] and it was shown that in a typical case the layers, in order to be fully effective, should have a permeability at least 10^6 times that of the embankment soil: an acceptable efficiency would be obtained with a permeability ratio of about 10^5.

Steady seepage

After the reservoir has been full for some time, conditions of steady seepage become established through the dam with the soil below the top flow line in the fully saturated state. This condition must be analysed in terms of effective stress with values of pore pressure being determined from the flow net. Values of r_u up to 0.45 are possible in homogeneous dams but much lower values can be achieved in dams having internal drainage. The factor of safety for this condition should be at least 1.5. Internal erosion is a particular danger when the reservoir is full because it can arise and develop within a relatively short time, seriously impairing the safety of the dam.

Rapid drawdown

After a condition of steady seepage has become established, a drawdown of the reservoir level will result in a change in the pore water pressure distribution. If the permeability of the soil is low, a drawdown period measured in weeks may be 'rapid' in relation to dissipation time and the change in pore water pressure can be assumed to take place under undrained conditions. Referring to Figure 9.13, the pore water pressure before drawdown at a typical point P on a potential failure surface is given by

$$u_0 = \gamma_w(h + h_w - h')$$

where h' is the loss in total head due to seepage between the upstream slope surface and the point P. It is again assumed that the total major principal stress at P is equal to the fill pressure. The change in total major principal stress is due to the total or partial removal of water above the slope on the vertical through P. For a drawdown depth exceeding h_w:

$$\Delta\sigma_1 = -\gamma_w h_w$$

and the change in pore water pressure is then given by

$$\Delta u = \overline{B}\,\Delta\sigma_1$$
$$= -\overline{B}\gamma_w h_w$$

Therefore the pore water pressure at P immediately after drawdown is

$$u = u_0 + \Delta u$$
$$= \gamma_w\{h + h_w(1 - \overline{B}) - h'\}$$

Hence

$$r_u = \frac{u}{\gamma_{sat}h}$$
$$= \frac{\gamma_w}{\gamma_{sat}}\left\{1 + \frac{h_w}{h}(1 - \overline{B}) - \frac{h'}{h}\right\} \qquad (9.17)$$

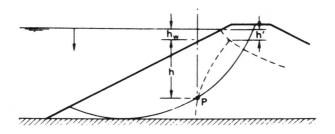

Figure 9.13 Rapid drawdown conditions. (Reproduced from A.W. Bishop and L. Bjerrum (1960) *Proceedings ASCE Research Conference on Shear Strength of Cohesive Soils*, Boulder, Colorado, by permission of the American Society of Civil Engineers.)

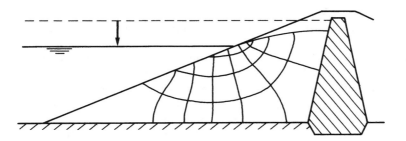

Figure 9.14 Drawdown flow net.

For total stress decreases the value of $\overline{B}$ is slightly greater than 1. An upper bound value of r_u could be obtained by assuming $\overline{B} = 1$ and neglecting h'. Typical values of r_u immediately after drawdown are within the range 0.3–0.4. A minimum factor of safety of 1.2 may be acceptable after rapid drawdown.

Morgenstern [11] published stability coefficients for the approximate analysis of homogeneous slopes after rapid drawdown.

The pore water pressure distribution after drawdown in soils of high permeability decreases as pore water drains out of the soil above the drawdown level. The saturation line moves downwards at a rate depending on the permeability of the soil. A series of flow nets can be drawn for different positions of the saturation line and values of pore water pressure obtained. The factor of safety can thus be determined, using an effective stress analysis, for any position of the saturation line. An example of a drawdown flow net is shown in Figure 9.14.

PROBLEMS

9.1 For the given failure surface, determine the factor of safety in terms of total stress for the slope detailed in Figure 9.15. The unit weight for both soils is

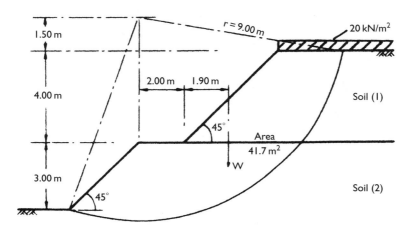

Figure 9.15

$19\,\mathrm{kN/m^3}$. The characteristic undrained strength (c_{uk}) is $30\,\mathrm{kN/m^2}$ for soil 1 and $45\,\mathrm{kN/m^2}$ for soil 2. What is the factor of safety if allowance is made for the development of a tension crack? Check the stability of the slope using the limit state method.

9.2   A cutting 9 m deep is to be excavated in a saturated clay of unit weight $19\,\mathrm{kN/m^3}$. The design shear strength parameters are $c_u = 30\,\mathrm{kN/m^2}$ and $\phi_u = 0$. A hard stratum underlies the clay at a depth of 11 m below ground level. Using Taylor's stability coefficients, determine the slope angle at which failure would occur. What is the allowable slope angle if a factor of safety of 1.2 is specified?

9.3   For the given failure surface, determine the factor of safety in terms of effective stress for the slope detailed in Figure 9.16, using the Fellenius method of slices. The unit weight of the soil is $21\,\mathrm{kN/m^3}$ and the characteristic shear strength parameters are $c' = 8\,\mathrm{kN/m^2}$ and $\phi' = 32°$. According to the limit state method, will a slip failure occur?

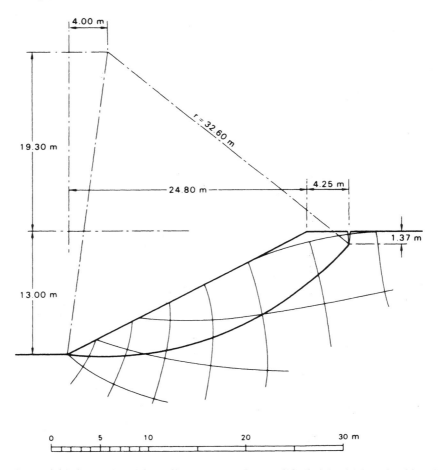

Figure 9.16 (Reproduced from Skempton and Brown (1961) A landslide in boulder clay at Selset, Yorkshire, *Geotechnique*, **11**, p. 280, by permission of the Council of the Institution of Civil Engineers.)

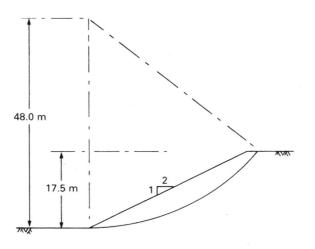

Figure 9.17

9.4   Repeat the analysis of the slope detailed in Problem 9.3 using the Bishop routine method of slices.

9.5   Using the Bishop routine method of slices, determine the factor of safety in terms of effective stress for the slope detailed in Figure 9.17 for the specified failure surface. The value of r_u is 0.20 and the unit weight of the soil is $20 \, \text{kN/m}^3$. Characteristic values of the shear strength parameters are $c' = 0$ and $\phi' = 33°$.

9.6   A long slope is to be formed in a soil of unit weight $19 \, \text{kN/m}^3$ for which the characteristic shear strength parameters are $c' = 0$ and $\phi' = 36°$. A firm stratum lies below the slope. It is to be assumed that the water table may occasionally rise to the surface, with seepage taking place parallel to the slope. (a) Using the traditional method, determine the maximum slope angle to ensure a factor of safety of 1.5, assuming a potential failure surface parallel to the slope. What would be the factor of safety of the slope, formed at this angle, if the water table were well below the surface? (b) Using the limit state method, check the stability for the slope angle determined above when the water table is at the surface.

REFERENCES

1   Bell, J.M. (1968) General slope stability analysis, *Journal ASCE*, **94** (SM6), 1253–70.

2   Bishop, A.W. (1955) The use of the slip circle in the stability analysis of slopes, *Geotechnique*, **5** (1), 7–17.

3   Bishop, A.W. and Bjerrum, L. (1960) The relevance of the triaxial test to the solution of stability problems, in *Proceedings ASCE Research Conference on Shear Strength of Cohesive Soils*, Boulder, CO, ASCE, New York, pp. 437–501.

4   Bishop, A.W. and Morgenstern, N.R. (1960) Stability coefficients for earth slopes, *Geotechnique*, **10** (4), 129–47.

5   British Standard 6031 (1981) *Code of Practice for Earthworks*, British Standards Institution, London.

  6 Building Research Establishment (1990) *An Engineering Guide to the Safety of Embankment Dams in the United Kingdom*, Watford.
  7 Gens, A., Hutchinson, J.N. and Cavounidis, S. (1988) Three-dimensional analysis of slides in cohesive soils, *Geotechnique*, **38** (1), 1–23.
  8 Gibson, R.E. and Morgenstern, N.R. (1962) A note on the stability of cuttings in normally consolidated clays, *Geotechnique*, **12** (3), 212–16.
  9 Gibson, R.E. and Shefford, G.C. (1968) The efficiency of horizontal drainage layers for accelerating consolidation of clay embankments, *Geotechnique*, **18** (3), 327–35.
 10 Lo, K.Y. (1965) Stability of slopes in anisotropic soils, *Journal ASCE*, **91** (SM4), 85–106.
 11 Morgenstern, N.R. (1963) Stability charts for earth slopes during rapid drawdown, *Geotechnique*, **13**, 121–31.
 12 Morgenstern, N.R. and Price, V.E. (1965) The analysis of the stability of general slip surfaces, *Geotechnique*, **15**, 79–93.
 13 Morgenstern, N.R. and Price, V.E. (1967) A numerical method for solving the equations of stability of general slip surfaces, *Computer Journal*, **9**, 388–93.
 14 Penman, A.D.M. (1986) On the embankment dam, *Geotechnique*, **36**, 301–48.
 15 Sarma, S.K. (1973) Stability analysis of embankments and slopes, *Geotechnique*, **23**, 423–33.
 16 Skempton, A.W. (1964) Long-term stability of clay slopes, *Geotechnique*, **14**, 75–102.
 17 Skempton, A.W. (1970) First-time slides in overconsolidated clays (Technical Note), *Geotechnique*, **20**, 320–4.
 18 Spencer, E. (1967) A method of analysis of the stability of embankments assuming parallel inter-slice forces, *Geotechnique*, **17**, 11–26.
 19 Taylor, D.W. (1937) Stability of earth slopes, *Journal of the Boston Society of Civil Engineers*, **24** (3), 337–86.
 20 Whitman, R.V. and Bailey, W.A. (1967) Use of computers for slope stability analysis, *Journal ASCE*, **93** (SM4), 475–98.

Chapter 10

Ground investigation

10.1 INTRODUCTION

An adequate ground investigation is an essential preliminary to the execution of a civil engineering project. Sufficient information must be obtained to enable a safe and economic design to be made and to avoid any difficulties during construction. The principal objects of the investigation are: (1) to determine the sequence, thicknesses and lateral extent of the soil strata and, where appropriate, the level of bedrock; (2) to obtain representative samples of the soils (and rock) for identification and classification and, if necessary, for use in laboratory tests to determine relevant soil parameters; (3) to identify the groundwater conditions. The investigation may also include the performance of *in-situ* tests to assess appropriate soil characteristics. Additional considerations arise if it is suspected that the ground may be contaminated. The results of a ground investigation should provide adequate information, for example, to enable the most suitable type of foundation for a proposed structure to be selected and to indicate if special problems are likely to arise during excavation.

A study of geological maps and memoirs, if available, should give an indication of the probable soil conditions of the site in question. If the site is extensive and if no existing information is available, the use of aerial photographs can be useful in identifying features of geological significance. Before the start of fieldwork an inspection of the site and the surrounding area should be made on foot. River banks, existing excavations, quarries and road or railway cuttings, for example, can yield valuable information regarding the nature of the strata and groundwater conditions: existing structures should be examined for signs of settlement damage. Previous experience of conditions in the area may have been obtained by adjacent owners or local authorities. All information obtained in advance enables the most suitable type of investigation to be decided.

The actual investigation procedure depends on the nature of the strata and the type of project but will normally involve the excavation of boreholes or trial pits. The number and location of boreholes or pits should be planned to enable the basic geological structure of the site to be determined and significant irregularities in the subsurface conditions to be detected. The greater the degree of variability of the ground conditions, the greater the number of boreholes or pits required. The locations should be offset from areas on which it is known that foundations are to be sited. A preliminary investigation on a modest scale may be carried out to obtain the general characteristics of the strata, followed by a more extensive and carefully planned investigation including sampling and *in-situ* testing.

It is essential that the investigation is taken to an adequate depth. This depth depends on the type and size of the project but must include all strata liable to be significantly affected by the structure and its construction. The investigation must extend below all strata which might have inadequate shear strength for the support of foundations or which would give rise to significant settlement. If the use of piles is anticipated the investigation will thus have to extend to a considerable depth below the surface. A general rule often applied in the case of foundations is that the ground conditions must be known within the significant depth (Section 8.1) provided there is no weak stratum below this depth which would cause unacceptable settlement. If rock is encountered it should be penetrated at least 3 m in more than one location to confirm that bedrock, and not a large boulder has been reached, unless geological knowledge indicates otherwise. The investigation may have to be taken to depths greater than normal in areas of old mine workings or other underground cavities.

Boreholes and trial pits should be backfilled after use. Backfilling with compacted soil may be adequate in many cases but if the groundwater conditions are altered by a borehole and the resultant flow could produce adverse effects then it is necessary to use a cement-based grout to seal the hole.

The cost of an investigation depends on the location and extent of the site, the nature of the strata and the type of project under consideration. In general, the larger the project, and the less critical the ground conditions are to the design and construction of the project, the lower the cost of the ground investigation as a percentage of the total cost. The cost is generally within the range 0.1–2% of the project cost. To reduce the scope of an investigation for financial reasons alone is never justified.

10.2 METHODS OF INVESTIGATION

Trial pits

The excavation of trial pits is a simple and reliable method of investigation but is limited to a maximum depth of 4–5 m. The soil is generally removed by means of the back-shovel of a mechanical excavator. Before any person enters the pit, the sides must always be supported unless they are sloped at a safe angle or are stepped: the excavated soil should be placed at least 1 m from the edge of the pit. If the pit is to extend below the water table, some form of dewatering is necessary in the more permeable soils, resulting in increased costs. The use of trial pits enables the *in-situ* soil conditions to be examined visually, and thus the boundaries between strata and the nature of any macro-fabric can be accurately determined. It is relatively easy to obtain disturbed or undisturbed soil samples: in cohesive soils block samples can be cut by hand from the sides or bottom of the pit and tube samples can be obtained below the bottom of the pit. Trial pits are suitable for investigations in all types of soil, including those containing cobbles or boulders.

Shafts and headings

Deep pits or shafts are usually advanced by hand excavation, the sides being supported by timbering. Headings or adits are excavated laterally from the bottom of shafts or from the surface into hillsides, both the sides and roof being supported. It is unlikely that shafts or headings would be excavated below the water table. Shafts and headings are very costly and their use would be justified only in investigations for very large structures, such as dams, if the ground conditions could not be ascertained adequately by other means.

Percussion boring

The boring rig (Figure 10.1) consists of a derrick, a power unit and a winch carrying a light steel cable which passes through a pulley on top of the derrick. Most rigs are fitted with road wheels and when folded down can be towed behind a vehicle. Various boring tools can be attached to the cable. The borehole is advanced by the percussive action of the tool which is alternately raised and dropped (usually over a distance of 1–2 m) by means of the winch unit. The two most widely used tools are the *shell (or baler)* and the *clay cutter*. If necessary a heavy steel element called a sinker bar can be fitted immediately above the tool to increase the impact energy.

The shell, which is used in sands and other coarse soils, is a heavy steel tube fitted with a flap or clack valve at the lower end. Below the water table the percussive action of the shell loosens the soil and produces a slurry in the borehole. Above the water table a slurry is produced by introducing water into the borehole. The slurry passes through the clack valve during the downward movement of the shell and is retained by the valve during the upward movement. When full, the shell is raised to the surface to be emptied. In cohesionless soils the borehole must be cased to prevent collapse. The casing, which consists of lengths of steel tubing screwed together, is lowered into the borehole and will normally slide down under its own weight; however, if necessary, installation of the casing can be aided by driving. On completion of the investigation

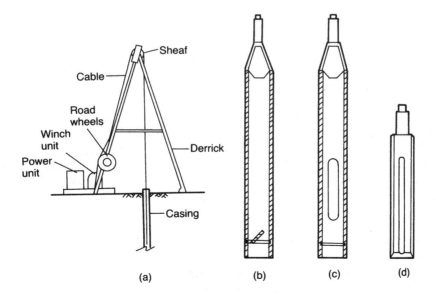

Figure 10.1 (a) Percussion boring rig, (b) shell, (c) clay cutter and (d) chisel.

the casing is recovered by means of the winch or by the use of jacks: excessive driving during installation may make recovery of the casing difficult.

The clay cutter, which is used in cohesive soils, is an open steel tube with a cutting shoe and a retaining ring at the lower end: the tool is used in a dry borehole. The percussive action of the tool cuts a plug of soil which eventually fractures near its base due to the presence of the retaining ring. The ring also ensures that the soil is retained inside the cutter when it is raised to the surface to be emptied.

Small boulders, cobbles and hard strata can be broken up by means of a *chisel*, aided by the additional weight of a sinker bar if necessary.

Borehole diameters can range from 150 to 300 mm. The maximum borehole depth is generally between 50 and 60 m. Percussion boring can be employed in most types of soil, including those containing cobbles and boulders. However, there is generally some disturbance of the soil below the bottom of the borehole, from which samples are taken, and it is extremely difficult to detect thin soil layers and minor geological features with this method. The rig is extremely versatile and can normally be fitted with a hydraulic power unit and attachments for mechanical augering, rotary core drilling and cone penetration testing.

Mechanical augers

Power-operated augers are generally mounted on vehicles or in the form of attachments to the derrick used for percussion boring. The power required to rotate the auger depends on the type and size of the auger itself and the type of soil to be penetrated. Downward pressure on the auger can be applied hydraulically, mechanically or by dead weight. The types of tool generally used are the *flight auger* and the *bucket auger*. The diameter of a flight auger is usually between 75 and 300 mm, although diameters as large as 1 m are available: the diameter of a bucket auger can range from 300 mm to 2 m. However, the larger sizes are used principally for excavating shafts for bored piles. Augers are used mainly in soils in which the borehole requires no support and remains dry, i.e. mainly in stiffer, overconsolidated clays. The use of casing would be inconvenient because of the necessity of removing the auger before driving the casing; however, it is possible to use bentonite slurry (Section 6.9) to support the sides of unstable holes. The presence of cobbles or boulders creates difficulties with the smaller-sized augers.

Short-flight augers (Figure 10.2(a)) consist of a helix of limited length, with cutters below the helix. The auger is attached to a steel shaft, known as a kelly bar, which passes through the rotary head of the rig. The auger is advanced until it is full of soil, then it is raised to the surface where the soil is ejected by rotating the auger in the reverse direction. Clearly, the shorter the helix, the more often the auger must be raised and lowered for a given borehole depth. The depth of the hole is limited by the length of the kelly bar.

Continuous-flight augers (Figure 10.2(b)) consist of rods with a helix covering the entire length. The soil rises to the surface along the helix, obviating the necessity for withdrawal: additional lengths of auger are added as the hole is advanced. Borehole depths up to 50 m are possible with continuous-flight augers, but there is a possibility that different soil types may become mixed as they rise to the surface and it may be difficult to determine the depths at which changes of strata occur.

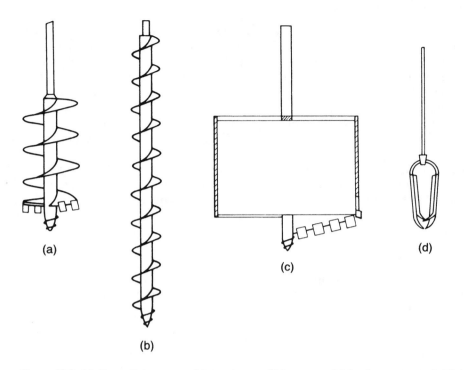

Figure 10.2 (a) Short-flight auger, (b) continuous-flight auger, (c) bucket auger and (d) Iwan
(hand) auger.

Continuous-fight augers with hollow stems are also used. When boring is in progress the hollow stem is closed at the lower end by a plug fitted to a rod running inside
the stem. Additional lengths of auger (and internal rod) are again added as the hole is
advanced. At any depth the rod and plug may be withdrawn from the hollow stem to
allow undisturbed samples to be taken, a sample tube mounted on rods being lowered
down the stem and driven into the soil below the auger. If bedrock is reached, drilling
can also take place through the hollow stem. The internal diameter of the stem can
range from 75 to 150 mm. As the auger performs the function of a casing it can be used
in sands below the water table, although difficulty may be experienced with sand being
forced upwards into the stem by hydrostatic pressure: this can be avoided by filling the
stem with water up to water table level.

Bucket augers (Figure 10.2(c)) consist of a steel cylinder, open at the top but fitted
with a base plate on which cutters are mounted, adjacent to slots in the plate: the auger
is attached to a kelly bar. When the auger is rotated and pressed downwards the soil
removed by the cutters passes through the slots into the bucket. When the bucket is full
it is raised to the surface to be emptied by releasing the hinged base plate.

Augered holes of 1 m diameter and larger can be used for the examination of the soil
strata *in situ*, the person carrying out the examination being lowered down the hole in
a special cage. The hole must be cased when used for this purpose and adequate
ventilation is essential.

Hand and portable augers

Hand augers can be used to excavate boreholes to depths of around 5 m using a set of extension rods. The auger is rotated and pressed down into the soil by means of a T-handle on the upper rod. The two common types are the Iwan or post-hole auger (Figure 10.2(d)) with diameters up to 200 mm, and the small helical auger, with diameters of about 50 mm. Hand augers are generally used only if the sides of the hole require no support and if particles of coarse gravel size and above are absent. The auger must be withdrawn at frequent intervals for the removal of soil. Undisturbed samples can be obtained by driving small-diameter tubes below the bottom of the borehole.

Small portable power augers, generally transported and operated by two persons, are suitable for boring to depths of 10–15 m; the hole diameter may range from 75 to 300 mm. The borehole may be cased if necessary and therefore the auger can be used in most soil types provided the larger particle sizes are absent.

Wash boring

In this method, water is pumped through a string of hollow boring rods and is released under pressure through narrow holes in a chisel attached to the lower end of the rods (Figure 10.3). The soil is loosened and broken up by the water jets and the up-and-down movement of the chisel. There is also provision for the manual rotation of the chisel by means of a tiller attached to the boring rods above the surface. The soil particles are washed to the surface between the rods and the side of the borehole and are allowed to settle out in a sump. The rig consists of a derrick with a power unit, a winch and a water pump. The winch carries a light steel cable which passes through the sheaf of the derrick and is attached to the top of the boring rods. The string of rods is raised and dropped by means of the winch unit, producing the chopping action of the chisel. The borehole is generally cased but the method can be used in uncased holes. Drilling fluid may be used as an alternative to water in the method, eliminating the need for casing.

Wash boring can be used in most types of soil but progress becomes slow if particles of coarse gravel size and larger are present. The accurate identification of soil types is difficult due to particles being broken up by the chisel and to mixing as the material is washed to the surface: in addition, segregation of particles takes place as they settle out in the sump. However, a change in the feel of the boring tool can sometimes be detected, and there may be a change in the colour of the water rising to the surface, when the boundaries between different strata are reached. The method is unacceptable as a means of obtaining soil samples. It is used only as a means of advancing a borehole to enable tube samples to be taken or *in-situ* tests to be carried out below the bottom of the hole. An advantage of the method is that the soil immediately below the hole remains relatively undisturbed.

Rotary drilling

Although primarily intended for investigations in rock, the method is also used in soils. The drilling tool, which is attached to the lower end of a string of hollow drilling

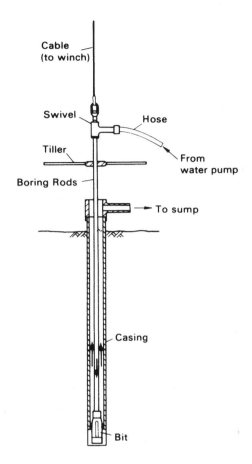

Figure 10.3 Wash boring.

rods (Figure 10.4), may be either a cutting bit or a coring bit: the coring bit is fixed to the lower end of a core barrel which in turn is carried by the drilling rods. Water or drilling fluid is pumped down the hollow rods and passes under pressure through narrow holes in the bit or barrel: this is the same principle as used in wash boring. The drilling fluid cools and lubricates the drilling tool and carries the loose debris to the surface between the rods and the side of the hole. The fluid also provides some support to the sides of the hole if no casing is used.

The rig consists of a derrick, power unit, winch, pump and a drill head to apply high-speed rotary drive and downward thrust to the drilling rods. A rotary head attachment can be supplied as an accessory to a percussion boring rig.

There are two forms of rotary drilling, open-hole drilling and core drilling. Open-hole drilling, which is generally used in soils and weak rock, uses a cutting bit to break down all the material within the diameter of the hole. Open-hole drilling can thus be used only as a means of advancing the hole: the drilling rods can then be removed to allow tube samples to be taken or *in-situ* tests to be carried out. In core drilling, which

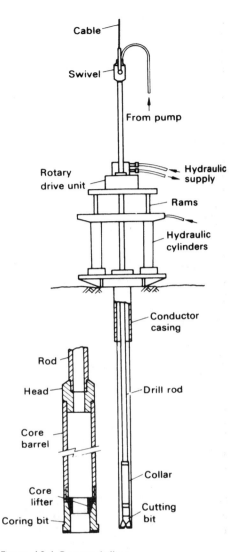

Figure 10.4 Rotary drilling.

is used in rocks and hard clays, the diamond or tungsten carbide bit cuts an annular hole in the material and an intact core enters the barrel, to be removed as a sample. However, the natural water content of the material is liable to be increased due to contact with the drilling fluid. Typical core diameters are 41, 54 and 76 mm, but can range up to 165 mm.

The advantage of rotary drilling in soils is that progress is much faster than with other investigation methods and disturbance of the soil below the borehole is slight. The method is not suitable if the soil contains a high percentage of gravel (or larger) particles as they tend to rotate beneath the bit and are not broken up.

Groundwater observations

An important part of any ground investigation is the determination of water table level and of any artesian pressure in particular strata. The variation of level or pressure over a given period of time may also require determination. Groundwater observations are of particular importance if deep excavations are to be carried out.

Water table level can be determined by measuring the depth to the water surface in a borehole. Water levels in boreholes may take a considerable time to stabilize; this time, known as the response time, depends on the permeability of the soil. Measurements, therefore, should be taken at regular intervals until the water level becomes constant. It is preferable that the level should be determined as soon as the borehole has reached water table level. If the borehole is further advanced it may penetrate a stratum under artesian pressure, resulting in the water level in the hole being above water table level. It is important that a stratum of low permeability below a perched water table should not be penetrated before the water level has been established. If a perched water table exists, the borehole must be cased in order that the main water table level is correctly determined: if the perched aquifer is not sealed, the water level in the borehole will be above the main water table level.

When it is required to determine the pore water pressure in a particular stratum, a piezometer (Section 11.2) should be used. The simplest type is the open standpipe piezometer (Figure 11.9) with the porous element sealed at the appropriate depth. However, this type has a long response time in soils of low permeability and in such cases it is preferable to install a hydraulic piezometer (Figure 11.7) having a relatively short response time.

Groundwater samples may be required for chemical analysis to determine if they contain sulphates (which may attack Portland cement concrete) or other corrosive constituents. It is important to ensure that samples are not contaminated or diluted. A sample should be taken immediately the water-bearing stratum is reached in boring. It is preferable to obtain samples from the standpipe piezometers if these have been installed.

10.3 SAMPLING

Soil samples are divided into two main categories, undisturbed and disturbed. Undisturbed samples, which are required mainly for shear strength and consolidation tests, are obtained by techniques which aim at preserving the *in-situ* structure and water content of the soil. In boreholes, undisturbed samples can be obtained by withdrawing the boring tools (except when hollow-stem continuous-flight augers are used) and driving or pushing a sample tube into the soil at the bottom of the hole. The sampler is normally attached to a length of boring rod which can be lowered and raised by the cable of the percussion rig. When the tube is brought to the surface, some soil is removed from each end and molten wax is applied, in thin layers, to form a seal approximately 25 mm thick: the ends of the tube are then covered by protective caps. Undisturbed block samples can be cut by hand from the bottom or sides of a trial pit. During cutting, the samples must be protected from water, wind and sun to avoid any change in water content: the samples should be covered with molten wax immediately they have been brought to the surface. It is impossible to obtain a sample that is completely undisturbed, no matter how elaborate or careful the ground investigation and sampling technique might be. In the case of clays, for example, swelling will take place adjacent to the bottom of a borehole due to the reduction in total stresses when soil is removed and structural disturbance may be caused by the action of the boring tools; subsequently, when a sample is removed from the ground the total stresses are reduced to zero.

Soft clays are extremely sensitive to sampling disturbance, the effects being more pronounced in clays of low plasticity than in those of high plasticity. The central core of a soft clay sample will be relatively less disturbed than the outer zone adjacent to the sampling tube. Immediately after sampling, the pore water pressure in the relatively undisturbed core will be negative due to the release of the *in-situ* total stresses. Swelling of the relatively undisturbed core will gradually take place due to water being drawn from the more disturbed outer zone and resulting in the dissipation of the negative excess pore water pressure: the outer zone of soil will consolidate due to the redistribution of water within the sample. The dissipation of the negative excess pore water pressure is accompanied by a corresponding reduction in effective stresses. The soil structure of the sample will thus offer less resistance to shear and will be less rigid than the *in-situ* soil.

A disturbed sample is one having the same particle size distribution as the *in-situ* soil but in which the soil structure has been significantly damaged or completely destroyed; in addition, the water content may be different from that of the *in-situ* soil. Disturbed samples, which are used mainly for soil classification tests, visual classification and compaction tests, can be excavated from trial pits or obtained from the tools used to advance boreholes (e.g. from augers and the clay cutter). The soil recovered from the shell in percussion boring will be deficient in fines and will be unsuitable for use as a disturbed sample. Samples in which the natural water content has been preserved should be placed in airtight, non-corrosive containers: all containers should be completely filled so that there is negligible air space above the sample.

All samples should be clearly labelled to show the project name, date, location, borehole number, depth and method of sampling; in addition, each sample should be given a serial number. Special care is required in the handling, transportation and storage of samples (particularly undisturbed samples) prior to testing.

The sampling method used should be related to the quality of sample required. Quality can be classified as follows, according to the uses to which the sample can be put:

Class 1: classification, water content, density, shear strength, deformation and consolidation tests.
Class 2: classification, water content and density tests.
Class 3: classification and water content tests.
Class 4: classification tests only.
Class 5: strata identification only.

For Classes 1 and 2 the sample must be undisturbed. Samples of Classes 3, 4 and 5 may be disturbed.

The principal types of tube samplers are described below.

Open drive sampler

An open drive sampler (Figure 10.5(a)) consists of a long steel tube with a screw thread at each end. A cutting shoe is attached to one end of the tube. The other end of

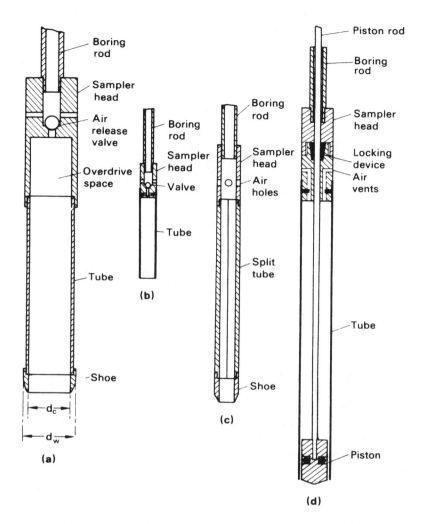

Figure 10.5 (a) Open drive sampler, (b) thin-walled sampler, (c) split-barrel sampler and (d) stationary piston sampler.

the tube screws into a sampler head to which, in turn, the boring rods are connected. The sampler head also incorporates a non-return valve to allow air and water to escape as the soil fills the tube and to help retain the sample as the tube is withdrawn. The inside of the tube should have a smooth surface and must be maintained in a clean condition.

The internal diameter of the cutting edge (d_c) should be approximately 1% smaller than that of the tube to reduce frictional resistance between the tube and the sample. This size difference also allows for slight elastic expansion of the sample on entering the tube and assists in sample retention. The external diameter of the cutting shoe (d_w) should be slightly greater than that of the tube to reduce the force required to withdraw the tube. The volume of soil displaced by the sampler as a proportion of the sample volume is represented by the *area ratio* (C_a) of the sampler, where

$$C_a = \frac{d_w^2 - d_c^2}{d_c^2} \tag{10.1}$$

The area ratio is generally expressed as a percentage. Other factors being equal, the lower the area ratio, the lower the degree of sample disturbance.

The sampler can be driven dynamically by means of a drop weight or sliding hammer, or statically by hydraulic or mechanical jacking. Prior to sampling all loose soil should be removed from the bottom of the borehole. Care should be taken to ensure that the sampler is not driven beyond its capacity, otherwise the sample will be compressed against the sampler head. Some types of sampler head have an overdrive space below the valve to reduce the risk of sample damage. After withdrawal, the cutting shoe and sampler head are detached and the ends of the sample are sealed.

The most widely used sample tube has an internal diameter of 100 mm and a length of 450 mm: the area ratio is approximately 30%. This sampler is suitable for all clay soils. When used to obtain samples of sand, a core-catcher, a short length of tube with spring-loaded flaps, should be fitted between the tube and cutting shoe to prevent loss of soil. The class of sample obtained depends on soil type.

Thin-walled sampler

Thin-walled samplers (Figure 10.5(b)) are used in soils which are sensitive to disturbance such as soft to firm clays and plastic silts. The sampler does not employ a separate cutting shoe, the lower end of the tube itself being machined to form a cutting edge. The internal diameter may range from 35 to 100 mm. The area ratio is approximately 10% and samples of first-class quality can be obtained provided the soil has not been disturbed in advancing the borehole. In trial pits and shallow boreholes the tube can often be driven manually.

Split-barrel sampler

Split-barrel samplers (Figure 10.5(c)) consist of a tube which is split longitudinally into two halves: a shoe and a sampler head incorporating air-release holes are screwed onto the ends. The two halves of the tube can be separated when the shoe and head are detached to allow the sample to be removed. The internal and external diameters are 35 and 50 mm, respectively, the area ratio being approximately 100%, with the result that there is considerable disturbance of the sample (Class 3 or 4). This sampler is used mainly in sands, being the tool specified in the standard penetration test.

Stationary piston sampler

Stationary piston samplers (Figure 10.5(d)) consist of a thin-walled tube fitted with a piston. The piston is attached to a long rod which passes through the sampler head and runs inside the hollow boring rods. The sampler is lowered into the borehole with the piston located at the lower end of the tube, the tube and piston being locked together by means of a clamping device at the top of the rods. The piston prevents water or loose soil from entering the tube. In soft soils the sampler can be pushed below the bottom of the borehole, bypassing any disturbed soil. The piston is held against the soil (generally by clamping the piston rod to the casing) and the tube is pushed past the piston (until the sampler head meets the top of the piston) to obtain the sample. The sampler is then withdrawn, a locking device in the sampler head holding the piston at the top of the tube as this takes place. The vacuum between the piston and the sample helps to retain the soil in the tube: the piston thus serves as a non-return valve.

Piston samplers should always be pushed down by hydraulic or mechanical jacking: they should never be driven. The diameter of the sampler is usually between 35 and 100 mm but can be as large as 250 mm. The samplers are generally used for soft clays and can produce samples of first-class quality up to 1 m in length.

Continuous sampler

The continuous sampler is a highly specialized type of sampler which is capable of obtaining undisturbed samples up to 25 m in length: the sampler is used mainly in soft clays. Details of the soil fabric can be determined more easily if a continuous sample is available. An essential requirement of continuous samplers is the elimination of frictional resistance between the sample and the inside of the sampler tube. In one type of sampler, developed in Sweden [7], this is achieved by superimposing thin strips of metal foil between the sample and the tube. The lower end of the sampler (Figure 10.6) has a sharp cutting edge above which the external diameter is enlarged to enable 16 rolls of foil to be housed in recesses within the wall of the sampler. The ends of the foil are attached to a piston which fits loosely inside the sampler: the piston is supported on a cable which is fixed at the surface. Lengths of sample tube (68 mm in diameter) are attached as required to the upper end of the sampler.

As the sampler is pushed into the soil the foil unrolls and encases the sample, the piston being held at a constant level by means of the cable. As the sampler is withdrawn the lengths of tube are uncoupled and a cut is made, between adjacent tubes, through the foil and sample. Sample quality is generally Class 1 or 2.

Another type is the Delft continuous sampler, of either 29 or 66 mm in diameter. The sample feeds into an impervious nylon stockinette sleeve. The sleeved sample, in turn, is fed into a fluid-supported thin-walled plastic tube.

Compressed air sampler

The compressed air sampler (Figure 10.7) is used to obtain undisturbed samples of sand (generally Class 2) below the water table. The sample tube, usually 60 mm in diameter, is attached to a sampler head having a relief valve which can be closed by a rubber diaphragm. Attached to the sampler head is a hollow guide rod surmounted by a guide head. An outer tube, or bell, surrounds the sample tube, the bell being attached to a weight which slides on the guide rod. The boring rods fit loosely into a plain socket in the top of the guide head, the weight of the bell and sampler being supported by means of a shackle which hooks over a peg in the lower length of boring rod: a light cable, leading to the surface, is fixed to the shackle. Compressed air, produced by a foot

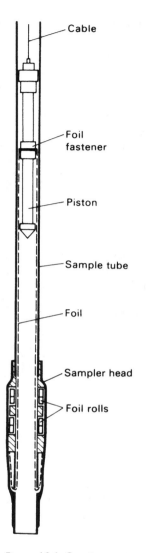

Figure 10.6 Continuous sampler.

pump, is supplied through a tube leading to the guide head, the air passing down the hollow guide rod to the bell.

The sampler is lowered on the boring rods to the bottom of the borehole, which will contain water below the level of the water table. When the sampler comes to rest at the bottom of the borehole the shackle springs off the peg, removing the connection between the sampler and the boring rods. The tube is pushed into the soil by means of the boring rods, a stop on the guide rod preventing overdriving: the boring rods are then withdrawn. Compressed air is now introduced to expel the water from the bell and to close the valve in the sampler head by pressing the diaphragm downwards. The

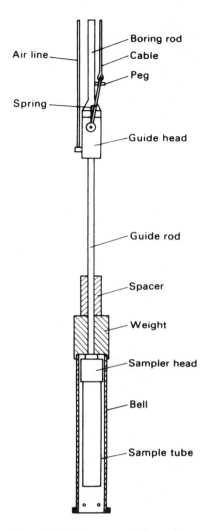

Figure 10.7 Compressed air sampler.

tube is withdrawn into the bell by means of the cable and then the tube and bell together are raised to the surface. The sand sample remains in the tube by virtue of arching and the slight negative pore water pressure in the soil. A plug is placed at the bottom of the tube before the suction is released and the tube is removed from the sampler head.

Window sampler

This sampler, which is most suited to dry cohesive soils, employs a series of tubes, usually 1 m in length and of different diameters (typically 80, 60, 50 and 36 mm). Tubes of the same diameter can be coupled together. A cutting shoe is attached to the end of the bottom tube. The tubes are driven into the soil by percussion using either a manual or rig-supported device and are extracted either manually or by means of the rig. The tube of largest diameter is the first to be driven and extracted with its sample inside. A tube of lesser diameter is then driven below the bottom of the open hole left by extraction of the larger tube. The operation is repeated using tubes of successively lower diameter and depths of up to 8 m can be reached. There are longitudinal slots or 'windows' in the walls at one side of the tubes to allow the soil to be examined and enable disturbed samples of Class 3 or 4 to be taken.

10.4 BOREHOLE LOGS

After an investigation has been completed and the results of any laboratory tests are available, the ground conditions discovered in each borehole (or trial pit) are summarized in the form of a borehole log. An example of such a log appears in Table 10.1, but details of the layout can vary. The last few columns are originally left without headings to allow for variations in the data presented. The method of investigation and details of the equipment used should be stated on each log. The location, ground level and diameter of the hole should be specified together with details of any casing used. The names of the client and the project should be stated.

The log should enable a rapid appraisal of the soil profile to be made. The log is prepared with reference to a vertical scale. A detailed description of each stratum is given and the levels of strata boundaries clearly shown: the level at which boring was terminated should be indicated. The different soil (and rock) types are represented by means of a legend using standard symbols. The depths, or ranges of depth, at which samples were taken or at which *in-situ* tests were performed are recorded: the type of sample is also specified. The results of certain laboratory or *in-situ* tests may be given in the log. The depths at which groundwater was encountered and subsequent changes in levels, with times, should be detailed.

The soil description should be based on particle size distribution and plasticity, generally using the rapid procedure in which these characteristics are assessed by means of visual inspection and feel: disturbed samples are generally used for this purpose. The description should include details of soil colour, particle shape and composition: if possible the geological formation and type of deposit should be given. The structural characteristics of the soil mass should also be described but this requires an examination of undisturbed samples or of the soil *in situ* (e.g. in a trial pit). Details should be given of the presence and spacing of bedding features, fissures and other relevant characteristics. The density index of sands (Table 8.3) and the stiffness of clays (Table 4.1) should be indicated.

10.5 GEOPHYSICAL METHODS

Under certain conditions geophysical methods may be useful in ground investigation, especially at the reconnaissance stage. However, the methods are not suitable for all ground conditions and there are limitations to the information that can be obtained:

Table 10.1

BOREHOLE LOG

Location: Barnhill
Client: RFC Consultants
Boring method: Shell and auger to 14.4 m
           Rotary core drilling to 17.8 m
  Diameter: 150 mm
               NX
  Casing: 150 mm to 5 m

Borehole No. 1
Sheet 1 of 1
Ground level: 36.30
Date: 30.7.77
Scale: 1:100

Description of strata	Level	Legend	Depth	Samples	N	c_u (kN/m^2)	
TOPSOIL	35.6		0.7				
Loose, light brown SAND				D	6		
	33.7		2.6				
Medium dense, brown gravelly SAND ▽	32.5			D	15		
	31.9		4.4				
				U		80	
				U		86	
Firm, yellowish-brown, closely fissured CLAY				U		97	
				U		105	
	24.1		12.2				
Very dense, red, silty SAND with decomposed SANDSTONE				D	50 for 210 mm		
	21.9		14.4				
Red, medium-grained, granular, fresh SANDSTONE, moderately weak, thickly bedded	18.5		17.8				

U: Undisturbed sample
D: Disturbed sample
B: Bulk disturbed sample
W: Water sample
▽: Water table

REMARKS

Water level (0930 h)
29.7.77   32.2 m
30.7.77   32.5 m
31.7.77   32.5 m

they must be considered mainly as supplementary methods. It is possible to locate strata boundaries only if the physical properties of the adjacent materials are significantly different. It is always necessary to check the results against data obtained by direct methods such as boring. Geophysical methods can produce rapid and economic results, e.g. for the filling in of detail between widely spaced boreholes or to indicate where additional boreholes may be required, provided such correlations are established. The methods can be useful in estimating the depth to bedrock or to the water table. There are several geophysical techniques, based on different physical principles. Two of these techniques are described below.

Seismic refraction method

The seismic refraction method depends on the fact that seismic waves have different velocities in different types of soil (or rock); in addition, the waves are refracted when they cross the boundary between different types of soil. The method enables the general soil types and the approximate depths to strata boundaries, or to bedrock, to be determined.

Waves are generated either by the detonation of explosives or by striking a metal plate with a large hammer. The equipment consists of one or more sensitive vibration transducers, called geophones, and an extremely accurate time-measuring device called a seismograph. A circuit between the detonator or hammer and the seismograph starts the timing mechanism at the instant of detonation or impact. The geophone is also connected electrically to the seismograph: when the first wave reaches the geophone the timing mechanism stops and the time interval is recorded in milliseconds.

When detonation or impact takes place, waves are emitted in every direction. One particular wave, called the direct wave, will travel parallel to the surface in the direction of the geophone. Other waves travel in a downward direction, at various angles to the horizontal, and will be refracted if they pass into a stratum of different seismic velocity. If the seismic velocity of the lower stratum is higher than that of the upper stratum, one particular wave will travel along the top of the lower stratum, parallel to the boundary, as shown in Figure 10.8(a): this wave continually 'leaks' energy back to the surface. Energy from this refracted wave can be detected by the geophone.

The procedure (Figure 10.8(a)) consists of installing a geophone in turn at a number of points in a straight line, at increasing distances from the source of wave generation. The length of the line of points should be 3–5 times the required depth of investigation. A series of detonations or impacts are produced and the arrival time of the first wave at each geophone position is recorded in turn. When the distance between the source and the geophone is short, the arrival time will be that of the direct wave. When the distance between the source and the geophone exceeds a certain value (depending on the thickness of the upper stratum) the refracted wave will be the first to be detected by the geophone. This is because the path of the refracted wave, although longer than that of the direct wave, is partly through a stratum of higher seismic velocity. The use of explosives is generally necessary if the source–geophone distance exceeds 30–50 m or if the upper soil stratum is loose.

An alternative procedure consists of using a single geophone position and producing a series of detonations or impacts at increasing distances from the geophone.

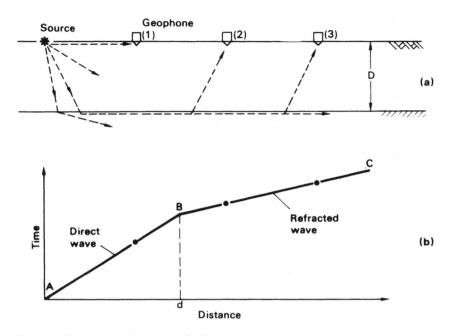

Figure 10.8 Seismic refraction method.

Arrival time is plotted against the distance between the source and the geophone, a typical plot being shown in Figure 10.8(b). If the source–geophone spacing is less than d, the direct wave reaches the geophone in advance of the refracted wave and the time–distance relationship is represented by a straight line (AB) through the origin. On the other hand, if the source–geophone distance is greater than d the refracted wave arrives in advance of the direct wave and the time–distance relationship is represented by a straight line (BC) at a different slope to that of AB. The slopes of the lines AB and BC are the seismic velocities (v_1 and v_2) of the upper and lower strata, respectively. The general types of soil or rock can be determined from a knowledge of these velocities. The depth (D) of the boundary between the two strata (provided the thickness of the upper stratum is constant) can be estimated from the formula

$$D = \frac{d}{2}\sqrt{\left(\frac{v_2 - v_1}{v_2 + v_1}\right)} \tag{10.2}$$

The method can also be used where there are more than two strata and procedures exist for the identification of inclined strata boundaries and vertical discontinuities.

The formulae used to estimate the depths of strata boundaries are based on the assumptions that each stratum is homogeneous and isotropic, the boundaries are plane, each stratum is thick enough to produce a change in slope on the time–distance plot and the seismic velocity increases in each successive stratum from the surface downwards. A layer of clay lying below a layer of compact gravel, for example, would not be detected.

Other difficulties arise if the velocity ranges of adjacent strata overlap, making it difficult to distinguish between them, and if the velocity increases with depth in a particular stratum. It is important that the results are correlated with data from borings.

Electrical resistivity method

The method depends on differences in the electrical resistance of different soil (and rock) types. The flow of current through a soil is mainly due to electrolytic action and therefore depends on the concentration of dissolved salts in the pore water: the mineral particles of a soil are poor conductors of current. The resistivity of a soil therefore decreases as both the water content and the concentration of salts increase. A dense, clean sand above the water table, for example, would exhibit a high resistivity due to its low degree of saturation and the virtual absence of dissolved salts. A saturated clay of high void ratio, on the other hand, would exhibit a low resistivity due to the relative abundance of pore water and the free ions in that water.

In its usual form (Figure 10.9(a)) the method involves driving four electrodes into the ground at equal distances (L) apart in a straight line. Current (I), from a battery, flows through the soil between the two outer electrodes, producing an electric field within the soil. The potential drop (E) is then measured between the two inner electrodes. The apparent resistivity (R) is given by the equation

$$R = \frac{2\pi LE}{I} \tag{10.3}$$

The apparent resistivity represents a weighted average of true resistivity in a large volume of soil, the soil close to the surface being more heavily weighted than the soil at depth. The presence of a stratum of soil of high resistivity lying below a stratum of low

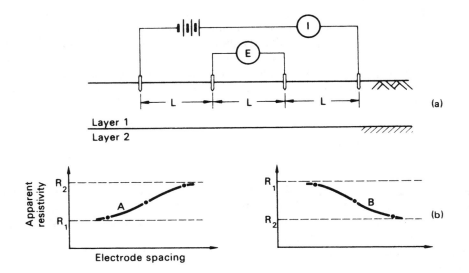

Figure 10.9 Electrical resistivity method.

resistivity forces the current to flow closer to the surface, resulting in a higher voltage drop and hence a higher value of apparent resistivity. The opposite is true if a stratum of low resistivity lies below a stratum of high resistivity.

The procedure known as *sounding* is used when the variation of resistivity with depth is required: this enables rough estimates to be made of the types and depths of strata. A series of readings are taken, the (equal) spacing of the electrodes being increased for each successive reading; however, the centre of the four electrodes remains at a fixed point. As the spacing is increased the apparent resistivity is influenced by a greater depth of soil. If the resistivity increases with increasing electrode spacing it can be concluded that an underlying stratum of higher resistivity is beginning to influence the readings. If increased separation produces decreasing resistivity, on the other hand, a stratum of lower resistivity is beginning to influence the readings. The greater the thickness of a layer, the greater the electrode spacing over which its influence will be observed, and vice versa.

Apparent resistivity is plotted against electrode spacing, preferably on log–log paper. Characteristic curves for a two-layer structure are illustrated in Figure 10.9(b). For curve A the resistivity of layer 1 is lower than that of layer 2: for curve B layer 1 has a higher resistivity than layer 2. The curves become asymptotic to lines representing the true resistivities R_1 and R_2 of the respective layers. Approximate layer thicknesses can be obtained by comparing the observed curve of resistivity versus electrode spacing with a set of standard curves. Other methods of interpretation have also been developed for two-layer and three-layer systems.

The procedure known as *profiling* is used in the investigation of lateral variation of soil types. A series of readings are taken, the four electrodes being moved laterally as a unit for each successive reading: the electrode spacing remains constant for each reading. Apparent resistivity is plotted against the centre position of the four electrodes, to natural scales: such plots can be used to locate the positions of soil of high or low resistivity. Contours of resistivity can be plotted over a given area.

The apparent resistivity for a particular soil or rock type can vary over a wide range of values; in addition, overlap occurs between the ranges for different types. This makes the identification of soil or rock type and the location of strata boundaries extremely uncertain. The presence of irregular features near the surface and of stray potentials can also cause difficulties in interpretation. It is essential, therefore, that the results obtained are correlated with borehole data. The method is not considered to be as reliable as the seismic method.

10.6   GROUND CONTAMINATION

The scope of an investigation must be extended if it is known or suspected that the ground in question has been contaminated. In such cases the soil and groundwater may contain potentially harmful substances such as organic or inorganic chemicals, fibrous materials such as asbestos, toxic or explosive gases, biological agents and radioactive elements. The contaminant may be in solid, liquid or gaseous form. Chemical contaminants may be adsorbed on the surfaces of fine soil particles. The presence of contamination influences all other aspects of ground investigation and may have consequences for foundation design and the general suitability of the site for the project under consideration. Adequate precautions must be taken to ensure the safety from health hazards of all personnel working on the site and when dealing with samples. During the investigation, precautions must be taken to prevent the spread of contaminants by personnel, by surface or groundwater flow and by wind.

At the outset, possible contamination may be predicted from information on previous uses of the site or adjacent areas, such as by certain types of industry, mining operations, reported leakage of hazardous liquids on the surface or from underground pipelines. The visual presence of contaminants and the presence of odours give direct evidence of potential problems. Remote sensing and geophysical techniques (e.g. infra-red photography and conductivity testing, respectively) can be useful in assessing possible contamination.

Soil and groundwater samples are normally obtained from shallow trial pits or boreholes. The depths at which samples are taken depend on the probable source of contamination and details of the types and structures of the strata. Experience and judgement are thus required in formulating the sampling programme. Solid samples, which would normally be taken at depth intervals of 100–150 mm, are obtained by means of stainless steel tools, which are easily cleaned and are not contaminated, or in driven steel tubes. Samples should be sealed in water-tight containers made of material that will not react with the sample. Care must be taken to avoid the loss of volatile contaminants to the atmosphere. Groundwater samples can be taken directly from trial pits, from observation pipes in boreholes or by means of specially designed sampling probes. Water samples should be taken over a suitable period to determine if properties are constant or variable. Gas samples can be obtained from tubes suspended in a perforated standpipe in a borehole, or from special probes. There are several types of receptacle suitable for collecting the gas. Details of the sampling process should be based on the advice of the analysts who will undertake the testing programme and report on the results.

REFERENCES

1   Bishop, A.W. (1948) A new sampling tool for use in cohesionless sands below groundwater level, *Geotechnique*, **1**, 125–31.
2   British Standard 5930 (1999) *Code of Practice for Site Investigations*, British Standards Institution, London.
3   Cairney, T. (ed.) (1987) *Reclaiming Contaminated Land*, Blackie, London.
4   Dobrin, M.D. (1988) *Introduction to Geophysical Prospecting*, 4th edn, McGraw-Hill, New York.
5   Griffiths, D.H. and King, R.F. (1965) *Applied Geophysics for Engineers and Geologists*, Pergamon Press, London.
6   Hvorslev, M.J. (1949) *Subsurface Exploration and Sampling of Soils for Civil Engineering Purposes*, US Waterways Experimental Station, Vicksburg, Mississippi.
7   Kjellman, W., Kallstenius, T. and Wager, O. (1950) Soil sampler with metal foils, *Proceedings Royal Swedish Geotechnical Institute*, No. 1.
8   Lerner, D.N. and Walter, R.G. (eds) (1998) Contaminated land and groundwater: future directions, *Geological Society, London, Engineering Geology Special Publication*, **14**, 37–43.
9   Rowe, P.W. (1972) The relevance of soil fabric to site investigation practice, *Geotechnique*, **22**, 193–300.
10 Weltman, A.J. and Head, J.M. (1983) *Site Investigation Manual*, CIRIA/PSA, London.

Chapter 11

Case studies

11.1 INTRODUCTION

There can be many uncertainties in the application of soil mechanics in geotechnical engineering practice. Soil is a natural (not a manufactured) material, therefore some degree of heterogeneity can be expected within a deposit. A ground investigation may not detect all the variations and geological detail within soil strata so that the risk of encountering unexpected conditions during construction is always possible. Specimens of relatively small size, and subject to some degree of disturbance even with the most careful sampling technique, are tested to model the behaviour of large *in-situ* masses. Results obtained from *in-situ* tests can reflect uncertainties due to heterogeneity. Consequently, judgements must be made regarding the accuracy of soil parameters obtained for use in design. In clays the scatter normally apparent in plots of undrained shear strength against depth is an illustration of the problem of selecting parameters. A geotechnical design is based on an appropriate theory which inevitably involves simplification of real soil behaviour and a simplified soil profile. In general, however, such simplifications are of lesser significance than uncertainties in the values of the soil parameters necessary for the calculation of quantitative results. Details of construction procedure and the standard of workmanship can result in further uncertainties in the prediction of soil behaviour.

In most cases of simple, routine construction the design is based on precedent and serious difficulties seldom arise. In larger or unusual projects, however, it may be desirable, or even essential, to compare actual and predicted behaviour. Lambe [9] classified the different types of prediction. Type A predictions are those made before the event. Predictions made during the event are classed as Type B and those made after the event are Type C: in both these cases no results from observations are known before predictions are made but if observational data are available at the time of prediction these types are classified as B1 and C1, respectively.

Studies of particular projects, as well as showing whether or not a safe and economic design has been achieved, provide the raw material for advances in the theory and application of soil mechanics. Case studies normally involve the monitoring over a period of time of quantities such as ground movement, pore water pressure and stress. Comparisons are then made with the theoretical or predicted values, e.g. the measured settlement of a foundation could be compared with the calculated value. If failure of a soil mass has occurred and the profile of the slip surface has been determined, e.g. in the slope of a cutting or embankment, the mobilized shear strength parameters could be back-calculated and compared with the test values. Empirical design procedures are based on *in-situ* measurements, e.g. the design of braced excavations is based on measurements of strut loads in different soil types.

The measurements required in case studies depend on the availability of suitable instrumentation, the role of which is to monitor soil response as construction proceeds so that decisions made at the design stage can be evaluated and if necessary revised. Instrumentation can also be used at the ground investigation stage to obtain information for use in design, e.g. details of groundwater conditions. However, instrumentation is only justified if it can lead to the answer to a specific question: it cannot by itself ensure a safe and economic design and the elimination of unpredicted problems during construction. The observational method, described in Section 11.3, is a set of procedures which makes use of instrumentation as part of a design and construction strategy or as a means of progressing a project in which unexpected difficulties have arisen. It should be appreciated that a sound understanding of the basic principles of soil mechanics is essential if the data obtained from field instrumentation are to be correctly interpreted. A selection of studies of various construction projects is presented later in this chapter.

11.2   FIELD INSTRUMENTATION

The most important requirements of a geotechnical instrument are reliability and sensitivity. In general the greater the simplicity of an instrument, the more reliable it is likely to be. On the other hand, the simplest instrument may not be sensitive enough to ensure that measurements are obtained to the required degree of accuracy and a compromise may have to be made between sensitivity and reliability. Instrumentation can be based on optical, mechanical, hydraulic, pneumatic and electrical principles: these principles are listed in order of decreasing simplicity and reliability. It should be appreciated, however, that the reliability of modern instruments of all types is of a high standard. The most widely used instruments for the various types of measurement are described below.

Vertical movement

The most straightforward technique for measuring surface settlement or heave is precise levelling. A stable bench mark must be established as a datum and in some cases it may be necessary to anchor a datum rod, separated by a sleeve from the surrounding soil, in rock or a firm stratum at depth. For settlement observations of the foundations of structures, durable levelling stations should be established in foundation slabs or near the bottom of columns or walls. A convenient form of station, illustrated in Figure 11.1, consists of a stainless steel socket into which a round-headed plug is screwed prior to levelling. After levelling the plug is removed and the socket is sealed with a perspex screw. To measure the settlement due to placement of an overlying fill, a horizontal plate, to which a vertical rod or tube is attached, is located on the ground surface before the fill is placed, as shown in Figure 11.2(a). The level of the top of the rod or tube is then determined. The settlement of the fill itself could be determined from surface levels, generally using levelling stations embedded in

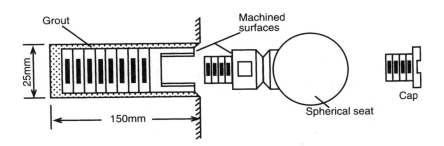

Figure 11.1 Levelling plug.

concrete. Vertical movements in an underlying stratum can be determined by means of a deep settlement probe. One type of probe, illustrated in Figure 11.2(b), consists of a screw auger attached to a rod which is surrounded by a sleeve to isolate it from the surrounding soil. The auger is located at the bottom of a borehole and anchored at the required level by rotating the inner rod. The borehole is backfilled after installation of the probe.

The rod extensometer, shown in Figure 11.2(c), is a simple and accurate device for measuring movement. The rods used are generally aluminium alloy tubes, typically 14 mm in diameter. Various lengths of rod can be coupled together as required. The lower end of the rod is grouted into the soil at the bottom of a borehole, a ragged anchor being threaded onto the rod, and the upper end passes into a reference tube grouted into the top of the borehole. The rod is isolated inside a plastic sleeve. The relative movement between the bottom anchor and the reference tube is measured with a dial gauge or displacement transducer operating against the top of the rod. An adjustment screw is fitted into a threaded collar at the upper end of the rod to extend the range of measurement. A multiple rod installation with rods anchored at different levels in the borehole enables settlement over different depths to be determined. The borehole is backfilled after installation of the extensometer. The use of rod extensometers is not limited to the measurement of vertical movement; they can be used in boreholes inclined in any direction.

Settlement at various depths within a soil mass can also be determined by means of a multi-point extensometer, one such device being the magnetic extensometer, shown in Figure 11.2(d), designed for use in boreholes in clay. The equipment consists of permanent ring magnets, axially magnetized, mounted in plastic holders which are supported at the required levels in the borehole by springs. The magnets, which are coated with epoxy resin as a guard against corrosion, are inserted around a plastic guide tube placed down the centre of the borehole. If necessary for stability, the borehole is filled with bentonite slurry. The levels of the magnets are determined by lowering a sensor incorporating a reed switch down the central plastic tube. When the reed switch moves into the field of a magnet it snaps shut and activates an indicator light or buzzer. A steel tape attached to the sensor enables the level of the station to be obtained to an accuracy of 1–2 mm. Greater accuracy can be obtained by locating, inside the guide tube, a measuring rod to which separate reed switches are attached at the level of each magnet, each switch operating on a separate electrical circuit.

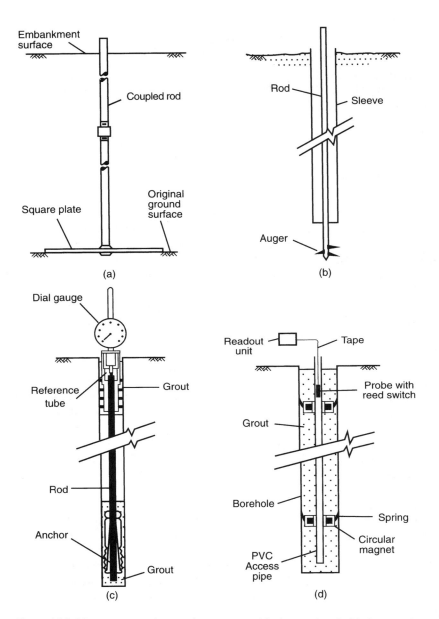

Figure 11.2 Measurement of vertical movement: (a) plate and rod, (b) deep settlement probe, (c) rod extensometer and (d) magnetic extensometer.

The measuring rod, which consists of hollow stainless steel tubing with the electrical wiring running inside, is drawn upwards by means of a measuring head incorporating a screw micrometer, the level of each magnet being determined in turn. As a precaution against failure, two switches may be mounted at each level.

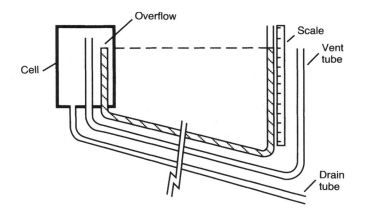

Figure 11.3 Hydraulic settlement cell.

The settlement of embankments can be measured by means of steel plates with central holes which are threaded over a length of vertical plastic tubing and laid, at various levels, on the surface of the fill as it is placed. Subsequently, the levels of the plates are determined by means of an induction coil which is lowered down the inside of the tubing.

Hydraulic settlement gauges, which are used mainly in embankments, provide another means of determining vertical movement. In principle the hydraulic overflow settlement cell (Figure 11.3) consists of a U-tube with limbs of unequal height: water overflows from the lower limb when settlement occurs causing a fall in water level, equal to the settlement, in the higher limb. The cell, which is cast into a concrete block at the chosen location, consists of a sealed rigid plastic cylindrical container housing a vertical tube acting as an overflow weir. Polythene-coated nylon tubing leads from the weir to a vertical graduated standpipe remote from the cell. A drain tube leading from the base of the cell allows overflow water to be removed and another tube acting as an air vent ensures that the interior of the cell is maintained at atmospheric pressure. The water tube is filled with de-aired water until it overflows at the weir. The standpipe is also filled and, when connected to the water tube from the cell, the water level in the standpipe will fall until it is equal to that of the weir. De-aired water is used to prevent the formation of air locks which would affect the accuracy of the gauge. The use of back air pressure applied to the cell enables it to be located below the level of the manometer. The system gives the settlement at one point only but can be used at locations which are inaccessible to other devices and is free of rods and tubes which might interfere with construction.

Horizontal movement

In principle the horizontal movement of stations relative to a fixed datum can be measured by means of a theodolite, using precision surveying techniques. However in many situations this method is impractical due to site conditions. Movement in one particular direction can be measured by means of an extensometer, of which there are several types.

The tape extensometer (Figure 11.4(b)) is used to measure the relative movement between two reference studs with spherical heads, which may be permanent or demountable as shown in Figure 11.4(a). The extensometer consists of a stainless steel measuring tape with punched holes at equal spacings. The free end of the tape is fixed to a spring-loaded connector which locates onto one of the spherical reference studs. The tape reel is housed in a cylindrical body incorporating a spring-tensioning device and a digital readout display. The end of the body locates onto the second spherical stud. The tape is locked by engaging a pin in one of the punched holes and tension applied to the spring by rotating the front section of the body, an indicator showing when the required tension has been applied. The distance is obtained from the reading on the tape at the pinned hole and the reading on the digital display. An accuracy of 0.1 mm is possible.

For relatively short measurements a rod extensometer (Figure 11.4(c)) incorporating a micrometer can be used. The instrument consists of a micrometer head with extension rods of different lengths and two end pieces, one with a conical seating and another with a flat surface, which locate against spherical reference studs. The rods have precision connectors to minimize errors when assembling the device. An appropriate gauge length is selected and the micrometer is adjusted until the end pieces make contact with the reference studs. The distance between the studs is read from the graduated micrometer barrel. A typical rod extensometer can measure a maximum movement of 25 mm.

The tube extensometer (Figure 11.4(d)) operates on a similar principle. One end piece has a conical seating as above, the other consists of a slotted tube with a coil spring inside. The slotted end piece is placed over one of the spherical reference studs, compressing the spring and ensuring firm contact between the second stud and the conical end piece at the other end of the extensometer. The distance between the outside end of the slotted tube and the reference stud is measured by means of a dial gauge enabling the movement between the two studs to be deduced. Movements up to 100 mm can be measured using a tube extensometer.

Various extensometers using transducers have also been developed. One such instrument uses a vibrating wire strain gauge, the diagram of which is shown in Figure 11.5. The wire is made to oscillate by means of an electromagnet situated approximately 1 mm from the wire. The oscillation induces alternating current, of frequency equal to that of the wire, in a second magnet. A change in the tension of the wire causes a change in its frequency of vibration and a corresponding change in the frequency of the induced current. The meter recording the output frequency can be calibrated in units of length. One end of the device is fixed to the first reference point by a pinned connector, the other end consisting of a sliding head which moves with the other reference point. The vibrating wire is fixed at one end and is connected to the moving head through a tensioned spring. When a displacement occurs there is a change in spring tension resulting, in turn, in a change in the frequency of vibration of the wire which is a function of the movement between the reference points. Movements of 200 mm can be measured to an accuracy of 0.1 mm. A number of extensometer units can be joined together by lengths of connecting rod to enable movements to be measured over a considerable length.

The potentiometric extensometer consists of a rectilinear resistance potentiometer inside a cylindrical housing filled with oil. Contacts attached to the piston of the

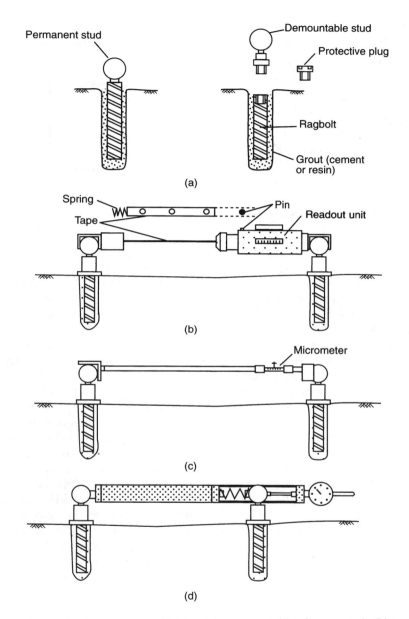

Figure 11.4 Measurement of horizontal movement: (a) reference studs, (b) tape extensometer, (c) rod extensometer and (d) tube extensometer.

extensometer slide along the resistance wire. The piston position is determined by means of a null balance readout unit calibrated directly in displacement units. Extension rods coupled to the instrument are used to give the required gauge length. Movements of up to 300 mm can be measured to an accuracy of 0.2 mm.

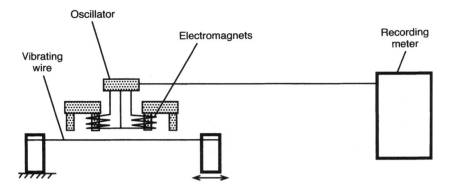

Figure 11.5 Vibrating wire strain gauge.

Horizontal movements at depth within a soil mass can be determined by means of the inclinometer, illustrated in Figure 11.6(a). The instrument operates inside a vertical (or nearly vertical) access tube which is grouted into a borehole or embedded within a fill, enabling the displacement profile along the length of the tube to be determined. The tube has four internal keyways at 90° centres. The inclinometer probe (known as the torpedo) consists of a stainless steel casing which houses a force balance acceler-ometer. The casing is fitted with two diametrically opposite spring-loaded wheels at each end, enabling the probe to travel along the access tube, the wheels being located in one pair of keyways, the accelerometer sensing the inclination of the tube in the plane of the keyways. The principle of the force balance accelerometer is shown in Figure 11.6(c). The device consists of a mass suspended between two electromagnets – a detector coil and a restoring coil. A lateral movement of the mass induces a current in the detector coil. The current is fed back through a servo-amplifier to the restoring coil which imparts an electromotive force to the mass equal and opposite to the component of gravitational force which causes the initial movement. Thus the forces are balanced and the mass does not, in fact, move. The voltage across a resistor in the restoring circuit is proportional to the restoring force and, in turn, to the angle of tilt of the probe. This voltage is measured and the voltmeter can be calibrated to give both angular and horizontal displacements. The vertical position of the probe is obtained from graduations on the cable attached to the device. Use of the other pair of keyways enables movements in the orthogonal direction to be determined. Readings are taken at intervals (δ) along the casing and horizontal movement is calculated as shown in Figure 11.6(b). Depending on the readout equipment, movements can be determined to an accuracy of 0.1 mm.

The slip indicator is a simple device to enable the location of a slip surface to be detected. A flexible plastic tube, to which a baseplate is attached, is placed in a bore-hole. The tube, typically 25 mm in diameter, is surrounded by a stiff sleeve. The tube is surrounded with sand, the sleeve being withdrawn as the sand is introduced. A probe at the end of a cable is lowered to the bottom of the tube. If a slip movement takes place the tube will deform. The position of the slip zone is determined both by raising the probe and by lowering a second probe from the surface until they meet resistance.

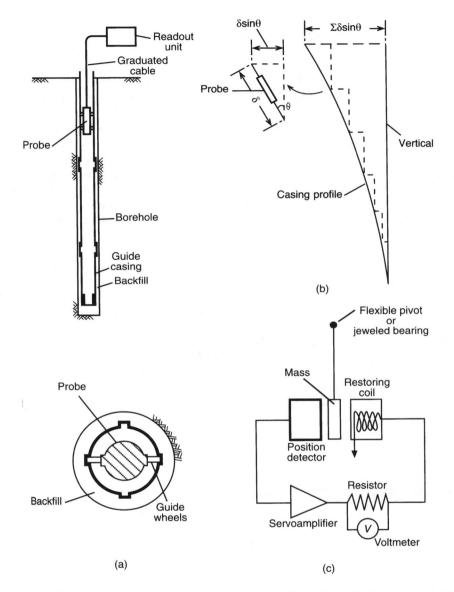

Figure 11.6 Inclinometer: (a) probe and guide tube, (b) method of calculation and (c) force balance accelerometer.

Pore water pressure

Predicted values of pore water pressure can be verified by *in-situ* measurements. *In-situ* pore pressures may be used, for example, to check the stability of slopes both during and after construction by monitoring the dissipation of excess pore water pressure. Pressure is measured by means of piezometers placed in the ground.

A piezometer consists of an element, filled with de-aired water, and incorporating a porous tip which provides continuity between the pore water in the soil and the water within the element. The element is connected to a pressure-measuring system. A high air-entry ceramic tip is essential for the measurement of pore water pressure in partially saturated soils (e.g. compacted fills), the air-entry value being the pressure difference at which air would bubble through a saturated filter. Therefore the air-entry value must exceed the difference between the pore air and pore water pressures, otherwise pore air pressure will be recorded. A coarse porous tip can only be used if it is known that the soil is fully saturated. If the pore water pressure is different from the pressure of the water in the measuring system, a flow of water into or out of the element will take place. This, in turn, results in a change in pressure adjacent to the tip and consequent seepage of pore water towards or away from the tip. Measurement involves balancing the pressure in the measuring system with the pore water pressure in the vicinity of the tip. However the time taken for the pressures to equalize, known as the response time, depends on the permeability of the soil and the flexibility of the measuring system. The response time of a piezometer should be as short as possible. Factors governing flexibility are: the volume change required to actuate the measuring device, the expansion of the connections and the presence of entrapped air. A de-airing unit forms an essential part of the equipment: efficient de-airing during installation is essential if errors in pressure measurement are to be avoided. To achieve a rapid response time in soils of low permeability the measuring system must be as stiff as possible, requiring the use of a closed hydraulic system in which virtually no flow of water is required to operate the measuring device.

Three types of piezometer for use with a closed hydraulic system are illustrated in Figure 11.7. The piezometers consist of a brass or plastic body into which a porous tip of ceramic, bronze or stone is sealed. Two tubes lead from the device to the measuring instrument, which may be a Bourdon gauge, a mercury manometer or a transducer. Use of a transducer enables results to be recorded automatically. The tubes are of nylon coated with polythene, nylon being impermeable to air and polythene to water.

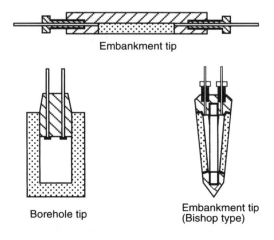

Figure 11.7 Piezometer tips.

The two tubes enable the system to be kept air-free by the periodic circulation of de-aired water. Allowance must be made for the difference in level between the tip and the measuring instrument, which should be sited below the tip whenever possible.

The pneumatic piezometer, represented diagramatically in Figure 11.8, incorporates a flexible diaphragm within the body of the unit, close to the porous tip, and has a rapid response time due to its negligible flexibility. The water pressure on one side of the diaphragm is balanced by pneumatic pressure, generally applied by air or nitrogen, on the other side. Gas inlet and outlet tubes lead through the body of the transducer, terminating at the diaphragm. Initially the water presses the diaphragm against the body, closing the ends of the inlet and outlet tubes. Gas is introduced at a low rate of flow and when its pressure just exceeds that of the water the diaphragm is pushed outwards allowing gas to flow into the outlet tube. The pressures are then in balance, the value being read from a pressure gauge attached to the inlet tube.

The vibrating-wire piezometer also employs a diaphragm housed within the unit. A steel wire in tension is fixed to the back of the diaphragm. Deflection of the diaphragm due to a change in water pressure causes a change in the tension of the wire and a corresponding change in the frequency of vibration. The frequency meter can be calibrated in units of pressure. Again, because the sensor is adjacent to the porous tip, the flexibility of the system is negligible and the response time is rapid.

If the soil is fully saturated and the permeability is relatively high, pore water pressure can be determined by measuring the water level in an open standpipe sealed in a cased borehole, as shown in Figure 11.9. The water level is normally determined by means of an electrical dipper, a probe with two conductors on the end of a measuring tape: the battery-operated circuit closes, triggering an indicator, when the conductors come into contact with the water. The standpipe is normally a plastic tube of 50 mm diameter or smaller, the lower end of which is either perforated or fitted with a porous element. A relatively large volume of water must pass through the porous element to change the standpipe level, therefore a short response time will only be obtained in soils of relatively high permeability. Sand or fine gravel is packed around the lower end and the standpipe is sealed in the borehole with clay (generally by the use of bentonite pellets) immediately above the level at which pore pressure is to be measured. The remainder of the borehole is backfilled with sand except near the

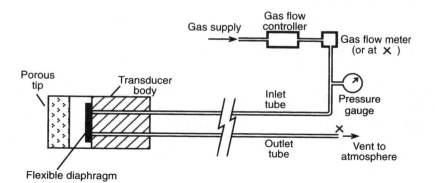

Figure 11.8 Pneumatic piezometer.

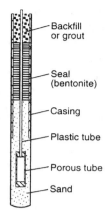

Backfill
or grout

Seal
(bentonite)

Casing

Plastic tube

Porous tube

Sand

Figure 11.9 Open standpipe piezometer.

surface where a second seal is placed to prevent the inflow of surface water. The top of the standpipe is fitted with a cap, again to prevent ingress of water. The piezometer illustrated in Figure 11.9 is a type first developed by Casagrande. Open standpipe piezometers which can be pushed or driven into the ground have also been developed.

Total normal stress

Total stress in any direction can be measured by means of pressure cells placed in the ground. Cells are normally disc-shaped and house a relatively stiff membrane in contact with the soil. Such cells are also used for the measurement of soil pressure on structures. It should be appreciated that the presence of a cell, as well as the process of installation, modifies the *in-situ* stresses due to arching and stress redistribution. Cells must be designed to reduce stress modification to within acceptable limits. Theoretical studies have indicated that the extent of stress modification depends on a complex relationship between the aspect ratio (the thickness/diameter ratio of the cell), the flexibility factor (depending on the soil/cell stiffness ratio and the thickness/ diameter ratio of the diaphragm) and the soil stress ratio (the ratio of stresses normal and parallel to the cell). The central deflection of the diaphragm should not exceed 1/5000 of its diameter and the cell should have a stiff outer ring. Cell diameter should also be related to the maximum particle size of the soil.

There are two broad categories of instrument, known as the diaphragm cell and the hydraulic cell, represented in Figure 11.10. In the first category the deflection of the diaphragm is measured by means of sensing devices such as electrical resistance strain gauges, linear transducers and vibrating wires, the readout in each case being calibrated in stress units. In the hydraulic cell, which normally consists of two stainless steel plates welded together around their periphery, the interior of the cell being filled with oil. The oil pressure, which is equal to the total normal stress acting on the outside of the cell, is measured by a transducer (generally of the pneumatic or hydraulic type) connected to the cell by a short length of steel tubing. Thin spade-shaped cells, which can be jacked into the soil, are useful for the measurement of *in-situ* horizontal stress.

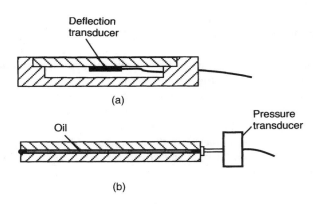

Figure 11.10 (a) Diaphragm pressure cell and (b) hydraulic pressure cell.

Load

The cells described above can be calibrated to measure load, the hydraulic type being the most widely used for this purpose. Such cells are used, e.g., in the load testing of piles and the measurement of strut loads in braced excavations. Cells with a central hole can be used to measure tensile loads in tie-backs. The use of surface-mounted strain gauges (e.g. electrical resistance or vibrating wire) is an alternative method for the measurement of load in struts and tie-backs. Load cells based on the principle of photo-elasticity are also used. The load is applied to a cylinder of optical glass, the corresponding strains producing interference fringe patterns which are visible under polarized light. The number of fringes is observed by means of a fringe counter, the load being obtained by multiplying the fringe count by a calibration factor. Another method of determining the load in a structural member such as a tie-back is by using a tell-tale. A tell-tale is an unstressed rod attached to the member at one end and running parallel to it. The change in length of the member under load is measured using the free end of the tell-tale as datum. The load can then be determined from the change in length provided the elastic modulus of the member is known.

11.3 THE OBSERVATIONAL METHOD

The main uncertainties in geotechnical prediction are the degree of variation and continuity of strata, pore water pressures (which may be dependent on macro-fabric features) and the values of soil parameters. Standard methods of coping with these uncertainties are the use of an excessive factor of safety, which is uneconomic, and the making of assumptions related to general experience, which ignores the danger of the unexpected. The observational method, as described by Peck [11], offers an alternative approach. The method is one of the design approaches listed in Eurocode 7 by which it can be verified that no relevant limit state is exceeded.

The philosophy of the observational method is to base the design initially on whatever information can be obtained, then to set out all conceivable differences between assumptions and reality. Calculations are then made, on the basis of the original assumptions, of relevant quantities which can be measured reliably in the field, e.g. settlement, lateral movement, pore water pressure. Predicted and measured values are compared as construction proceeds and the design or the construction procedure is modified if necessary. The engineer must have a plan of action, prepared in advance, for every unfavourable situation that observations might disclose. It is essential, therefore, that the appropriate instrumentation to detect all such situations should be installed. If, however, the nature of the project is such that the design cannot be changed during construction if certain unfavourable conditions arise, the method is not applicable. In such circumstances a design based on the least favourable conditions conceived must be adopted even if the probability of their occurrence is very low. If an unforeseen event were to arise for which no strategy existed, then the project could be in serious trouble. The potential advantages of the method are: savings in cost, savings in time and assurance of safety.

There are two distinct situations in which the method may be appropriate. (1) Projects in which use of the method is envisaged from conception, known as *ab initio* applications. It may be that an acceptable working hypothesis is not possible and that use of the observational method offers the only hope of achieving the objective. (2) Situations in which circumstances arise such that the method is necessary to ensure successful completion of the project, known as best-way-out applications. It should be appreciated that projects do go wrong and the observational method may then offer the only satisfactory way out of the difficulties encountered.

The comprehensive application of the method involves the following steps. The extent to which all steps can be followed depends on the type and complexity of the project.

1   Investigation to establish the general nature, sequence, extent and properties of the strata, but not necessarily in detail.
2   Assessment of the most probable conditions and the most unfavourable conceivable deviations from these conditions, geological conditions usually being of major importance.
3   Design process based on a working hypothesis of behaviour under the most probable conditions.
4   Selection of quantities to be observed as construction proceeds and calculation of their anticipated values on the basis of the working hypothesis.
5   Calculation of values of the same quantities under the most unfavourable conditions compatible with the available data on ground conditions.
6   Decision in advance on a course of action or design modification for every conceivable deviation between observations and predictions based on the working hypothesis.
7   Measurement of quantities to be observed and evaluation of actual conditions.
8   Modification of design or construction procedure to suit actual conditions.

A simple example, given by Peck, of the use of the method planned from the design stage is the case of a braced excavation in clay, with struts at three levels. The excavation, to a depth of 14 m, was for the construction of a building in Chicago. The design of the struts was based on the trapezoidal diagram (Figure 6.33(d)), which is an envelope of strut load measurements reported from a variety of sites. For a particular site, therefore, most of the struts could be expected to carry loads significantly less than those predicted from the diagram. For the site in question the struts were designed for loads equal to two-thirds of those given by the diagram and the load factor used was relatively low. On this basis a total of 39 struts was required. The probability then existed that a few struts would be overloaded (reaching loads around the values given by the diagram) and would fail. The procedure adopted was to measure the loads in every strut at critical stages of construction and to have additional struts available for immediate insertion if necessary. Only three additional struts were required. The extra cost of monitoring the loads and in having additional struts available was low compared to the overall saving in using struts of smaller section. The procedure also ensured that no strut became overloaded. If a much higher number of additional struts had been required the overall cost might have been higher and construction might have been delayed but such a risk was judged to be minimal.

11.4 ILLUSTRATIVE CASES

Bearing capacity failure

This case concerns the collapse in 1955, during initial filling, of a reinforced concrete grain elevator near Fargo, North Dakota, reported by Nordlund and Deere [10]. The structure consisted of 20 cylindrical storage bins 37 m high, in two rows of 10, constructed on a raft foundation 66×16 m located at a depth of 1.8 m. The thickness of the raft was 0.7 m except for the outer 0.9 m which was 1.3 m thick. The profile of soil heave which occurred at the south side of the elevator during collapse is shown in Figure 11.11(a). The pattern of heave indicated that soil displacement and settlement were greater at the west end of the structure than at the east end. It was established from loading records that the average net foundation pressure at collapse was 228 kN/m^2 but due to unequal loading in the bins the total load was applied at an eccentricity of around 1.0 m from the centroid of the raft. Because of the relative rigidity of the structure, the pressures at the corners varied between 206 and 250 kN/m^2 as a result of this eccentricity: the two highest pressures occurred at the west end of the structure where the soil displacement was greater. Settlement measurements had been taken by levelling but only after approximately 20% of the live load at failure had been applied. The recorded settlements of the centre of the raft are shown in Figure 11.11(b), the last observation being taken 4 days before collapse.

Subsequent to collapse the ground conditions were investigated by means of three boreholes at the locations shown in Figure 11.11(a). The investigation showed the presence of four distinct strata. The strata, denoted A to D respectively in Figure 11.11(a), were as follows: silty clay (within which shrinkage cracks were evident); stratified clay and silt; fine sand (interbedded with silt and silty clay in its lower reaches); and a deep deposit of clay (with a weathered crust over the upper 1.5 m). Tube samples of 47 mm diameter were taken at frequent intervals in the cohesive strata and standard penetration tests were performed in the sand. *In-situ* vane tests were also commissioned. The probable failure surface, shown in Figure 11.11(a), was identified from the extent of the heave area and from the position of a zone of remoulded clay

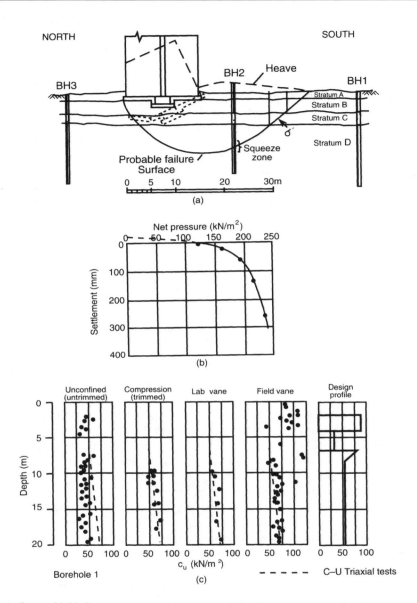

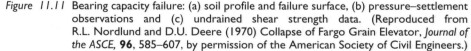

Figure 11.11 Bearing capacity failure: (a) soil profile and failure surface, (b) pressure–settlement observations and (c) undrained shear strength data. (Reproduced from R.L. Nordlund and D.U. Deere (1970) Collapse of Fargo Grain Elevator, *Journal of the ASCE*, **96**, 585–607, by permission of the American Society of Civil Engineers.)

encountered in the second borehole. This borehole was uncased below a depth of 7 m and difficulty was experienced in advancing the boring at a depth of around 15 m due to soft soil being squeezed into the hole: it was assumed that this soil had been remoulded by the slip movement.

In the laboratory, unconfined compression and vane tests (on samples retained within the sampling tube) were performed. Consolidated undrained triaxial tests were performed on specimens from borehole 1 only. Values of undrained shear strength obtained by the different test methods for samples from borehole 1 are given in Figure 11.11(c), the dotted line representing the results from consolidated undrained triaxial tests. Unconfined compression tests on specimens that had not been trimmed prior to testing showed significantly lower strength than that for trimmed specimens, the difference probably being the result of sample disturbance and swelling (although Nordlund and Deere were uncertain that this was the case). The strains at failure in the various laboratory tests varied between 1.25 and 8%: strength values mobilized at 1.25% strain were used in the analysis because the maximum shear strength of each stratum would not be mobilized simultaneously along a failure surface. Although the test results indicated a slight increase in shear strength with depth, a constant distribution was used in analysis. Nowadays it is unlikely that unconfined compression tests would be used in such an investigation – unconsolidated-undrained triaxial or *in-situ* vane tests would be employed.

The application of bearing capacity theory to this case is not straightforward due to the presence of the layer of sand (stratum C). Based on the standard penetration resistance (N) within this layer, which varied within the range 5–13, Nordlund and Deere assumed a value of shear strength parameter ϕ' of 25°; however, based on Figure 8.9, a value of 30° would have been more realistic. The shear strength within the sand layer was calculated as the average value of $\sigma' \tan \phi'$ over the depth of sand intersected by the failure surface, as shown in Figure 11.11(a), adopting the same principle as is used in the method of slices in slope stability analysis. Using this procedure a shear strength of 30 kN/m^2 was obtained and this was conceived to be an equivalent c_u value. The strength profile used in the analysis of the failure is shown in Figure 11.11(c), the average value of c_u over a depth of $\frac{2}{3}$ B (where B is the width of the raft) being 46 kN/m^2. The exercise of considerable judgement was required in deciding the appropriate shear strength distribution.

In stratified deposits the use of Skempton's values of bearing capacity factor (N_c) is considered to be appropriate only if the value of c_u of any stratum is within 50% of the average value of all strata within the significant depth of the foundation. Such a condition is not met in this case. An effect of the above condition is to reduce the influence of progressive failure due to different failure strains in the individual strata. However, because differences in stress–strain characteristics of the strata were considered in deciding average strength values, the use of Skempton's approach was considered to be justified. For a foundation of dimensions 66 × 16 m at a depth of 1.8 m ($B/L = 0.24$ and $D/B = 0.12$) the value of N_c from Figure 8.5 is 5.4. Thus the net ultimate bearing capacity ($c_u N_c - \gamma D$) is 250 kN/m^2. This corresponds exactly to the highest net foundation pressure at a corner of the elevator but is greater than the average pressure of 228 kN/m^2.

Thus an analysis in advance of construction, using the average foundation pressure, would have indicated a factor of safety of only 1.1. Even with the most optimistic interpretation of the shear strength data, it is clear that the factor of safety would have been inadequate and that an alternative foundation design would have been required, perhaps involving partial loading for a period long enough to allow some consolidation, and a corresponding increase in shear strength, to take place. Piling would have been a satisfactory solution but would have increased the cost of the structure. The settlement data clearly indicated that failure was imminent but no action was taken: immediate unloading would have prevented collapse. Fortunately bearing capacity failures are rare and it is difficult to imagine that the circumstances of this failure would occur in present times: even the simplest investigation and testing programme would have shown that the factor of safety was dangerously low.

Foundation settlement in clay

This case, reported by Somerville and Shelton [13], concerns the settlement performance of six 15-storey blocks of flats at two sites in Glasgow. The sites lie over a buried river channel, the ground investigations showing alluvial deposits of firm laminated clays and silts, underlain by glacial deposits of stiff sandy clay with gravel, cobbles and boulders. Piled foundations would normally have been used in such conditions but, because of the presence at a deeper level of coal measures which had been extensively mined, it was considered inadvisable to impose concentrated loading on these strata by piles terminating a short distance above. The solution adopted was the use of semi-buoyant cellular raft foundations to reduce the net foundation pressure on the clays and silts to a value which would give an adequate factor of safety against shear failure and which would limit long-term settlement to 65–75 mm, a limit imposed by the structural engineer on the basis of tolerable differential settlement. Provision was also made for the jacking of each structure from the cellular raft to correct any distortion which might result from future mining subsidence. Settlement observations were made during construction and over the subsequent 3 years to compare predicted and actual values.

The soil profile for one of the sites (Parkhead) is shown in Figure 11.12(a). Values of undrained shear strength (c_u) were determined by means of triaxial tests on undisturbed samples and *in-situ* vane tests. Values of coefficient of volume compressibility (m_v) were obtained from oedometer tests. Distributions of c_u and m_v with depth are also shown in Figure 11.12(a), the scatter of c_u values being particularly pronounced.

The raft was located at a depth of 4.3 m and the gross foundation pressure imposed by the structure was 135 kN/m^2. Taking the average unit weight of the soils as 19 kN/m^3 the net foundation pressure, therefore, is 53.5 kN/m^2. The plan of the raft, detailed in Figure 11.12(b), can be approximated to a rectangle of the same area having dimensions of 33.5 and 22.2 m. Consolidation settlements at the centre of the raft and at the mid-point of the longer side, calculated by the one-dimensional method, are 57 and 30 mm, respectively. For a foundation breadth of 22.2 m and a depth of consolidating soil of 13.5 m, it would be appropriate to multiply the above settlements by a settlement coefficient (μ): in the absence of a value of pore pressure coefficient A, a settlement coefficient of 0.85 was judged to be appropriate, reducing the calculated settlements to 48 and 25 mm, respectively. However, these values imply that the foundation is flexible, whereas the cellular raft in this case is stiff. Thus the settlement distribution across the breadth of the structure would be expected to be less extreme, being less than the calculated value at the centre and greater than the calculated value at the edge. In the absence of realistic values of undrained modulus (E_u), Somerville and Shelton assumed that immediate settlement would be 20% of consolidation settlement, i.e. 10 and 5 mm, respectively. Consideration was also given to the rate of

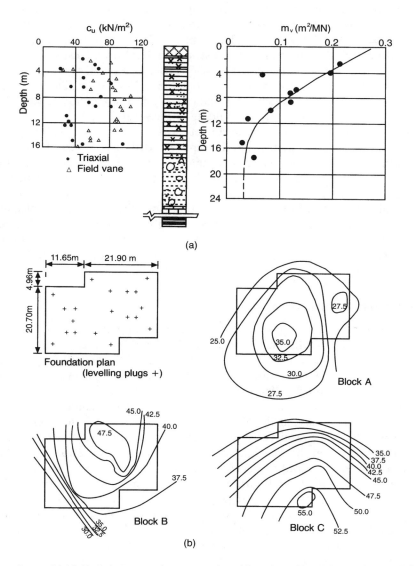

Figure 11.12 Foundation settlement in clay: (a) soil profile and test data and (b) settlement contours. (Reproduced from S.H. Somerville and J.C. Shelton (1972) Observed settlement of multi-storey buildings on laminated clays and silts in Glasgow, *Geotechnique*, **22**, 513–20, by permission of Thomas Telford Ltd.)

settlement of the foundation but because radial drainage would be significant due to the fact that the clay and silt were laminated, use of Terzaghi's theory was inappropriate. However a value of coefficient of consolidation (c_v) was deduced from published settlement–time observations of a storage tank foundation on similar soils in Kent. Using this value (780 m^2/year) it was estimated that 90% of consolidation settlement would have taken place after 9 months.

A check was made on the factor of safety against shear failure. The relevant foundation dimensions are depth (D) 4.3 m, breadth (B) 22.2 m and length (L) 33.5 m; thus $D/B = 0.19$ and $B/L = 0.66$. The appropriate value of bearing capacity factor N_c, therefore, is 5.85. Referring to the distribution of undrained shear strength (c_u) with depth shown in Figure 11.12(a) and selecting a worst-conceivable value of 30 kN/m^2, the net ultimate bearing capacity is $30 \times 5.85 = 175$ kN/m^2. Therefore the factor of safety is $175/53.5 = 3.3$ which is satisfactory. Using a more realistic value for c_u of, say, 60 kN/m^2, the factor of safety would be in excess of 6.

Settlement observations were made by means of levelling, using plugs screwed into sockets set into a number of columns. Levels were recorded at intervals of approximately 3 months. The observations showed that settlements did not exceed the predicted values and that 90% consolidation settlement occurred after approximately 12 months. Maximum settlements of the three blocks at the site in question were 35, 46 and 51 mm; minimum settlements were 23, 23 and 34 mm, respectively. Settlement contours are shown in Figure 11.12(b) and indicate slight tilting of the foundations: maximum angular distortions of 1/1280, 1/350 and 1/550, respectively, can be deduced. The study showed that acceptable settlement predictions were made.

Foundation settlement in sand

The settlement performance of a raft foundation supporting a 13-storey building in Belfast was observed by Stuart and Graham [14]. The raft, which carries a net foundation pressure of 161 kN/m^2 (the net load being 121 MN), has a length (L) of 55 m, a breadth (B) of 20 m and is located at a depth of 6 m below final ground level. It was established from boreholes adjacent to each end of the structure that the soil conditions consisted of approximately 25 m of glacio/lacustrine sands and gravels underlain by hard boulder clay, as detailed in Figure 11.13(a), the water table being at a depth of around 9 m. The results of standard penetration tests are also shown. The original design was based on the Terzaghi and Peck method, the criterion being that the maximum settlement for a raft should not exceed 50 mm. This criterion is met by doubling the pressure derived from Figure 8.10, the resultant value then being multiplied by the water table correction factor (Equation 8.16). Conversely, if the above foundation pressure is halved and divided by the water table correction factor, calculated to be 0.67, a value of 120 kN/m^2 is obtained and it is apparent from Figure 8.10 that an average N value of 14 over a depth of 20 m (B) below foundation level would result in an acceptable design. Loose sand, with N values as low as 6, was located at foundation level, especially at one end of the raft, and it was decided to excavate this material and replace it with lean concrete over depths varying between 2.3 and 0.5 m.

The average values of standard penetration resistance (N), corrected for overburden pressure (using Figure 8.8), for the two boreholes are 19 and 21 (although values were not available over the full depth of B below foundation level). Using the lower average value of 19 in Figure 8.10 and extrapolating to $B = 20$ m, a pressure of 180 kN/m^2 is obtained. Doubling this value for a raft and multiplying by the water table correction factor yields an allowable bearing capacity of 240 kN/m^2. The settlement under an actual pressure of 161 kN/m^2, therefore, should be well within the design criterion.

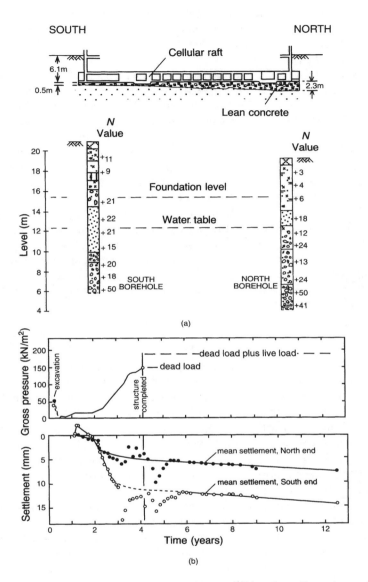

Figure 11.13 Foundation settlement in sand: (a) soil profile and test data and (b) settlement observations. (Reproduced from J.G. Stuart and J. Graham (1975) Settlement performance of a raft foundation on sand, in *Settlement of Structures*, by permission of Thomas Telford Ltd.)

It should be recalled that the Terzaghi and Peck method is not intended to yield actual settlement values.

Using the Burland and Burbidge method (which, of course, post-dates the project in question) the value of influence depth (z_I) for $B = 20\,\text{m}$ is $10\,\text{m}$ (Figure 8.12). The average (uncorrected) value of N between depths of 6 and 16 m is 23, giving a value of

I_c (Equation 8.18) of 0.021 and a settlement (Equation 8.19c) of 9 mm, it being assumed that the lower strata are overconsolidated. However this value must be multiplied by the shape factor f_s (Equation 8.20), the value of which is 1.31. Thus the calculated settlement is 12 mm.

Settlement points were established by casting stainless steel pegs into the basement floor and observations were made by means of a level fitted with a parallel plate micrometer. However access to the basement became difficult and pegs were inadvertently displaced with the result that there was considerable uncertainty about the accuracy of the readings towards and shortly after the end of construction, as is apparent from the settlement–time plot. Later observations were taken on pegs at ground level. The datum for levelling was a bench mark on a nearby building constructed 4 years earlier on piled foundations. This datum, in turn, was checked at regular intervals against Ordnance Survey benchmarks and was found to be rising at the rate of about 1.5 mm per year: the settlement observations were adjusted accordingly. The average settlements recorded for the two ends of the building over a period of 12 years, together with the loading details, are shown in Figure 11.13(b). At the end of construction the settlements at the two ends of the foundation were 5 and 12 mm (a difference of 7 mm). After the end of construction, settlement continued at a constant rate, probably due to creep. The observations confirm that the Terzaghi and Peck method is excessively conservative.

Pre-loading

It may be necessary to locate groups of oil storage tanks on sites underlain by soft compressible soils. Large long-term consolidation settlements would be unacceptable and the use of piled foundations would normally be ruled out on economic grounds. One possible solution would be to pre-load the newly constructed tanks for an appropriate period by filling them with water and allowing most of the settlement to take place before the tanks are brought into service. (Tanks are usually water-tested for leakage in any event.) The pre-load may have to be applied in increments to avoid undrained shear failure of the supporting soil. As consolidation proceeds, the shear strength of the soil increases, allowing further load increments to be applied. The rate of dissipation of excess pore water pressure can be monitored by installing piezometers below the tanks, thus providing a means of controlling the rate of loading. It should be appreciated that differential settlement will occur if non-uniform soil conditions exist and re-levelling of the tanks may then be necessary at the end of the pre-loading period. If the compressible soil were of limited depth, another possible solution to the settlement problem would be to excavate and replace it with compacted fill. If the level of the site were to be raised, the fill used might itself provide sufficient pre-load prior to construction of the tanks.

An example of water pre-loading, reported by Darragh [2], occurred during the construction of over 100 tanks at the Pascagoula refinery on the coast of Mississippi. The soil conditions at the site consisted of 3.0 m of silty sand underlain by 4.8 m of soft clay of very high plasticity and 12.0 m of firm sandy clay, dense clayey sand and stiff clay. Further deposits of clay occurred below this level. Parameters determined for the soft clay were $c_u = 20{-}28 \, \text{kN/m}^2$, $C_c = 1.0$ and $c_v = 0.93 \, \text{m}^2/\text{year}$. The section in

Figure 11.14(a) shows a tank 38 m in diameter and 14.5 m in height constructed on a berm of silty sand. The final settlement under the centre of this tank was calculated as 900 mm (and approximately half this value under the edge), two-thirds of which was due to consolidation of the soft clay. Initially three tanks were pre-loaded and observations of settlement, lateral movement and pore water pressure were made. Levelling points were established on the sides of the tanks and on the adjacent ground surface, the ground points also being used for measurements of lateral movement. Lateral movements within the soft clay were determined by means of four inclinometers below the edge of each tank. Piezometers were installed at different levels in the soft clay to measure pore water pressure below four points around the rim of each tank.

The tanks were filled with water incrementally over a period of about 10 months, as indicated in Figure 11.14(b). Based on observations during the first two increments, the rates of settlement, lateral movement and pore water pressure dissipation were estimated, by extrapolation, for subsequent increments. Each increment was held for a period long enough to ensure that pre-determined values of settlement, lateral movement and pore water pressure were not exceeded. During the test period, however, settlements were well below the permitted values and differential settlements were small. The maximum and minimum settlements recorded for one of the tanks are shown in Figure 11.14(b) along with the predicted settlement–time curve, the predicted curve being developed by extrapolating the observed settlements during the early stages of loading. In developing this curve it was assumed (1) that the rate of settlement was proportional to the square root of time as fitted to the observations during the first increment and the early observations of the second increment and (2) that settlements under subsequent increments were directly proportional to load. Subsequent back-calculation indicated that the observed settlements could have been predicted by elastic theory using an undrained modulus of $200c_u$ and the rate of settlement by a value of c_v of approximately twice the average laboratory value. Readings from one of the piezometers at the centre of the soft clay, showing the development and dissipation of excess pore water pressure, are also shown in Figure 11.14(b). Using observed increases in pore water pressure under load increments, values of pore pressure coefficient A were calculated from Equation 4.25. At the higher loads the results indicated that A approached the failure value obtained in undrained triaxial tests on specimens of the soft clay. Although there are uncertainties in such comparisons (e.g. in the calculation of $\Delta\sigma_3$ in Equation 4.25), the observed pore water pressures indicated that the factor of safety against shear failure was relatively low. The inclinometer readings (Figure 11.14(c)) indicated maximum lateral movements of around 175 and 50 mm, respectively, at the top and bottom of the soft clay layer. Under a given load, lateral movement consisted of both elastic and plastic components. The rate of plastic movement decreased at the higher loads, indicating increased shear strength due to consolidation of the clay under previous load increments and that shear failure was unlikely.

Based on the results from the initial three tanks, the loading period for all other tanks was reduced to 6 months, the initial load increment was increased by 33% and settlement observations alone were employed as the means of loading control.

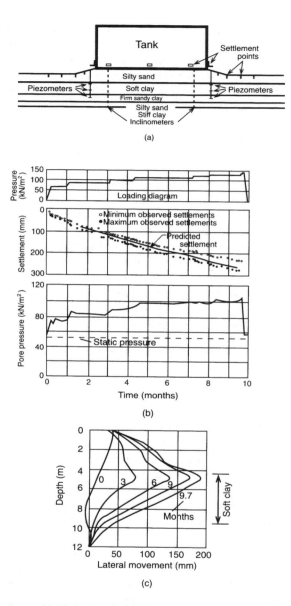

(a)

(b)

(c)

Figure 11.14 Pre-loading of storage tanks: (a) soil profile, (b) settlement and pore water pressure observations and (c) observed lateral movement. (Reproduced from R.D. Darragh (1964) Controlled water tests to pre-load tank foundations, in *Proceedings ASCE*, **90**, 303–29, by permission of the American Society of Civil Engineers.)

The principle of pre-loading can also be applied to accelerate the settlement of embankments by surcharging with an additional depth of fill. At the end of an appropriate period, surcharge is removed down to formation level.

Pile load test

This case concerns the choice of pile foundations for an office building in Perth, reported by Whitworth and Thomson [16], and the difficulties in predicting their ultimate load capacity. Due to the use of high-quality architectural finishes it was a design requirement that total and differential settlements were to be minimal. Column loads of up to 1.65 MN were to be carried on a total of 123 piles, each required to support a working load of 550 kN. For a load factor of 2.5, this would necessitate an ultimate load capacity of 1375 kN. The ground investigation showed the presence of relatively recent deposits consisting of soft clay (plasticity index 25), a thin layer of peat, medium dense to dense sand and gravel, and soft to firm silty clay (plasticity index 18), the soil profile used in design being shown in Figure 11.15(a). These soils were underlain at depth by till and sandstone. Groundwater conditions were investigated by observation of water levels in boreholes and by means of piezometers. However groundwater level could not be determined with certainty and a level of 9.0 m above Ordnance Datum was used in the initial design calculations. Undrained shear strength in the two clay layers was measured by *in-situ* vane tests and by triaxial tests on specimens from 100 mm diameter tube samples. The results are shown in Figure 11.15(a), it being apparent that the laboratory values were significantly less than those from the *in-situ* tests. From Figure 4.12 the correction factor for vane strength is 1.0 for a plasticity index of 18. Standard penetration tests in the sand and gravel gave (uncorrected) N values varying between 11 and in excess of 50, with an average value of 25.

Initially, various pile types were considered. Driven piles were ruled out because of the effects during installation of vibration and noise on adjacent buildings and their occupants. The use of cast-*in-situ* piles in shafts excavated by conventional continuous-flight augers was also rejected because of doubts about the ability of the auger to penetrate the sand and gravel layer: if rotation of the auger were then to continue without penetration of the layer, soil could be drawn laterally towards the auger, resulting in adjacent settlement. Bored piles cast under bentonite slurry, however, were considered to be a viable option. Consideration was given to the possibility of locating the piles in the sand and gravel layer but uncertainty about the continuity of the layer ruled out this option. The solution eventually adopted was the use of cast-*in-situ* concrete piles of 620 mm diameter, the shafts being excavated by a special type of continuous-flight auger incorporating an integral rock drilling bit and a tremie pipe inside the stem, through which the concrete was pumped (these being known as *Starsol* piles). Initial calculations indicated that piles installed to a level of −14.5 m OD would be satisfactory but subsequently this was increased to −16.5 m in view of the limited data available on the performance of this form of pile in the above types of strata.

In the initial design, shaft resistance in the upper layer of soft clay was neglected. For the soft to firm clay an average value of undrained shear strength of 40 kN/m² and an adhesion coefficient (α) of 0.85 were used, giving a resistance of 827 kN on a shaft length of 12.5 m. A bearing capacity coefficient (N_c) of 9 was used for base resistance, giving a value of 108 kN for $c_u = 40$ kN/m². The calculation of shaft resistance in the sand and gravel stratum was based on Equation 8.29 (with $\delta = \phi'$) taking $K_s = 0.9$, a typical value for continuous auger piles, and $\phi' = 35°$, based on the average N value. The effective overburden pressure at the centre of the stratum is $8.0\gamma' = 65$ kN/m², assuming a total unit weight of 18 kN/m³, therefore the resistance on a shaft length of

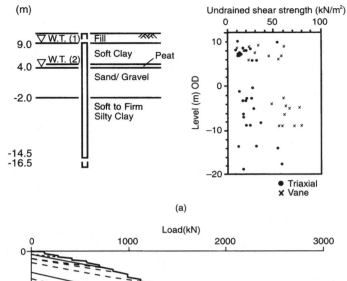

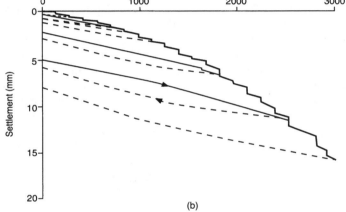

Figure 11.15 Pile load test: (a) soil profile and undrained shear strength and (b) load–settlement observations. (Reproduced from L.J. Whitworth and N. Thomson (1995) The design, construction and load testing of Starsol piles, Perth, Scotland, in *Proceedings Institution of Civil Engineers, Geotechnical Engineering*, **113**, 233–41, by permission of Thomas Telford Ltd.)

6.0 m is 479 kN. Thus the ultimate pile load is 1414 kN, giving a load factor of 2.57. The shaft resistance in the clay was also calculated by the effective stress method, assuming a conservative value of β of 0.25. At the centre of the clay stratum the effective overburden pressure is $18.25\gamma' = 149\,kN/m^2$, therefore the shaft resistance is 900 kN. Using this value the ultimate pile load becomes 1487 kN, giving a load factor of 2.70.

For the test pile the maintained load procedure was adopted, it being envisaged that five increments would be applied up to 1100 kN, twice the working load, followed by a further load increase to ultimate failure. However it became apparent that ultimate failure would not be achieved and, in the event, eight cycles were applied up to a load of 3000 kN, the maximum possible for the test equipment. The test results are shown in

Figure 11.15(b). Based on various extrapolation procedures (described in Ref. [20], Chapter 8) it was concluded that a reasonable estimate of ultimate load for the test pile was 3300 kN, i.e. almost $2\frac{1}{2}$ times the predicted value.

The large discrepancy between prediction and observation led to a re-evaluation of design assumptions. A groundwater level of 4.0 m OD was considered to be more realistic. The working level of 10.8 m used for the pile test was taken as ground level rather than 9.0 m, the length of the test pile thus being 27.3 m. Both these revisions influenced the value of overburden pressure in the effective stress calculations (values of effective overburden pressure at the centres of the successive strata being 21, 147 and 231 kN/m^2). An average value of c_u of 25 kN/m^2 was assumed for the upper layer of soft clay, the contribution to pile resistance of this layer initially having been neglected. The design value of c_u for the soft to firm silty clay was re-assessed as 65 kN/m^2, based only on the results of the *in-situ* vane tests, these not being subject to sampling disturbance. An adhesion coefficient (α) of 1.0 was adopted, this value being considered more appropriate for a pile cast by concrete injection. Using correlations between ϕ' and plasticity index, and taking $K_s = K_0$ in Equation 8.34, a value of β of 0.29 was obtained for both clays. Using these revised parameters and the increased shaft length of 14.5 m in the lower clay stratum, ultimate loads of around 3350 kN were obtained using the two methods of calculation, consistent with the extrapolated test values.

The case illustrates how the appraisal of soil conditions and parameters is the most crucial factor in predicting pile performance and the importance of pile loading tests in calibrating and refining design calculations. In situations where little or no previous performance data is available, it is inevitable that conservative values of design parameters initially will be used.

Piled raft

The behaviour of a piled raft foundation in London clay, supporting a building 90 m high with a basement 8.8 m deep, was monitored by Hooper [7]. The raft, cast on a blinding layer in contact with the clay, is 1.5 m thick and is supported by 51 under-reamed bored piles 25 m in length, the shaft and base diameters being 0.9 and 2.4 m, respectively. A section through the foundation is shown in Figure 11.16(a). The gross pressure on the raft is 368 kN/m^2 (corresponding to a gross building load of 228 MN) but overburden pressure of 172 kN/m^2 was removed by excavation, resulting in a net foundation pressure of 196 kN/m^2 (the corresponding net load being 121 MN). The soil profile and values of parameters c_u and m_v, determined from laboratory tests on specimens obtained from various depths, are shown in Figure 11.16(b). These distributions show the degree of scatter typical in practice, due mainly to sampling disturbance and to natural variations in soil structure. Evidence from a nearby borehole indicated that the clay extended to a depth of around 60 m. Strata below this depth would have no significant effect on the behaviour of the foundation.

Settlement was monitored by means of 14 levelling sockets installed in the columns and walls at ground floor level. The loads in three of the piles were measured by means of load cells, located 2 m below the tops of the piles. Each cell consisted of six photo-elastic load gauges located between two circular steel plates of the same diameter as the piles. The stiffness of the cells was approximately equal to that of the piles to ensure that the load distribution in the piles was not affected. The pressure between the raft

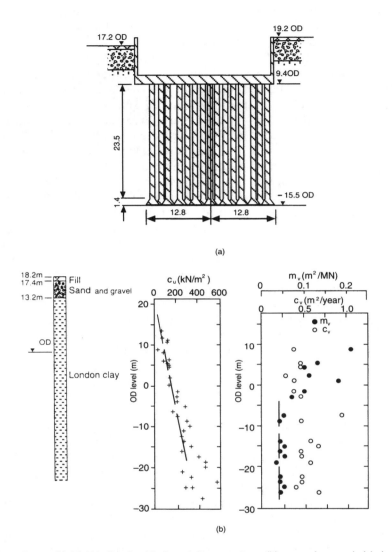

Figure 11.16 Piled raft: (a) foundation section, (b) test data and (c) load and settlement observations. (Reproduced from J.A. Hooper (1973) Observations on the behaviour of a piled-raft foundation on London Clay, *Proceedings Institution of Civil Engineers*, **55** (2), 855–77, by permission of Thomas Telford Ltd.)

and the clay, adjacent to the instrumented piles, was measured by single photo-elastic load gauges on top of a concrete piston in contact with the soil (the pressure being corrected for the upward movement of the piston). The load gauges were read by means of a polarized light unit which was lowered down access tubes, the fringe patterns being viewed by means of an inclined mirror within the unit.

The cells enabled the load distribution between the raft and the piles to be determined, the results being shown in Figure 11.16(c). During the initial stages of loading it is

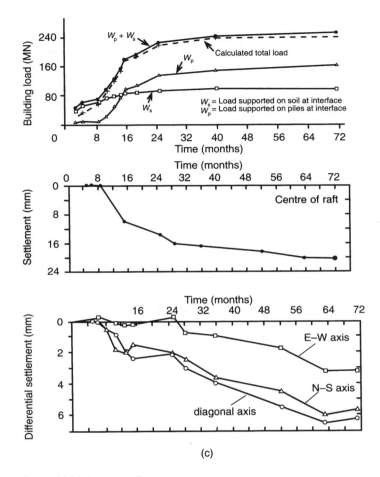

Figure 11.16 (continued)

apparent that most of the load was supported directly by the soil in contact with the raft but most of the subsequent loading was supported by the piles. At the end of construction approximately 60% of the structural load was supported by the piles and 40% by the soil in contact with the raft. As consolidation proceeded it would be expected that the proportion of load supported by the piles would increase and the raft contact pressure would decrease. The settlement observations are also shown in Figure 11.16(c), the data being obtained using a precision level incorporating a parallel plate micrometer capable of reading to an accuracy of 0.025 mm. The temporary bench marks employed for levelling were checked periodically against a permanent deep bench mark based in the chalk bedrock. The observations show that consolidation was virtually complete 6 years after the start of construction and that the maximum settlement at the centre of the raft was 22 mm, an acceptable value. Differential settlement varied between 7 mm across one diagonal axis and 3 mm along one longitudinal axis. There were no signs of cracking in the concrete structure and the architectural finishes.

The calculated load factor for the pile group assuming a group efficiency of 1.0 is 6.4 and the final consolidation settlement, calculated using the equivalent raft concept, is 40 mm. In these calculations no allowance was made for the proportion of load supported at the soil–raft interface and the above values, therefore, are very conservative. At the end of construction, the load supported by the piles is around 130 MN, as indicated in Figure 11.16(c), whereas the ultimate load for the pile group was calculated to be 770 MN. Calculation of rate of settlement using values of c_v determined from small oedometer specimens would be unrealistic due to the presence of fissures and laminations in the clay. Subsequently, the behaviour of the raft was analysed by means of a linearly elastic finite element model in which the soil modulus was assumed to increase linearly with depth, the objective being to reproduce the measured loads and settlements. A trial-and-error approach was used to establish the distribution of undrained modulus E_u with depth z which would match the calculated to the observed values of load and settlement at the soil–raft interface. It was established by back calculation that

$$E_u = 10 + 5.2z$$

the units of E_u and z being MN/m^2 and m, respectively. In the analysis the value of Poisson's ratio for undrained conditions (ν_u) was assumed to be constant with depth and was taken as 0.47, rather than the theoretical value of 0.5, to avoid computational difficulties. The value of Poisson's ratio for drained conditions (ν') was assumed to be 0.10 and if it is accepted that the shear modulus (G) is independent of drainage conditions then from Equation 5.3 it is apparent that

$$E' = 0.75E_u$$

Settlements were computed for both undrained and drained conditions, the maximum (drained) settlement being 20 mm. Further analysis showed that the under-reamed bases contributed very little to the performance of the group and that without piles the maximum settlement would be 48 mm. The piles effectively act as settlement reducers and provide a reserve capacity against local instability and tilting.

Slope failure

This case concerns the failure of a slope in clay near Oslo which was the subject of an extensive investigation by Sevaldson [12]. Initially, natural slopes existed at an inclination of around 1 : 2½ on each side of an old river bed and there was a history of slides in the area. However, about 30 years before the failure occurred, the soil was trimmed back to a slope of 1:2 on one side of the area to accommodate a railway marshalling yard. Approximately 5 years before the failure the ground was trimmed back further at the same slope. The case, therefore, is an example of long-term failure. A section through the centre of the slip area is shown in Figure 11.17(a) and a plan of the area (the lateral extent of which was about 40 m) in Figure 11.17(b).

The ground conditions in the area had been determined prior to the failure from borehole data. Below a drying crust, the soil consisted of a firm, homogeneous, intact, lightly overconsolidated clay with some thin layers of silt and sand. A further ground investigation took place subsequent to the slide including borings from which

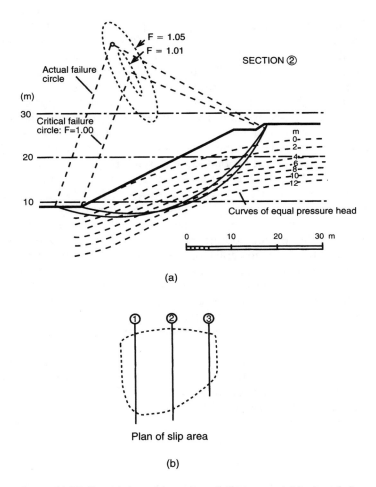

(a)

Plan of slip area

(b)

Figure 11.17 Slope failure: (a) section of slip area and (b) plan of slip area. (Reproduced from R.A. Sevaldson (1956) The slide in Lodalen, October 6th 1954, *Geotechnique, 6,* 167–82, by permission of Thomas Telford Ltd.)

undisturbed samples, 54 mm in diameter and 800 mm in length, were obtained both within and adjacent to the slide area by means of thin-walled piston samplers. In addition, samples of 100 mm in diameter and 600 mm in length were obtained outside the slide area only. The samples from within the slide area were examined carefully in the laboratory to identify the position of the slip surface. However, direct evidence of the slip surface was found only in samples from two of the boreholes. This evidence consisted of thin zones of clay which had been partially remoulded and which had a greater water content than the adjacent soil. Two samples also exhibited fissures, indicating proximity to the failure surface. The evidence described and the visible profile of the upper section of the slip surface enabled the surface to be identified, the shape of which could be represented in section by a circular arc. Pore water pressure was also measured at different depths in four boreholes, one inside and three outside the slide area, using open standpipe piezometers. The piezometers consisted of 300 mm lengths of brass tube with radially bored holes and covered by filters of porous sintered bronze, the lower ends being fitted with solid cones to give protection during installation. Plastic tubes were connected to the filter points and protected by casing screwed into the points, pressure head being indicated by the water levels in the tubes. Contours of pressure head determined from the piezometer observations are shown in Figure 11.17(a).

Values of the shear strength parameters in terms of effective stress (the appropriate parameters for analysis of a long-term failure) were determined from consolidated undrained triaxial tests with pore water pressure measurement. Three or four specimens of clay obtained from the same sample were tested at different values of all-round pressure and c' and ϕ' determined for that sample. Ten separate determinations were made in this way, giving the parameters at various depths in different boreholes. The average values of the parameters were $c' = 10\,kN/m^2 (\pm 2.2\,kN/m^2)$ and $\phi' = 27° (\tan\phi' = 0.512 \pm 0.038)$, i.e. ranges of ± 22 and $\pm 7\%$, respectively.

The stability of the slope, as represented by the three cross-sections shown in Figure 11.17(b), was analysed using the Bishop routine method, i.e. assuming failure along circular arcs. It was assumed that a tension crack opened to a depth of 1.0 m below the upper ground surface. The minimum values of factor of safety calculated for Sections 1, 2 and 3, respectively, were 1.10, 1.00 and 1.19, giving a weighted average for the overall slide of 1.05. The critical failure arc, corresponding to the minimum factor of safety, corresponded closely to the actual slip surface, as shown in Figure 11.17(a). Sevaldson stated that if c' were taken to be zero with the same value of ϕ', then the factor of safety would be 0.73.

It is now recognized that the parameter ϕ'_{cv} (c' being zero) should be used in slope stability analysis. A study of Sevaldson's test results indicates that the probable value of ϕ'_{cv} is around 31°. Using that value, a rough calculation would indicate a factor of safety of approximately 0.86. The case shows that failure of the slope would have been predicted using a simple analytical procedure and the available shear strength data.

Vertical drains

A section of a motorway link road in Belfast is carried on embankments up to 8 m in height over approximately 10 m of soft alluvial soils, the case being reported by Davies and Humpheson [3]. The ground investigation, using boreholes supplemented by continuous sampling and Dutch cone tests, showed that below 0.5–1.0 m of fill there were two distinct layers. Typically, the upper layer was 4–5 m in thickness, the soil varying from silty fine sand to silty clay. The lower layer, also 4–5 m thick, consisted of soft silty clay of high plasticity. These layers were underlain by 0.5–1.0 m of peat, below which there were depths of gravel and glacial till.

Piston samplers of 100 mm diameter were used to obtain material for a comprehensive laboratory programme of oedometer and undrained triaxial tests. *In-situ* vane tests were also included as part of the ground investigation. Undrained shear strength (c_u) varied over depth between 10 and 25 kN/m², as shown in Figure 11.18(a). The oedometer tests enabled the values of compression index (C_c) and coefficient of consolidation (c_v) to be determined. Using the C_c/depth profiles for appropriate boreholes, consolidation settlements were calculated for a range of embankment pressures. The resulting pressure–settlement relationship is shown in Figure 11.18(b), the division of

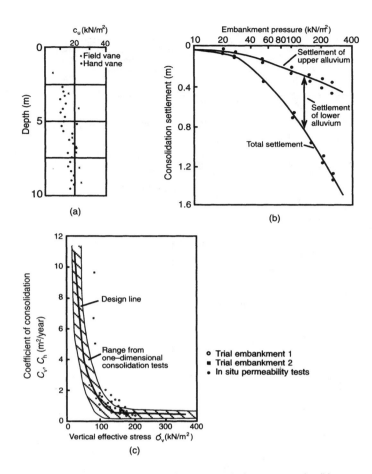

Figure 11.18 Vertical drains: (a) undrained shear strength, (b) pressure–settlement relationships, (c) coefficient of consolidation, (d) field instrumentation and (e) piezometer observations. (Reproduced from J.A. Davies and C. Humpheson (1981) A comparison between the performance of two types of vertical drain beneath a trial embankment in Belfast, *Geotechnique*, **31**, 19–31, by permission of Thomas Telford Ltd.)

settlement between the upper and lower layers also being indicated: for an embankment of 8 m high (pressure around $160 \, \text{kN/m}^2$), settlements of the order of 1 m are predicted. For the upper layer, values of c_v, determined from specimens obtained at different depths in four boreholes, varied from around 5 to over $500 \, \text{m}^2/\text{year}$. For the lower layer, c_v varied between 0.25 and $0.50 \, \text{m}^2/\text{year}$ at vertical effective stress levels above $100 \, \text{kN/m}^2$; however at lower stress levels values as high as $14 \, \text{m}^2/\text{year}$ were obtained. The range of values obtained for the lower layer is shown in Figure 11.18(c) together with the variation used in design. Some oedometer specimens were cut horizontally in order that values of c_h could be determined. Values of c_h for the lower layer were very close to those of c_v, indicating the absence of preferred drainage paths

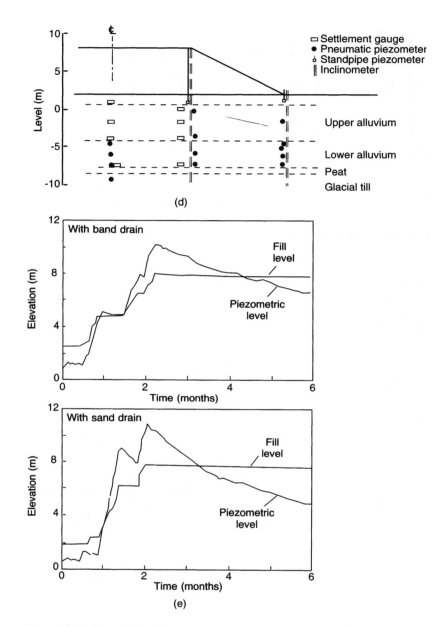

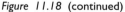

Figure 11.18 (continued)

within the clay. Values of coefficient of consolidation were also obtained from the results of *in-situ* permeability tests using piezometers installed in some of the boreholes.

Bearing capacity calculations based on the profile of undrained shear strength showed that for an embankment 8.5 m in height the factor of safety would be at least

1.25: therefore because the factor of safety would increase as the clay consolidated, stability was not considered to be a problem. Settlement, however, did present a problem. In the upper layer it was assumed that consolidation settlement would keep pace with the application of fill pressure because of the relatively high values of c_v. However, calculations based on the values of c_v in the lower layer indicated that significant consolidation settlement would continue well beyond the construction period of the road. It was decided, therefore, that vertical drains would be installed to ensure that post-construction settlement would be reduced to an acceptable value. The design of a system of vertical drains depends crucially on the values of coefficient of consolidation in the vertical and horizontal directions. Even if a careful and comprehensive investigation and testing programme is carried out, there is always uncertainty regarding the reliability of parameters determined from small oedometer specimens when applied to an *in-situ* soil mass. Accordingly, to ensure a reliable and economic design, two trial embankments, incorporating comprehensive instrumentation, were constructed. Two types of drain were assessed: backfilled sand drains of 200 mm diameter were installed below one embankment and band drains of dimensions 100×7 mm (assumed to have an equivalent diameter of 50 mm) below the other. In the trial the spacings of the two types of drain were designed to produce the same rate of dissipation of excess pore water pressure. The design value of c_v was taken as $0.50 \, m^2$/year, based on the final value of vertical effective stress in the lower layer. The spacings were 1.3 and 0.9 m for the sand and band drains, respectively. The instrumentation comprised pneumatic and standpipe piezometers, pneumatic settlement gauges and inclinometers, as shown in Figure 11.18(d).

The sequences of construction of the two trial embankments and the corresponding responses of pore water pressure below the centre lines, recorded by piezometers at mid-depth of the lower layer, are shown in Figure 11.18(e). The responses of piezometers in the upper layer indicated very rapid dissipation of excess pore water pressure. The rate of dissipation was greater in the trial with sand drains than in the one with band drains. If vertical drainage is neglected the coefficient of consolidation can be back-calculated, using consolidation theory, from the dissipation rates recorded by the piezometers. Values of c_h at mid-depth in the lower layer, under the two trial embankments, are superimposed in Figure 11.18(c). Values from the *in-situ* permeability tests are also shown. At low stress levels the *in-situ* values are higher than those determined in the laboratory but at higher levels the *in-situ* and laboratory values are relatively close. The design value ($0.50 \, m^2$/year), therefore, would result in the rate of dissipation being significantly underestimated during the early stages of embankment construction. Use of a constant value of c_v does not represent the observed rate of dissipation: the coefficient should be varied according to the value of vertical effective stress. It was shown that satisfactory predictions of dissipation rate could be made by varying c_v at short time intervals, generally five days. Comparison of the dissipation rates shows that the sand drains appeared to be the more efficient type. This could be due to the smear effect being less pronounced than in the case of the band drains which were installed by means of a mandrel, or to the equivalent diameter assumed for the band drains being incorrect.

Laboratory determination of coefficient of consolidation is normally expected to result in the significant underestimation of rate of settlement because of the presence of macro-fabric features in the soil mass. In this case, however, it is apparent that there is relatively close agreement between predicted and observed rates, due largely to the absence of a preferred drainage direction within the clay.

Embankment dam

This case concerns the performance of Balderhead dam in the north of England, constructed between 1961 and 1965, as reported by Kennard *et al.* [8] and Vaughan *et al.* [15]. The dam has a maximum height of 48 m and a crest length of 925 m. The foundation material is mainly shale with occasional thin horizontal layers of sandstone and limestone, but deposits of boulder clay occur on the lower valley slopes. The cross-section of the dam is shown in Figure 11.19(a). The core consists of boulder clay, watered before and after placement and compacted in 180 mm layers by a sheepsfoot roller. Subsequent measurements showed that the placement water content varied between 10 and 15%, being 2–3% below the optimum value at some levels. Around 20% of the core material was of clay size and particles larger than 150 mm were removed. The shoulders of the dam consist of shale, watered and compacted in 200 mm layers by a grid roller over the lower 18 m and in 800 mm layers by a vibrating roller above that height. The internal drainage layers consisted of crushed limestone, the ratio of D_{15} for the filter to D_{85} for the fill (Equation 2.35) being specified as 3. Cut-off in the foundation material was effected by means of a concrete wall 1.8 m thick and 25 m deep below the centre-line of the dam and a grout curtain extending to a depth of about 30 m below the base of the wall. Piezometers and water-overflow settlement gauges were installed in the shale fill, mainly adjacent to the clay core. Two plate settlement gauges, five pressure cells and three horizontal deformation gauges were located in the clay core. The settlement gauges in the core consisted of steel plates, laid at 3 m intervals as the fill was placed, threaded over plastic tubing set into the concrete cut-off wall; the subsequent levels of the plates were determined by means of an induction coil lowered inside the tubing. The horizontal gauges measured deformation over a 1.8 m gauge length by means of vibrating wire sensors, each end of the device being attached to a steel beam encased in concrete. Provision was also made for the measurement of the quantity of seepage through the dam. The layout of instruments in the core at the highest section of the dam is shown in Figure 11.19(b). Results from the instruments at this section are shown in Figures 11.19(c) and (d).

Shortly before maximum reservoir level was reached on impounding, a large increase in seepage flow, measured at the outlets to the underfilter, occurred through the dam, as indicated in Figure 11.19(c). A few months later a small depression was discovered in the crest above the chimney drain. Later, a sink hole, about 3 m in diameter and 2.5 m in depth, developed above the upstream edge of the core. In addition, the flow from the underfilter, which had previously been clear, became cloudy, indicating that the filter had failed to block particles from the core. In response to these developments the water level in the reservoir was drawn down by 9 m, resulting in a marked decrease in the downstream flow (shown in Figure 11.19(c)) and the water from the underfilter becoming clear. However during drawdown a second sink hole developed in the crest.

A trial pit was excavated to a depth of 12 m in the chimney drain to expose the downstream face of the core and no significant softening of the clay or other damage was evident. Further investigations by wash borings advanced from the crest of the

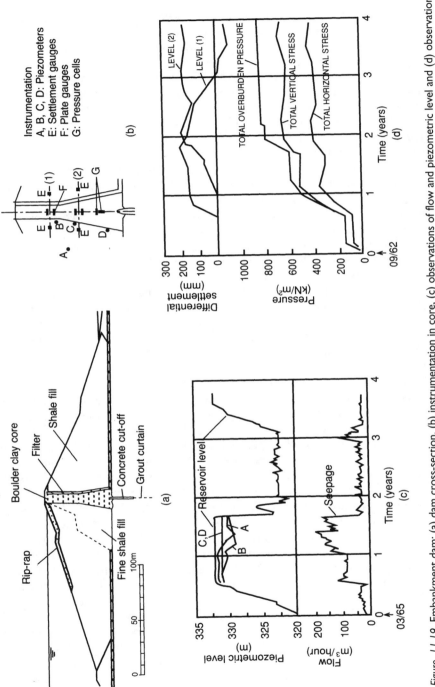

Figure 11.19 Embankment dam: (a) dam cross-section, (b) instrumentation in core, (c) observations of flow and piezometric level and (d) observations of stress and differential settlement. (Reproduced from M.F. Kennard et al. (1967) Stress and strain measurements in the clay core of Balderhead dam, Transactions 9th International Congress on Large Dams, 3, 129–51, by permission of ICOLD.)

dam to a depth of 18 m showed that wash water pumped through the boring rods was lost through the core. Further boreholes were advanced by means of a flight auger and at a depth of 14 m in one hole softened clay and water, which rose rapidly to reservoir level, were encountered. From all the evidence it was concluded that leakage had occurred through the core below drawdown level but that the seepage paths had closed up when the pore water pressure had decreased on drawdown. Readings from two of the piezometers upstream of the core (A and B in Figure 11.19(c)) confirmed this interpretation. The pressure heads indicated by these piezometers were in phase with the increase in reservoir level until the final stages of impounding when the heads decreased below reservoir level, matching the increase in seepage, showing that leakage was occurring at around the level of these piezometers. After drawdown the heads again corresponded to reservoir level. On the other hand, the pressure heads indicated by piezometers at a lower level (C and D in Figure 11.19(c)) remained in step with the increase in reservoir level throughout.

Differential settlements between the core and the shoulders, at two levels where water overflow gauges were located close to the core, are plotted in Figure 11.19(d), the maximum value being about 200 mm. Readings from two of the pressure gauges, indicating the values of total vertical and horizontal stress, are also plotted in Figure 11.19(d). (It should be noted that Figures 11.19(c) and (d) cover different time spans.) It is apparent that the total vertical stress became less than the overburden pressure indicating that load was being transferred from the core to the shoulders as a result of differential settlement and arching. However the total vertical stress was not low enough to cause hydraulic fracture. On the other hand, the total horizontal stress could have been lower than the local pore water pressure, producing the conditions necessary for fracture. Vertical and horizontal strains at two points in the core were derived from readings of the plate settlement gauges and the horizontal deformation gauges, respectively. The maximum vertical and horizontal strains in the part of the core over a shale foundation are 3 and $\frac{1}{2}$%, respectively. The corresponding values in the part of the core founded on boulder clay are $5\frac{1}{2}$% and 1%.

It was concluded that the leakage through the dam was due to hydraulic fracture. The fact that the water content in the core was less than the optimum value was a contributory factor. The remedial measures adopted consisted of grouting the core to increase the total horizontal stresses and prevent further hydraulic fracture and the construction of a diaphragm wall within damaged sections of the core. It was judged that the filter would be adequate provided that the core remained intact. The grout consisted of equal proportions of cement and bentonite with a water/solids ratio of 4. It was important to ensure that the grout pressure was less than the overburden pressure at the points of injection. Subsequent borehole samples showed that the grout had flowed along approximately vertical planes within the core. The diaphragm wall was considered to be necessary because of segregation of core material under the flow during leakage: complete permeation by the grout could not be assured and the risk of ungrouted silt and fine sand adjacent to the filter was unacceptable. The diaphragm wall, 200 m in length, was constructed on top of the cut-off wall using plastic concrete with 2.5% of bentonite in the mix. Extra piezometers were installed in inclined boreholes just upstream from the core to detect any further leakage. After refilling no significant flow was apparent, although, 2 years later, piezometer readings indicated decreasing pressure head at one cross-section and additional grouting was undertaken.

Observational method

An example of the use of the observational method occurred during construction of the cut and cover section of the Limehouse Link, a highway between the City of London and the Docklands area, described by Glass and Powderham [5]. The soil conditions and a cross-section of the tunnel at one location are shown in Figure 11.20; however soil conditions were variable along the line of construction. During excavation, more details of the soil conditions were obtained to supplement the original ground investigation data. Work proceeded at nine construction fronts along the 1.8 km length of the tunnel. Over most of the length, diaphragm walls were used to support the sides of the excavation during construction and to form the permanent sides of the tunnel. As well as ensuring an adequate factor of safety with respect to passive resistance for the walls, the design requirements were that ground movements around the tunnel should be low enough to avoid damage to adjacent structures. Prior to construction, parametric studies using the finite element method were performed to obtain an indicative range of wall displacements. Due to uncertainties regarding the variable soil conditions the original design was based on the use of props, both above the level of the roof slab and midway between the roof slab and formation level, during excavation.

The observational method was introduced with the objective of eliminating the need for props at mid-height, thus providing a clear working space for excavation below the roof slab and leading to a consequent increase in the rate of construction. Calculations based on the most probable soil conditions indicated that excavation could be undertaken without the use of mid-height props. However there was insufficient confidence in the data to justify starting construction without the use of these props. The procedure adopted was to start construction at each front using mid-height props then to introduce, in stages, planned modifications in technique as excavation proceeded, with each stage being shown to be safe by means of *in-situ* monitoring. A pre-determined course of action was available if monitoring were to show unacceptable risk to the integrity of the walls or adjacent structures. The monitoring of wall movement was the principal means of construction control. This approach was intermediate between the *ab initio* and best-way-out applications of the observational method described by Peck.

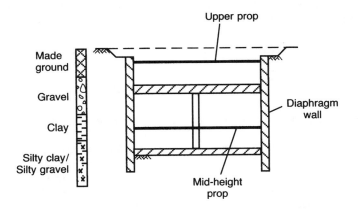

Figure 11.20 Limehouse Link: soil profile and tunnel cross-section.

The first stage of construction involved the installation of mid-height props in accordance with the original design, the props being prestressed to 10% of the design load. The prop loads were monitored during excavation by means of vibrating wire strain gauges and it was found that the loads were considerably less than expected, being within the range of only 6–15% of the design values. After excavation to formation level the concrete base slab was cast on top of a 100 mm blinding layer. The second stage consisted of placing the props as above, followed by the placing of a 300 mm blinding layer to act as a strut at formation level. When this layer had gained sufficient strength, the mid-height props were de-stressed and inward wall movement was monitored. The third stage involved positioning the mid-height props but leaving a small gap at one end between wall and prop, allowing a movement of 20 mm to take place before load acted on the prop, the size of the gap being monitored to indicate whether or not the prop was being loaded. The fourth stage was to dispense with the mid-height props but to limit the length of excavation at formation level to 5 m, the blinding layer being cast immediately. Wall movement was monitored at frequent intervals by means of a tape extensometer and using surveying techniques. To guard against the possibility of excessive wall movement a number of reserve props were readily available.

Wall convergence (the sum of the two wall movements), at formation level, was plotted against time and compared with pre-determined 'warning' values. Movements of less than 20 mm (0.1% of the typical tunnel depth of 20 m) were of no concern and confirmed that dispensing with mid-height props was justified. Increased frequency of monitoring would be initiated if movements between 20 and 25 mm were recorded. If movements exceeded 25 mm during the fourth stage then the reserve props would be used. Movements exceeding 25 mm during the third stage would have resulted in the reintroduction of prestressed props. The maximum allowable convergence was 70 mm (35 mm for each wall). In the event the maximum recorded wall convergence was 11 mm with most values being less than 7 mm. For all nine construction fronts, no further propping was necessary after the first three stages described above, saving the requirement for most of the steelwork initially envisaged.

A further stage concerned an assessment of the action of the blinding strut. Lengths of blinding were cast with a 15 mm gap at one end. The closure of this gap was monitored and the readings showed that the struts did not become active before the base slab itself was cast. Values of wall movement were essentially the same as those during the earlier construction stages. Accordingly, the thickness of the blinding layer was reduced to 100 mm.

REFERENCES

1   British Geotechnical Society (1974) *Field Instrumentation in Geotechnical Engineering*, Butterworths, London.
2   Darragh, R.D. (1964) Controlled water tests to preload tank foundations, *Proceedings ASCE*, **90** (SM5), 303–29.
3   Davies, J.A. and Humpheson, C. (1981) A comparison between the performance of two types of vertical drain beneath a trial embankment in Belfast, *Geotechnique*, **31**, 19–31.
4   Dunnicliff, J. (1988) *Geotechnical Instrumentation for Monitoring Field Performance*, John Wiley & Sons, New York.

5   Glass, P.R. and Powderham, A.J. (1994) Application of the observational method at the Limehouse Link, *Geotechnique*, **44**, 665–79.

6   Hanna, T.H. (1985) *Field Instrumentation in Geotechnical Engineering*, Trans Tech Publications, Germany.

7   Hooper, J.A. (1973) Observations on the behaviour of a piled-raft foundation on London clay, *Proceedings Institution of Civil Engineers*, **55** (2), 855–77.

8   Kennard, M.F., Penman, A.D.M. and Vaughan, P.R. (1967) Stress and strain measurements in the clay core at Balderhead dam, *Transactions 9th International Congress on Large Dams*, **3**, 129–51.

9   Lambe, T.W. (1973) Predictions in soil engineering, *Geotechnique*, **23**, 149–202.

10 Nordlund, R.L. and Deere, D.U. (1970) Collapse of Fargo Grain Elevator, *Journal ASCE*, **96** (SM2), 585–607.

11 Peck, R.B. (1969) Advantages and limitations of the observational method in applied soil mechanics, *Geotechnique*, **19**, 171–87.

12 Sevaldson, R.A. (1956) The slide in Lodalen, October 6th, 1954, *Geotechnique*, **6**, 167–82.

13 Somerville, S.H. and Shelton, J.C. (1972) Observed settlement of multi-storey buildings on laminated clays and silts in Glasgow, *Geotechnique*, **22**, 513–20.

14 Stuart, J.G. and Graham, J. (1975) Settlement performance of a raft foundation on sand, *Settlement of Structures*, Pentech Press, London.

15 Vaughan, P.R., Kluth, D.J., Leonard, M.W. and Pradoura, H.H.M. (1970) Cracking and erosion of the rolled clay core of Balderhead dam and the remedial works adopted for its repair, *Transactions 10th International Congress on Large Dams*, **1**, 73–93.

16 Whitworth, L.J. and Thomson, N. (1995) The design, construction and load testing of Starsol piles, Perth, Scotland, *Proceedings Institution of Civil Engineers, Geotechnical Engineering*, **113**, 233–41.

Principal symbols

A, a	Area
A	Air content
$A, \overline{A}$	Pore pressure coefficients
a'	Modified shear strength parameter (effective stress)
a	Dial gauge reading in oedometer test
B	Width of footing
$B, \overline{B}$	Pore pressure coefficients
C_U	Coefficient of uniformity
C_N	Correction factor for overburden pressure
C_Z	Coefficient of curvature
C_a	Area ratio
C_c	Compression index
C_s	Isotropic compressibility of soil skeleton
C_{s0}	Uniaxial compressibility of soil skeleton
C_v	Compressibility of pore fluid
C_w	Correction factor for water table position
C_α	Rate of secondary compression
c	Shear strength parameter
c_u	Undrained (total stress) shear strength parameter
c'	Drained (effective stress) shear strength parameter
c_w	Wall adhesion
c_v	Coefficient of consolidation (vertical drainage)
c_h	Coefficient of consolidation (horizontal drainage)
D	Depth of footing; depth of excavation
D	Depth factor
D	Particle size
D_b	Depth of embedment of pile in bearing stratum
d	Length of drainage path
d	Diameter; depth
d	Depth factor
E	Young's modulus
e	Void ratio
e	Eccentricity
F	Factor of safety
F	Axial load on pile

f_s	Sleeve resistance of static penetrometer
G	Shear modulus
G_s	Specific gravity of soil particles
g	$9.8 \, \text{m/s}^2$
H, h	Height
H	Horizontal component of load
H	Layer or specimen thickness
h	Total head
I	Influence factor
I_C	Compressibility index
I_D	Density index
I_P	Plasticity index
I_L	Liquidity index
i	Hydraulic gradient
i	Inclination factor
J	Seepage force
j	Seepage pressure
K	Lateral pressure coefficient
K_a	Active pressure coefficient
K_p	Passive pressure coefficient
K_0	Coefficient of earth pressure at-rest
K	Absolute permeability
k	Coefficient of permeability
L, l	Length
M	Mass
M	Moment
M	Slope of critical-state line
m_v	Coefficient of volume compressibility
N	Normal force
N	Standard penetration resistance
N_d	Number of equipotential drops
N_f	Number of flow channels
N_γ	Bearing capacity factor
N_c	Bearing capacity factor
N_q	Bearing capacity factor
N_s	Stability coefficient
n	Porosity
n_d	Equipotential number
P_a	Total active thrust
P_p	Total passive resistance
p	Pressure
p_a	Active pressure
p_p	Passive pressure
p_0	At-rest pressure
p_1	Limit pressure
Q	Surface load
Q_f	Ultimate load

q	Flow per unit time
q	Total foundation pressure
q	Surface pressure
q_a	Allowable bearing capacity
q_b	Base resistance of pile per unit area
q_c	Cone penetration resistance
q_f	Ultimate bearing capacity
q_{nf}	Net ultimate bearing capacity
q_n	Net foundation pressure
q_s	Shaft resistance of pile per unit area
R, r	Radius
R	Bearing resistance
R_{oc}	Overconsolidation ratio
r	Compression ratio
r_u	Pore pressure ratio
S	Shearing or sliding resistance
S_r	Degree of saturation
s	Shape factor
s	Settlement
s_c	Consolidation settlement
s_i	Immediate settlement
T_v	Time factor (vertical drainage)
T_r	Time factor (radial drainage)
t	Time
U	Boundary water force
U	Degree of consolidation
u, u_w	Pore water pressure
u_a	Pore air pressure
u_e	Excess pore water pressure
u_i	Initial excess pore water pressure
u_s	Static pore water pressure
u_{ss}	Steady seepage pore water pressure
V	Volume
V	Vertical component of load
v	Specific volume
v	Discharge velocity
v'	Seepage velocity
W	Weight
w	Water content
w_L	Liquid limit
w_P	Plastic limit
z	Depth coordinate
z	Elevation head
α'	Modified shear strength parameter (effective stress)
α	Angle of wall inclination
α	Skin friction coefficient
β	Skin friction coefficient

β	Slope angle
γ	Partial factor
γ	Shear strain
γ	Unit weight
γ_d	Dry unit weight
γ_{sat}	Saturated unit weight
γ'	Buoyant unit weight
γ_w	Unit weight of water
δ	Angle of wall friction
ε	Normal strain
ε_v	Volumetric strain
η	Dynamic viscosity
κ	Slope of isotropic swelling/recompression line
λ	Slope of isotropic normal consolidation line
μ	Settlement coefficient
ν	Poisson's ratio
ξ	Correlation (or reduction) factor for pile load tests
ρ	Bulk density
ρ_d	Dry density
ρ_s	Particle density
ρ_{sat}	Saturated density
ρ_w	Density of water
σ	Total normal stress
σ'	Effective normal stress
$\sigma_1, \sigma_2, \sigma_3$	Total principal stresses
$\sigma'_1, \sigma'_2, \sigma'_3$	Effective principal stresses
τ	Shear stress
τ_f	Shear strength; peak shear strength
τ_r	Residual shear strength
ϕ	Potential function
ϕ	Shear strength parameter
ϕ_u	Undrained (total stress) shear strength parameter
ϕ'	Drained (effective stress) shear strength parameter
ϕ'_r	Drained residual shear strength parameter
ϕ'_{max}	Maximum (peak) angle of shearing resistance
ϕ'_{cv}	Angle of shearing resistance at constant volume (i.e. at the critical state)
ϕ_μ	True angle of friction
χ	Parameter in effective stress equation for partially saturated soil
ψ	Flow function
ψ	Angle of dilation
k (subscript)	Characteristic
d (subscript)	Design
c (subscript)	Compression
t (subscript)	Tension

Answers to problems

CHAPTER 1

1.1 SW, MS, ML, CV
1.2 0.55, 46.6%, 2.10 Mg/m^3, 20.4%
1.3 15.7 kN/m^3, 19.7 kN/m^3, 9.9 kN/m^3, 18.7 kN/m^3, 19.3%
1.4 98%
1.5 1.92 Mg/m^3, 0.38, 83.7%, 4.5%; no
1.6 15%, 1.83 Mg/m^3, 3.5%
1.7 70%

CHAPTER 2

2.1 4.9×10^{-8} m/s
2.2 1.3×10^{-6} m^3/s (per m)
2.3 5.8×10^{-5} m^3/s (per m), 316 kN/m
2.4 2.0×10^{-6} m^3/s (per m)
2.5 4.7×10^{-6} m^3/s (per m)
2.6 1.1×10^{-6} m^3/s (per m)
2.7 1.8×10^{-5} m^3/s (per m)
2.8 5.9×10^{-5} m^3/s (per m)
2.9 1.0×10^{-5} m^3/s (per m)

CHAPTER 3

3.1 51 kN/m^2
3.2 51 kN/m^2
3.3 51.4 kN/m^2, 33.4 kN/m^2
3.4 105.9 kN/m^2
3.5 (a) 94.0 kN/m^2, 154.2 kN/m^2; (b) 94.0 kN/m^2, 133.8 kN/m^2
3.6 9.9 kN, 73° below horizontal
3.7 30.2 kN/m^2, 10.6 kN/m^2
3.8 1.5, 14 kN/m^2, 90 kN/m^2
3.9 2.0, 0.65 m

CHAPTER 4

4.1 $113\,kN/m^2$
4.2 $44°$
4.3 $110\,kN/m^2$, $0°$
4.4 $36\,kN/m^2$
4.5 0, $25\frac{1}{2}°$, $170\,kN/m^2$
4.6 $34°$, $31.5°$, $29°$; $20\,kN/m^2$, $31°$
4.7 $76\,kN/m^2$
4.8 0.73
4.9 0.96, 0.23

CHAPTER 5

5.1 $96\,kN/m^2$
5.2 $277\,kN/m^2$
5.3 $45\,kN/m^2$
5.4 $68\,kN/m^2$
5.5 $76\,kN/m$
5.6 $7\,mm$

CHAPTER 6

6.1 $76.5\,kN/m$, $122\,kN/m$
6.2 $571\,kN/m$, $8.57\,m$
6.3 $11.5\,kN/m^2$, $202\,kN/m^2$; $18.7\,kN/m^2$, $198\,kN/m^2$
6.4   (a) $195\,kN/m^2$, $49\,kN/m^2$, 1.7; (b) Yes: $1054\,kN\,m > 345\,kN\,m$, $227\,kN > 163\,kN$
6.5 $3.95\,m$
6.6   Yes: $1307\,kN\,m > 512\,kN\,m$, $250\,kN/m^2 > 228\,kN/m^2$, $245\,kN > 205\,kN$
6.7 $5.60\,m$, $226\,kN$
6.8   (a) 1.8, $354\,kN$; (b) $5.44\,m$
6.9 2.25
6.10 $110\,kN$; $154\,kN$
6.11 1.7
6.12 (a) $340\,N/mm^2 > 71\,N/mm^2$, $21.3\,kN > 13.8\,kN$; (b) $340\,N/mm^2 > 87\,N/mm^2$, $21.4\,kN > 17.0\,kN$

CHAPTER 7

7.1 $c_v = 2.7$, $2.6\,m^2/year$, $m_v = 0.98\,m^2/MN$, $k = 8.1 \times 10^{-10}\,m/s$
7.2 $318\,mm$, $38\,mm$ (four sublayers)
7.3 2.6 years, 0.95 years
7.4 $35.2\,kN/m^2$
7.5 0.65

7.6   130 mm, 95 mm
7.7   124 mm, 38 mm, 72 mm, 65 mm
7.8   280 mm (six sublayers)
7.9   0.80
7.10   8.8 years, 0.7 years

CHAPTER 8

8.1   2.8, 2.9
8.2   4.8; Yes: 970 kN/m > 500 kN/m
8.3   4100 kN; Yes: 7241 kN > 4100 kN
8.4   7.0, 5.3, 3.3
8.5   Yes: 6777 kN > 5950 kN
8.6   3600 kN
8.7   1.8
8.8   No: 4037 kN < 4125 kN; Yes: 13 mm < 20 mm
8.9   120 kN/m^2, 135 kN/m^2, 150 kN/m^2
8.10   25 mm
8.11   (a) 9200 kN; (b) 9650 kN
8.12   (a) 2.7; (b) Yes: 30.27 MN > 22.80 MN; (c) 30 mm
8.13   (a) 2.1; (b) Yes: 3348 kN > 3300 kN; (c) Yes: 15 mm < 20 mm
8.14   230 kN

CHAPTER 9

9.1   1.67, 1.49, 4175 kN/m > 3628 kN/m
9.2   50°, 27°
9.3   0.91; Yes: 734 kN/m < 1075 kN/m
9.4   1.01
9.5   1.08
9.6   (a) 13°, 3.1; (b) Yes: 5.04z kN > 4.16z kN

Index